Don't try this at home!

Greg

A Signal Integrity Engineer's Companion

Prentice Hall Modern Semiconductor Design Series

James R. Armstrong and F. Gail Gray
VHDL Design Representation and Synthesis

Mark Gordon Arnold
Verilog Digital Computer Design: Algorithms into Hardware

Jayaram Bhasker
A VHDL Primer, Third Edition

Mark D. Birnbaum
Essential Electronic Design Automation (EDA)

Eric Bogatin
Signal Integrity: Simplified

Douglas Brooks
Signal Integrity Issues and Printed Circuit Board Design

Ken Coffman
Real World FPGA Design with Verilog

Alfred Crouch
Design-for-Test for Digital IC's and Embedded Core Systems

Dennis Derickson and Marcus Müller (Editors)
Digital Communications Test and Measurement

Greg Edlund
Timing Analysis and Simulation for Signal Integrity Engineers

Daniel P. Foty
MOSFET Modeling with SPICE: Principles and Practice

Tom Granberg
Handbook of Digital Techniques for High-Speed Design

Nigel Horspool and Peter Gorman
The ASIC Handbook

William K. Lam
Hardware Design Verification: Simulation and Formal Method-Based Approaches

Mike Peng Li
Jitter, Noise, and Signal Integrity at High-Speed

Farzad Nekoogar and Faranak Nekoogar
From ASICs to SOCs: A Practical Approach

Farzad Nekoogar
Timing Verification of Application-Specific Integrated Circuits (ASICs)

Samir Palnitkar
Design Verification with

David Pellerin and Scott Thibault
Practical FPGA Programming in C

Christopher T. Robertson
Printed Circuit Board Designer's Reference: Basics

Chris Rowen
Engineering the Complex SOC

Madhavan Swaminathan and A. Ege Engin
Power Integrity Modeling and Design for Semiconductors and Systems

Wayne Wolf
FPGA-Based System Design

Wayne Wolf
Modern VLSI Design: System-on-Chip Design, Third Edition

Bob Zeidman
Verilog Designer's Library

A Signal Integrity Engineer's Companion

Real-Time Test and Measurement and Design Simulation

Geoff Lawday, David Ireland, and Greg Edlund

PRENTICE
HALL

An Imprint of Pearson Education

Upper Saddle River, NJ • Boston • Indianapolis • San Francisco
New York • Toronto • Montreal • London • Munich • Paris • Madrid
Cape Town • Sydney • Tokyo • Singapore • Mexico City

Many of the designations used by manufacturers and sellers to distinguish their products are claimed as trademarks. Where those designations appear in this book, and the publisher was aware of a trademark claim, the designations have been printed with initial capital letters or in all capitals.

The authors and publisher have taken care in the preparation of this book, but make no expressed or implied warranty of any kind and assume no responsibility for errors or omissions. No liability is assumed for incidental or consequential damages in connection with or arising out of the use of the information or programs contained herein.

The publisher offers excellent discounts on this book when ordered in quantity for bulk purchases or special sales, which may include electronic versions and/or custom covers and content particular to your business, training goals, marketing focus, and branding interests. For more information, please contact:

U.S. Corporate and Government Sales
(800) 382-3419
corpsales@pearsontechgroup.com

For sales outside the United States please contact:

International Sales
international@pearsoned.com

 This Book Is Safari Enabled

The Safari® Enabled icon on the cover of your favorite technology book means the book is available through Safari Bookshelf. When you buy this book, you get free access to the online edition for 45 days.

Safari Bookshelf is an electronic reference library that lets you easily search thousands of technical books, find code samples, download chapters, and access technical information whenever and wherever you need it.

To gain 45-day Safari Enabled access to this book:

- Go to http://www.prenhallprofessional.com/safarienabled
- Complete the brief registration form
- Enter the coupon code B7JD-PR1L-H1NS-BXK3-63AE

If you have difficulty registering on Safari Bookshelf or accessing the online edition, please e-mail customer-service@safaribooksonline.com.

Visit us on the Web: www.prenhallprofessional.com

Library of Congress Cataloging-in-Publication Data:

Lawday, Geoff, 1946-
 A signal integrity engineer's companion : real-time test and measurement and design simulation / Geoff Lawday, David Ireland, and Greg Edlund.
    p. cm.
  ISBN-10: 0-13-186006-2 (pbk. : alk. paper)
  ISBN-13: 978-0-13-186006-3
  1. Electronic apparatus and appliances—Testing. 2. Electronic apparatus and appliances—Design and construction. 3. Signal processing—Simulation methods. 4. Switching circuits—Reliability. 5. Oscillators, Electric—Testing. I. Ireland, David, 1957- II. Title.
  TK7870.23.L39 2008
  621.382'2—dc22

                              2008011981

ISBN-13: 978-0-131-86006-3
ISBN-10: 0-131-86006-2

Text printed in the United States on recycled paper at Courier in Westford, Massachusetts.
First printing June 2008

Credits

Editor-in-Chief

Mark Taub

Acquisitions Editor

Bernard Goodwin

Marketing Manager

Curt Johnson

Managing Editor

Kristy Hart

Project Editors

Lori Lyons, Andy Beaster

Copy Editors

Gayle Johnson, Geneil Breeze

Indexer

Brad Herriman

Proofreader

Linda K. Seifert

Technical Reviewers

Sam Shaw, James Peterson, Andy Martwick, Mike Resso

Publishing Coordinator

Michelle Housley

Multimedia Developer

Dan Scherf

Cover Designer

Louisa Adair

Composition

Bronkella Publishing

Graphics

Laura Robbins

To our wives: Linda, Jane, and Karen

Contents

2   Chip-to-Chip Timing and Simulation   31

Foreword

It is easy to be misled by the rhetoric of the day. Even the world of electronic engineering is not immune. Words and phrases such as "digitalization" and "digital convergence" carry the subtext that you only have to worry about ones and zeroes; that analog is being shown the door by a growing band of electronics designers in their quest to render more into binary logic. But it's an illusion. Far from being squeezed out, the word analog is seeping into all areas of electronic designs in increasingly subtle and potentially damaging ways.

Matters are made even worse by the ease with which digital control can be used to massage and reshape the form of signals. Pre-emphasis is almost trivial to implement in the circuits that precede a driver. Although this processing can improve the ability of a receiver to decode the signal, it can have deleterious effects on other receivers in the vicinity. Worse than that, the interference can depend heavily on the data being transmitted. From that, it is not hard to see how intermittent, apparently random "Heisenbugs" can pop up during operation and promptly disappear the moment you try to add instrumentation to work out what is going wrong.

Signal integrity has been a problem for many years but the issues were often isolated to small parts of a system design. Today, all the trends point to signal corruption getting worse and worse. Switching speeds are going up and the voltages provided on supply rails are going down.

But the trends are not all technical. As you can read in the introduction that follows, some of the biggest problems can result from commercial decisions: the pressure to reduce manufacturing costs are pushing designers to consider cheaper components, packages, and substrates; and designers have less time to get the job done.

In some markets, such as cellular handsets, companies want to be able to produce variants very quickly. They might be on the shelf for only six to nine months. Marketing may not know more than six months out whether the design will be a flyer or a dud: All they can do is extrapolate current trends and hope. The closer they define a product to when it is meant to go on sale, the better their chance of getting it right. But that is no help to an engineering team trying to make the phone work.

The core chips may have been designed for a different phone. But a new screen or keyboard, or a switch from a candy-bar to a flip-phone design means that the board layout has to change. And with a small rearrangement of the components on that board, you can suddenly find that the design decisions made by the chipmakers are at odds with the requirements of the new layout. What was meant to be a quick-turnaround project to create a simple variant of an existing phone for a new market suddenly looks a lot less tractable.

In this book, the authors take you through the methods available to digital designers to ensure that they are not vulnerable to the little tricks that the analog properties world can play on them. It is a comprehensive treatment that shows how working in the virtual and real worlds provide a combined methodology for avoiding signal-integrity problems. It is tempting to think of signal integrity as a subject dominated by "black magic" techniques. But there is plenty of science to help the time-starved engineer ensure that a high-speed, low-voltage bus will work in the final system.

This book demonstrates how modeling and behavioral simulation let the engineer make sensible decisions early on in the project. It covers tricky subjects such as the modeling of transmission lines—a skill that will prove vital in the coming years.

Equally important is the ability to work out where things are going wrong in the prototype, and to track down the source of the problem. Chapters on probing, oscilloscope use, and time-domain reflectometry provide practical advice on the best way to look beyond the ones and zeroes the logic is meant to see into the electromagnetic soup that the real world is made from.

It is not just about the wired world either: the last chapter concentrates on the wireless world and the challenges raised by new software-defined radio architectures.

This is a book that I am sure will be an essential addition to every electronic engineering lab as more people find they have to grapple with the analog infrastructure that underlies every software-driven digital system. In such a digital world, a book like this has never been so important.

Chris Edwards, Editor, *The IET Electronics Systems and Software* **magazine London, UK**

Chris Edwards reports on electronics, IT, and technology matters. He has more than 15 years of journalism experience as an editor and writer. He is currently the freelance editor of *Electronics Systems and Software*, published by the Institution of Engineering and Technology (IET), and a regular contributor to the magazines *Engineering and Technology* (formerly *IEE Review*), *Information Professional*, as well as the new tech magazine for teenagers, *Flipside*.

Preface

WHAT THIS BOOK IS ABOUT

We live in the high-speed digital age where embedded system developers in many cases must apply new design methodologies and perform complex signal integrity tests and measurements. This book guides the reader through the full life cycle of embedded system design from specification and simulation to test and measurement. A significant feature of the book is the explanations of the thorny issues of signal integrity engineering that are so often the cause of delay in a product's development—time to market is the signal integrity engineer's Achilles' heel. By considering the whole life cycle from simulation through to test and validation you can see the new interplay of simulation and real-time test, which drives the new design methodologies. A case in point is the design and implementation of a new high-speed serial bus where the simulation and real-time test are inextricably linked, especially where designs incorporate device driver pre-emphasis or receiver equalization.

The celebrated teacher and philosopher Archimedes said, "What we must learn to do is to learn by doing," and this book endeavors to do just that by presenting practical applications so that you can learn from the authors' experiences in embedded system design, simulation, test, and measurement. What's more, you are encouraged to consider and adopt good practices in signal integrity engineering throughout the entire life cycle of the design.

Of particular importance today is the need to meet regulatory compliance and interoperability requirements. Consequently this book treats the demands of compliance testing and interoperability design as a foremost topic. Alongside today's compliance testing is the migration to high-speed serial buses, where both topics lend themselves to some prime examples of good practice in signal integrity engineering. Therefore, the design and testing of high-speed serial buses with their associated low-voltage differential signaling are pivotal themes throughout several chapters of this book. Nevertheless, the fundamentals of signal integrity engineering underpin the understanding of compliance testing and interoperability design, and you are encouraged to refresh or learn the basics of signal acquisition, test, measurement, and device simulation, which for the most part is provided in this book.

Work as a team has become routine for engineers designing, developing, and testing digital systems. The members of the team build up companionships and regularly look to each other for advice, especially in today's complex high-speed digital world. Today, the task of developing or testing a digital system is complex, and even the design of what appears to be a simple circuit can be problematic. For example, an engineer can design a simple digital circuit that meets all the requirements, and then a year down the line a chip is changed because the new chip is cheaper and has the same functionality. What the engineer didn't know is that the new chip is now made with a smaller device size that reduces its cost and its switching time, leading to faster edge rates and signal integrity problems! This leads to the well-documented quote "There are two kinds of design engineers, those that have signal integrity problems, and those that will." As we have seen if a design is sensitive to edge rates, the component specification must make edge rate a formal product parameter since it is just not possible to anticipate the evolution of a silicon fabrication process. This book aims to be a companion to the engineer and part of the engineer's team by providing an understanding of design specification, simulation, test, and measurement along with some significant advice on maintaining signal integrity throughout the life cycle of a design.

Although we can take the big view, there are significant little problems that the engineer needs to know but, as they say, "is afraid to ask." For example, if a designer suspects ground bounce, or more accurately transients in the signal reference, and wants to measure the ground line with an oscilloscope, where is the probe ground connected? Actually, it is normally connected to a solid logic zero. Moreover, why is it that a state-of-the-art oscilloscope can give a 50% measurement error when measuring ground bounce? Well, the bandwidth of an oscilloscope is typically specified at the –3 dB point, and a voltage measured at the limit of the specified oscilloscope bandwidth will be shown as half the real voltage. Also it is important to measure the voltage but think in terms of current—since the current spike on a ground rail generates crosstalk and spurious switching. And we could go on; there are a myriad of signal integrity challenges from intermittent setup and hold violations, resulting in problems ranging from metastability to electromagnetically induced crosstalk. The ability to foresee signal integrity problems and how to avoid them is fundamental to this book.

A feature of this book is the blend of source material. Whereas a theoretical text on signal integrity is built on scientific laws and notable hypotheses, this book has sourced its applications from the authors' professional experiences, published papers, and the work of associates. This book is a blend of source material coherently assembled and expanded to provide an understanding of modern signal integrity applications. Practical issues concern us most in this book. Each chapter focuses on a day-to-day activity of the signal integrity engineer, giving advice and illustrations from the industry. Practicality forms the central theme throughout the book.

The twenty-first century is a digital era of media convergence where mobile telephony, computing, and digital broadcasting merge, and consumers expect the media to be transparent to the technology. Put simply, the sports fan expects to view an event, in real time, on his or her mobile phone, laptop, and personal music player or digital TV at a reasonable cost and with absolute reliability. We could have taken any number of other examples from industry, medicine, or the military. Today, the digital designer or maintenance engineer is expected to be accomplished at signal integrity engineering, which is at the heart of the provision of systems that will make tomorrow's innovative technologies happen. This book reflects this trend.

THE INTENDED AUDIENCE

Signal integrity engineering is a young and evolving science where few who proclaim to know it all. Writing this book has been a journey of discovery for the authors, and we have every reason to believe it will be a rewarding journey for you. We have no doubt that some topics discussed in this book will provoke debate as there is much to be standardized in this branch of engineering. However, indisputable principles are presented in this book that underlie signal integrity engineering. These principles give the emergent engineer a basis on which to build the knowledge and understanding necessary for good signal integrity engineering. Therefore, this book is recommended reading for the student signal integrity engineer and the practicing engineer whereby the authors present a wealth of applications that illustrate good practice and show the development, test, and validation of modern digital systems.

While the book is naturally partitioned into chapters of diverse topics a common thread runs throughout the book. Each chapter provides a guide for the reader by presenting the necessary prerequisites of a topic before detailing complex design or test applications. Consequently the experienced engineer can approach a topic by stepping through the beginning of a chapter and concentrating on the detail in the applications and advanced topics. No signal integrity book claims to be all encompassing, and this book is no exception. You may need to consolidate your understanding of the theory of signal integrity engineering via in-depth theoretical texts in the Prentice Hall SI series. Nevertheless, much of this book is self-contained in terms of addressing a wide audience in signal integrity engineering.

HOW THIS BOOK IS ORGANIZED

To guide you through the full life cycle of embedded system design, including specification, simulation, test, and measurement, the book is structured, where possible, chronologically to follow the development cycle. However to encompass the diverse aspects of signal integrity engineering and to provide a coherent thread as you read, chapter order is a compromise of product life cycle flow and a natural grouping of signal integrity engineering topics. Therefore both compliance and serial bus simulation are found toward the latter part of the book, whereas earlier topics are prerequisites for these more advanced subjects.

Chapter 1: Introduction: An Engineer's Companion

Chapter 1 takes you into the world of device and circuit simulation, which is a major phase in the successful development of a modern digital product. A thought-provoking example is given whereby a designer is under the intense pressure of time to market where it's easy to overlook a true understanding of operating margins—will a network continue to function reliably over the range of manufacturing and operating conditions it will encounter during the useful life of the product? What are the expected primary failure mechanisms, and how do they interact with one another? Some of these complex simulation questions will be considered in the body of the introduction, but more to the point, these discussions lead the way to the in-depth chapters that examine these concerns.

Following simulation, the introduction describes in detail a number of principal innovations in signal integrity engineering test and measurement. For example, to overcome some of the traditional signal integrity engineering problems, device manufacturers currently use novel integrated signal processing functions within device drivers and receivers to apply signal pre-emphasis and equalization. However, incorrectly applied pre-emphasis generates unwanted overshoot, crosstalk, and noise. This is one illustration of the complexity in solving today's signal integrity problems. This chapter throws light on such issues and presents a pathway for the solution of such problems by describing the basics of eye diagrams, which form the basis of many of today's automated compliance tests and signal analysis measurements.

Chapter 2: Chip-to-Chip Timing and Simulation

This chapter covers the circuits used to store information in a CMOS state machine and how they fail. A set of SPICE simulations and spreadsheet budgets introduces the common-clock architecture, the first of three paradigms for transferring digital signals between chips. Even though the source-synchronous and high-speed serial paradigms are more prevalent in contemporary systems, the common-clock architecture is not dead yet. A solid approach for timing common-clock transfers is a useful thing to have in the toolbox.

IO circuits play a pivotal role in signal integrity engineering, yet we seldom get to lay our eyes on a schematic for one of them. A handful of CMOS IO circuits get used time and again, and Chapter 2 examines their pertinent electrical characteristics. The chapter also covers the assumptions we make when using behavioral models for these circuits. Studying these circuits provides a basis for understanding the more esoteric circuits. This chapter and others make repeated references to the accuracy and quality of the models we use in signal integrity simulations.

Chapter 3: Signal Path Analysis as an Aid to Signal Integrity

This chapter describes signal path analysis based on intuitive time-domain reflectometry (TDR) techniques. TDR measurement theory and its application are described in detail, given that TDR is fundamental to the understanding of signal integrity effects, such as impedance mismatches and circuit board (PCB) issues. In particular, it provides an ideal vehicle for illustrating some principal signal integrity challenges and their solutions. Another facet of this chapter is the introduction to Vector Network Analysis (VNA), which is an important frequency domain measurement methodology used, among other things, to accurately characterize high-speed signal paths. A particular feature of this chapter is the introduction to the design, development, and test of Low Voltage Differential Signaling (LVDS) signal paths. However an understanding of basic transmission line theory underpins good practice in signal path design.

Today, with high-speed digital signal transmission, even the shortest passive PCB trace can exhibit transmission line effects. Transmission line theory encompasses electromagnetic field concepts and generally attracts complex mathematical analysis. However, using TDR, leads intuitively to a basic understanding of transmission line theory, even though some of the basic concepts require a few simple calculations.

Chapter 4: DDR2 Case Study

The DDR2 case study tackles the million-dollar question for a common source-synchronous bus: Will the interface operate with positive timing margin over the lifetime of the product without incurring the high costs associated with excessive conservatism? This approach involves picking the interface apart piece by piece—understanding how many mV of crosstalk a DIMM connector generates and how many ps of eye closure go along with it. Under the pressure of a project schedule, it is often tempting to gather a set of models, construct a simulation, and be done with the exercise. This chapter challenges the reader to take a deeper look.

Chapter 5: Real-Time Measurements: Probing

Central to the book and this chapter is the challenge of data acquisition whereby the problematic issues of the ideal nonintrusive probe are examined. How is signal fidelity achieved in a modern signal integrity measurement, and how are unwanted measurement artifacts avoided? Analog signal measurement is carefully investigated to demonstrate the importance of correctly connecting to a system under test because the analog features of a high-speed digital signal determine signal integrity. The chapter provides practical advice on probing and how to avoid probe problems. Special attention is given to the probing of today's fast-edge signals and Low Voltage Differential Signaling (LVDS).

Chapter 6: Testing and Debugging: Oscilloscopes and Logic Analyzers

This chapter reviews modern signal integrity testing from both an analog and digital viewpoint, since at the high frequencies encountered in today's designs the two are inextricably linked. The emphasis is on frequencies over 1 GHz, where the measurement tools of choice are the digital oscilloscope and the logic analyzer. The chapter provides practical examples to show how detailed observation of both analog and digital waveforms, side by side, can provide the data necessary for the understanding of the most challenging signal integrity timing budget issues, such as setup and hold violations, data skew, and metastable states. Real-world illustrations show these problems and how they can be detected and debugged.

Chapter 7: Replicating Real-World Signals with Signal Sources

A core theme throughout signal integrity engineering is the behavioral analysis of a digital circuit in terms of its analog properties and notably how the behavior of a digital circuit is determined. Sections of this book are devoted to the methods used to simulate models where computer-generated outputs show signals and data in a variety of formats in response to an array of simulated inputs. However, the fundamental method used to determine the real-time characteristics and operation of an actual circuit, or prototype, is to externally control and observe the circuit or device under test. Most of this book is about signal acquisition and measurement, and this chapter provides the balance; it is about the other half of the story—the signal source—which is used to control a circuit. Put simply, this chapter is about the externally provided signal that is used as a real-time stimulus for electronic measurements. This chapter describes and demystifies the complex issues of modern signal sources and shows how they can be used to stress a digital circuit to expose signal integrity faults.

Chapter 8: Signal Analysis and Compliance

The proliferation of digital systems, such as the new high-speed buses, has created numerous interoperability and compliance standards. This chapter explains how real-time test and measurement is the cornerstone of compliance testing. It describes the various standards, with particular emphasis on high-speed serial buses, and shows why, at frequencies of more than 2.5 GHz, real-time test and measurement is the only way to achieve compliance. The practical use of logic analyzers and oscilloscopes for compliance validation is examined, whereby techniques such as automated eye diagrams and statistical analysis are discussed in detail. This chapter is essential reading for the signal integrity engineer who needs to understand the challenges of meeting regulatory compliance tests.

Chapter 9: PCI Express Case Study

High-Speed Serial (HSS) is the last of the three major paradigms for transferring data between chips, and PCI Express is an excellent example of a high-speed serial interface. The PCI Special Interest Group published a set of guidelines that will keep designs out of trouble. However, situations often arise that force a designer to depart from the well-traveled path, either by breaking one of the guidelines or by trading one off for another. In these cases, it is helpful to acquire an understanding of how much each interconnect component contributes to the jitter budget and how many picoseconds remain after accounting for all relevant effects. Starting with a set of models and design rules, Chapter 9 examines the characteristics of each component in the time and frequency domains and then combines them one at a time to arrive at a total jitter budget.

Chapter 10: The Wireless Signal

The success of cellular technology and wireless data networks has caused the cost of basic radio frequency (RF) components to plummet. This has enabled manufacturers outside the traditional military and communications markets to embed relatively complex RF devices into all sorts of commodity products. RF transmitters have become so pervasive that they can be found in any number of embedded systems. Therefore this book introduces RF test and measurement for completeness in the understanding of signal integrity engineering. Moreover, given the challenge of characterizing the behavior of today's high-speed logic devices, this chapter provides an understanding of how radio frequency parameters such as jitter are measured. Although this chapter provides a detailed decision of the real-time spectrum analyzer (RTSA), the chapter also offers a detailed introduction to the Swept Spectrum Analyzer (SA) and the Vector Signal Analyzer (VSA).

THE WEBSITE THAT ACCOMPANIES THIS BOOK, WWW.INFORMIT.COM/TITLE/0131860062

Color Pictures and Illustrations

Illustrations and pictures are used throughout a number of chapters in this book to allow the reader to become involved with instrumentation applications. In particular the chapters describing test and measurement use logic analyzer, oscilloscope, and spectrum analyzer displays to quantify and simplify what would otherwise be difficult and wordy descriptions. However, in keeping with most other books, the illustrations are understandable in monochrome, but some of the detail or features of a picture can be lost. Therefore, downloading the figure files from the website, www.informit.com/title/0131860062, allows the reader to view the descriptive color images and relate them to the accompanying text contained in the following three chapters:

- Chapter 6, "Testing and Debugging: Oscilloscopes and Logic Analyzers"
- Chapter 8, "Signal Analysis and Compliance"
- Chapter 10, "The Wireless Signal"

Simulation Models

Chapters 2, 4, and 9 demonstrate the allocation of picoseconds using case studies of three interface paradigms: common clock, source synchronous, and high-speed serial.

The model kit for Chapter 2 includes SPICE transistor models for an ancient 3.3 V 0.5 um CMOS process, IO circuits, and some simple networks.

Chapter 4 features behavioral simulation of standard DDR2 IO circuits, lossy transmission lines, vias, and a DIMM connector. Although the chapter discusses DIMM connector crosstalk, the model kit only contains single-line models because the DIMM connector model is proprietary.

The PCI Express case study in Chapter 9 demands more accurate interconnect models: coupled s-parameter representation of a ball-grid array, the corresponding via field, and edge connector. The simulations utilize a simple 100 ohm de-emphasized driver model found in the Agilent ADS library. This model is not included in the kit.

All models are suitable for simulation in SPICE or a behavioral simulator.

Acknowledgments

As executive editor of this book, Peter Bush has made a unique and extensive contribution to its preparation and production. Peter has a B.Sc. in Physics from Leeds University, and after graduation he became a development engineer for Kodak. A career change to technical authoring led Peter to the position of Executive Editor for the journal *Electronics Power* and later the prestigious post of Managing Editor for the Institution of Electrical Engineers (IEE). Consequently, the authors were delighted when Peter agreed to take a significant role in the research and preparation of this book. Currently, Peter is the Managing Director of his public relations company Peter Bush Communications and is renowned for anonymously authoring distinguished articles and whitepapers on electronic instrumentation and its application. Above all, both David and Geoff are exceptionally grateful to Peter for his personal encouragement and professional thoroughness throughout the authoring of this book.

David and his colleagues at Tektronix contributed to a large part of this book, which would not have been written without their help and guidance. In particular, he would like to thank Dave Fink - MAPL, Mark Briscoe - VSPL, Mike Juliana - LAPL, Dmitry Smolyansky - EOPL, Bob Buxton - SSPL, Jon Mees and

Trevor Smith - EMEA Marketing, and John Marrinan - UKAE for their individual contributions, and any individual he has inadvertently overlooked. And, of course, he thanks the extensive community of Tektronix people in sales, marketing, and the product lines who freely gave of their time in the research and preparation of this book.

Special thanks are given to Bob Blake of Altera for his expert guidance and his significant contribution to the introduction of this book, and to Altera for generously giving copyright permission to publish their pre-emphasis and equalization illustrations.

Greg has been fortunate in that the path of his career has crossed those of many other talented engineers from whom he has learned so much. He is particularly grateful to Gene Sluss, who mentored him in the art of integrated circuit design at Supercomputer Systems, Inc. Greg is also indebted to Dale Becker and Wayne Vlasak, who guided his career at IBM and freely shared with him from their own experiences as engineers. Digital Equipment Corp. has a rich tradition of technical excellence, and Bob Haller passed some of it along to Greg during his two years there. Amanda Mikhail is a trusted colleague who has set an example of dedication and excellence for those around her.

Geoff, on his own behalf, and the authors profoundly thank the editorial team at Prentice Hall—especially Bernard Goodwin and Michelle Housley—for their help and patience on what was at times a rocky voyage and journey of discovery. Geoff also takes this opportunity to thank those who inspired his passion for engineering at school, university, and the workplace, who are too numerous to mention. However, he must thank his colleagues and students at Buckinghamshire New University for their encouragement and advice—in particular the Head of School, Peter Harding; also, Dawn and Jean at Peter Bush Communications, who were always cheerful and helpful.

We are deeply indebted to the reviewers from Honeywell, Intel, and Agilent, among others, who gave expert guidance, appraisals, and critiques. Wherever possible we have heeded their advice and amended the manuscript; however, any errors that remain are the sole responsibility of the authors.

This has been a joint authorship and the authors cannot, understandably, thank all those who have contributed to this book, and we apologize to anyone that we omitted to thank. We agreed to limit our acknowledgments to our professional acquaintances and colleagues, but we are all deeply indebted to our families for their patience and support without which this book would never have been written.

About the Authors

Dr. Geoffrey Lawday currently holds the Tektronix Chair in Measurement at Buckinghamshire New University where he teaches embedded system design and high performance computing in the School of Computing. Having gained a BSc in Physics and an MSc in Computer Engineering at Surrey University, he was awarded a PhD in Time-Frequency Signal Analysis from Brunel University. His research in signal integrity engineering is reflected in his publications, such as the critique on the introduction of the new serial buses published in the flagship journal of the Institution of Electrical Engineers.

David Ireland has more than thirty years experience in test and measurement ranging from an engineering apprenticeship with Racal, where he gained his formal electronic engineering qualifications, to his current position at Tektronix, where he is the marketing manager of design and manufacturing at Tektronix Europe. He is widely recognized by embedded system engineers in Europe for his signal integrity articles and collaborative workshops on high-speed digital system design, test, and measurement.

 **Greg Edlund**'s career in signal integrity began in 1988 at Supercomputer Systems, Inc., where he simulated and measured timing characteristics of bipolar embedded RAMs used in the computer's vector registers. Since then, he has participated in the development and testing of nine other high-performance computing platforms for Cray Research, Inc., Digital Equipment Corp., and IBM Corp. He has had the good fortune of learning from many talented engineers while focusing his attention on modeling, simulation, and measurement of IO circuits and interconnect components. A solid physical foundation and practical engineering experience combine to form a valuable perspective on optimizing performance, reliability, and cost.

1

Introduction: An Engineer's Companion

An engineer's companion is like any other companion: It's a fellow traveler and colleague who offers advice and support, sharing experiences with a friend, in what would otherwise be a solitary journey. Our journey will take us through the endeavors of embedded system design, simulation, prototype development, and test. The dangers are signal reflections, attenuation, crosstalk, unwanted ground currents, timing errors, electromagnetic radiation, and a host of other signal integrity (SI) issues. SI engineering is a relatively new branch of electronics engineering. For the most part, it relates to the analog factors that affect both the performance and reliability of modern high-speed digital signals and systems. In general, integrity has to do with truthfulness. When applied to digital electronics, such as communication and computer systems, it is specifically about signal accuracy and system reliability.

Although it is written for SI engineering professionals, this book is intended to support new engineers and students who have an interest in designing, simulating, and developing modern high-performance digital and embedded systems. Along with the central theme of how to think about SI engineering, this book includes practical guidance on how to achieve and interpret a simulation or real-time test and measurement. In many jobs within the SI industry, a technician,

engineer, or designer could be anyone who thinks about the reliability or operation of a modern embedded system. This book aims to address the concerns and
uncertainties faced by these people. Because this book was written with a wide
audience in mind, each topic is presented with a prerequisite theoretical or practical preamble that supports novice engineers and students and that can often be
omitted by practicing engineers. This chapter keeps with the format of this book,
in that it follows the development of an embedded system, from its simulation to
prototyping and real-time test.

We live in the digital age, in which providers of telephony, computing, and
broadcast systems are busily facilitating media convergence. Music, video, and
information systems must be transparent to the newly integrated communication
and computing systems. Consequently, consumers expect their modern high-
performance telephony and computing systems to interactively communicate the
latest news and entertainment while providing e-business transactions. Driving
forces such as media convergence are challenging digital designers to work in an
era of reasonably priced, high-performance, highly dependable digital systems
that typically are portable and generally are required to have worldwide compliance. Today the issues of interoperability are paramount where modern systems
and components from disparate manufacturers are often required to work together
seamlessly. Moreover it's widely documented that communication and computing
systems double in computational throughput every eighteen months. This implies
that any absolute frequency or data rate quoted in any SI book will be out of date
by the time the book is published—and this book is no exception. Nevertheless, it
is anticipated that, similar to most of the books in this SI engineering series, the
fundamental practical examples, guidance, and underlying theoretical concepts
will remain relevant for many years to come. Today's cutting-edge digital interfaces will become the bread and butter of tomorrow's digital systems.

1.1 LIFE CYCLE: THE MOTIVATION TO DEVELOP A SIMULATION STRATEGY

Most signal integrity engineers would agree that the primary motivation for simulating chip-to-chip networks is to maximize the probability that those networks
will function flawlessly on first power-up. Another compelling motivation is easy
to overlook under the intense pressure of time to market: understanding operating
margins (see Figure 1-1). It is tempting to stitch together IO circuit and interconnect models, run the simulations, check the results, and be done with the exercise.
This may prove that the network will function under a given set of conditions, but
will it continue to function reliably over the range of manufacturing and operating
conditions it will encounter during the product's useful life? What are the
expected primary failure mechanisms, and how do they interact with one another?

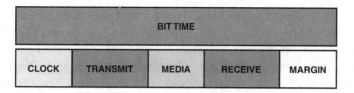

Figure 1-1   Operating margin.

These questions are indeed of primary importance, and it may be possible to answer them given unlimited resources and time. Unfortunately, most signal integrity engineers operate under somewhat different conditions. A contemporary PC board design may have upwards of a thousand nets that belong to two or three dozen different buses. The power spectrum will likely have significant content above 5 GHz. Supplying power and cooling to high-performance processors can place challenging constraints on layout and routing. On top of these technical challenges, the customer may require that the product be ready for manufacturing in a time period that severely stresses the team's ability to carry out the level of analysis required to ensure reliable operation. This is the irony of the business: the relevant physical effects become ever more difficult and expensive to analyze, while the market relentlessly exerts downward pressure on cost and schedule. These two freight trains are running full speed toward each other on the same track.

Given these technical and business challenges, is it still possible to achieve the goal of reliable operation of a system filled with dozens of digital IO buses over the product's lifetime? At times it may appear that the solution to this difficult problem is the empty set. In the heat of battle, the level of complexity can be so overwhelming that it seems impossible to satisfy all the constraints simultaneously. Nevertheless, it is the opinion of the authors that it is possible to successfully manage the signal integrity of a complex contemporary design if the lead engineers keep two important principles in mind. First, the signal integrity engineer must be involved at the very beginning of the design cycle, when the team is making critical architectural decisions and selecting component technology. Second, the team must develop a comprehensive simulation and measurement strategy that applies the appropriate level of analysis to each bus in the system.

1.1.1   The Benefits of Early Teamwork

It must be tempting for those who make weighty architectural decisions during the earliest stages of a new product to avoid addressing implementation details. After all, it is easier to define a product while cruising along at the top of the troposphere. The product must offer innovative, distinguishing features. It must also offer Earth-shattering performance at a lower cost than the competition. Most important, it must be ready for manufacturing before the infamous marketing window closes. If this is a consumer product, this means having the product shrink-wrapped and on the shelves for Christmas shoppers. These are all difficult goals to achieve, even without having to worry about how to manufacture the product and make it reliable.

As any experienced engineer will attest, usually a price must be paid for making architectural decisions without input from those whose job it is to implement the architecture. Probably the worst possible scenario is a product that is marginally functional, because the marginal part does not become apparent until production is in full swing. This is even more treacherous than a product that overruns its budget, misses its milestones, and never makes it to market. Assembly lines come down. Companies—more than one—lose large quantities of money each day. There may be recalls. There will certainly be redesigns under intense pressure. At the end of the whole experience lies a painful loss of reputation. This is certainly a scenario that no Vice President of Technology or Chief Financial Officer would choose to put in motion if the choice were made clear. One thing is crystal clear: the company greatly enhances it chances of success by building a cross-disciplinary team in the project's early architectural phases. A minimum team would include a board layout designer, firmware programmer, and engineers from the disciplines of logic, mechanical, thermal, power, manufacturing, electromagnetic compliance, and signal integrity. A large company may have separate engineers to represent each of these disciplines, while in most other companies one person wears several hats. In either case, it is important that each discipline be represented on the team.

As an illustration of the importance of early teamwork, consider the following fictional scenario that is close enough to reality to be disturbing. Project X took off like a rocket from the very beginning. Conceived in a boardroom behind closed doors, it was already well-defined before anyone in development engineering heard of it. The senior engineering staff voiced their opinions about the unrealistic schedule, especially during a time when other high-profile projects were consuming most of the company's resources. However, commitments to customers had already been made, a marketing plan was in the works, finances were allocated, and the wheels were in motion—the first domino in a sequence of related events.

The second domino was in place before PC board placement and routing began. The Vice President of Engineering had defined a budget that was consistent with the price point that Marketing deemed competitive. This budget called for a four-layer PC board: signal-ground-power-signal. Upon seeing the form factor for the first time, the PC board designer expressed concern about routability in certain areas he perceived to be bottlenecks. The lead signal integrity engineer expressed concern about high forward crosstalk due to microstrip transmission lines and high edge rates in the PCI Express (PCIe) signals. However, neither the board designer nor the signal integrity engineer could prove that a solution did not exist, so the design progressed as planned.

An exceptionally aggressive schedule became domino number three. To meet the schedule, the board design shop had two of their top designers work back-to-back twelve-hour shifts for three weeks, with one day off on the weekends. While it is true that an auto-router can save many hours of human labor, there is no substitute for an experienced designer when channels are fully allocated. The company's design process called for routing constraints to be in place before routing could begin. Again, schedule won the day. Routing began before the signal integrity team could assign constraints to the 800 nets on the board that required attention out of a total of 1,000 nets.

In a remarkable feat of skill and sweat, the design team generated Gerber files on schedule. Much to their credit, they reserved one day for a complete design review, with mandatory attendance by everyone on the team. They even invited a few seasoned veterans who had since gone on to jobs in other parts of the company. One of these veterans spotted the problem: a PCIe differential pair routed next to a Gigabit Media Independent Interface (GMII) clock signal. The edge rate of the 2.5 gigabits per second (Gbps) PCIe signal was clearly defined in the specification as 10 V/ns. The edge rate of the 125 MHz GMII clock was only 1 V/ns, making it vulnerable to crosstalk from aggressors with higher edge rates, but this information was unavailable in the component datasheet. The signal integrity engineer took an action item to acquire an IBIS datasheet for both the PHY chip and the IO controller. She was successful for the PHY chip, but the vendor for the IO controller required the company to sign a nondisclosure agreement before releasing the IBIS datasheet. This legal process took much longer than the one day allocated for reviewing the design, and the team sent the Gerber files to the PC board manufacturer. Domino number four.

The first set of boards came back from assembly and performed admirably on the benchtop. The software team did their job of loading new code, debugging, recompiling, and loading again. The hardware team did their job of measuring thermal characteristics, calculating power draw, dumping registers, and capturing traces for bus transactions on the logic analyzer. Knowing that there was exposure to crosstalk problems, the signal integrity engineer spent a lot of time probing the board, looking for them. She found some quiet line noise on the order of 200 to 300 mV, but the nets had enough noise and timing margin to tolerate it. After several weeks of intense work, the development team gave management the green light for production, and each team member spent a few days contributing to the final report before moving on to the next project.

Unbeknownst to the heroes of our story, the fifth and final domino was starting to fall at a semiconductor manufacturing plant on the other side of the planet. Although the IO controller chip had been in production for two years and the process had remained stable for most of this time, a recent drop in yield prompted the process engineers to make some tweaks that ultimately resulted in slightly lower edge rates. The IBIS datasheet for the IO controller contained edge rate information, but the component datasheet did not. The manufacturer did not feel a need to notify its customers, because all specified parameters remained within their limits.

Shortly after these new components hit the assembly floor, the boards began experiencing high levels of fallout in the form of intermittent failures that were associated with traffic on the PCI Express and Gigabit Ethernet buses. When a few of the boards made it to customer installations, the crisis was officially under way. Management told the team to drop what they were doing and focus on resolving the crisis. One sleepless night, the neurons in the brain of the signal integrity engineer rewired themselves to remind her of the comment during the design review about crosstalk between PCI Express data and the GMII clock. The next day she decided to probe these signals, using a logic analyzer to trigger the oscilloscope when both were active at the same time. This did not happen often, because the two buses were asynchronous to each other. After three days of persistent testing she captured a waveform that showed a slope reversal—double clocking—right in the threshold region of the GMII clock receiver on the PHY chip and coincident with a transaction on the PCI Express bus.

Not many pleasant alternatives presented themselves to the team. It was not possible to slow down a PCI Express signal and still expect the bus to function. While an FET switch would sharpen the edge of the GMII clock signal, the IO controller was in a BGA package, and there was no way to install the FET switch between the pin and the net. In the end, it was decided to stop production and rush a new six-layer PC board through design.

During the lull between releasing the new design and shipping the new boards, upper management held an all-day process review. The veteran engineer from the original design review retraced the trail of circumstances that led to the failure. First and most important, the signal integrity team did not have a representative at the table when the architecture was being defined. He pointed out that a cross-disciplinary design team is the cornerstone of a healthy, solid design process. Second, there was no cost-performance analysis of the PC board stack-up. This is admittedly one of most difficult things any engineer has to do. There was no excuse for the third contribution to the failure; schedule should never trump assigning design constraints unless the risk of nonfunctional hardware is deemed acceptable. Item number four was an unavailable model. This is understandable, because high-quality models are hard to come by, but it is possible to have the necessary models when they are needed by planning ahead. Finally, if a design is sensitive to edge rate, the component specification should call out edge rate as a parameter, because it is not possible to anticipate the evolution of a silicon fabrication process.

The recommendations that came from the design process review were to establish a cross-disciplinary team for all new designs and to develop a comprehensive process for applying the appropriate level of analysis to each net in a PC board or system design.

Most of the time, circumstances do not present themselves to us in such an obvious, logical progression. Only careful retrospection reveals the sequence of events that led to a particular conclusion. To some degree, the engineer's job is to play the role of the prophet who can foresee these circumstances and avoid them without becoming the Jeremiah to whom nobody pays much attention.

1.1.2   Defining the Boundaries of Simulation Space

Avoiding situations like the one just described requires a sound understanding of the physics involved. You also need the insight to know what level of analysis is required for a given net or bus. Although it is certainly possible to acquire models for a thousand nets and simulate every one of them before releasing a design to manufacturing, the company that practices such a philosophy may not be in business very long. It would appear that one of the more critical tasks of designing any digital IO interface is establishing the boundaries of simulation space. These are the criteria you use to decide whether a net needs to be simulated or whether some other method of analysis is more appropriate. Make no mistake: simulation is expensive and should be used only when there are strong economic and technical motivations for doing so. Once this critical question of whether to simulate is answered, you can go about the tasks of actually running the simulations and interpreting their results.

An excellent way to begin the decision-making process is to compile a comprehensive list of all nets in the design and some relevant information associated with each bus. This begs for a definition of a bus, which is a word that is frequently used but seldom defined. For the purposes of this discussion, a bus is defined as a collection of data and control signals that have a common functional purpose and are synchronized to the same clock or strobe signal.

One example of a bus is the traditional 33 MHz PCI bus. It is composed of 32 address-data signals and a set of control signals, each synchronized by a common clock signal that originates from a clock source chip. DDR memory is a source-synchronous bus in which the transmitting chip sends a clock or strobe signal along with the data. The source synchronous bus facilitates faster data rates by eliminating skew between multiple copies of the same clock and transmitter launch time from the timing budget (see Chapter 4, "DDR2 Case Study"). PCI Express is another example of a bus, although it is not necessary for each chip on a given PCI Express bus to share the same reference clock. PCI Express uses a clock-data recovery circuit. This means the receiving chip uses the same low-frequency reference clock as the transmitting chip. It boosts the clock to the data rate and infers the optimum sample point from the incoming data stream.

The analysis decision matrix shown in Table 1-1 allows the signal integrity engineer to view all the relevant electrical parameters of each bus at the same time. The engineer also can decide what level of analysis is necessary to ensure that each bus functions reliably over the product's lifetime. The simplest case might be a bus that a trusted colleague has analyzed in the past and that others have used successfully time and again. In this case no simulation is required—*provided* that all of the bus's electrical parameters are identical to the ones that were analyzed in the past. The next, more complicated case is the bus for which a designers' guide or specification exists. If a third party analyzed the bus and published a set of rules that, when followed, guarantee sufficient operating margins, simulation is not necessary. The job of the signal engineer defaults to describing design constraints to the CAD system and checking that they are met. Of course, the reliability of the source must be beyond reproach! Some buses may not require simulation but do require rudimentary hand calculation, such as the value of termination resistors, stub length as a function of rise time, or RC time constant of a heavily loaded reset net. Finally, if a bus passes through each of these three filters, it is time to assemble the models and fire up the simulator. The closer this process occurs to the beginning of the project, the higher the likelihood of success. The bus parameter spreadsheet should include the following items:

Table 1-1   Analysis Decision Matrix

Parameter	I2C	PCI-X	DDR2	PCIe	Units
Engineer					
Net count					
Data rate					Gbps
IO power supply voltage					V
IO circuit technology					
Input setup time					ps
Input hold time					ps
Input minimum edge rate					V/ns
Input high threshold					V
Input low threshold					V

Parameter	I2C	PCI-X	DDR2	PCIe	Units
Output rise time					ps
Output fall time					ps
Output maximum edge rate					V/ns
Output impedance					ohm
Output high level					V
Output low level					V
Pin capacitance					pF
System clock skew and jitter					ps
Net characteristic impedance					ohm
Termination					ohm
Maximum net length					in.
Number of loads					

It's helpful if the signal integrity engineer and the person who draws the schematics can agree on a naming convention that involves adding a prefix to the net name of each net in a bus. This will facilitate tracking coverage of all nets in a design. Someone who is good with programming can write a simple script that sorts and counts the nets from the "all nets" file, the best source of which is either the schematic entry or layout CAD tool. The goal is to make a decision about each net in each bus in the system: what level of analysis does it require? You can then keep track of the number of nets simulated and constrained and have an up-to-date measure of how close a project is to completion. Totaling the net count column will give you an excellent rough estimate of the work involved at the beginning of the project. Keeping track of who is analyzing each bus prevents unpleasant revelations toward the end of a project, such as "I thought so-and-so was working on that bus!"

1.2   PROTOTYPING: INTERCONNECTING HIGH-SPEED DIGITAL SIGNALS

Traditionally the professional engineer strives to design high-performance digital and embedded systems within tight time-to-market constraints, cost limitations, and quality demands while managing manufacturability requirements. Alongside the traditional concerns, the key challenge facing today's designer is the task of maintaining signal integrity in a modern high-performance digital system. A contemporary example of the changing landscape of embedded system design is the support given to the SI engineer by device manufacturers that integrate some ingenious circuitry within their devices to minimize or circumvent a number of SI issues. A particular case in point is the interplay between the SI engineer and modern programmable logic device manufacturers. The inclusion of programmable pre-emphasis, deskewing, edge rate control, and equalization provides a range of solutions to a number of key SI concerns.

As system speeds increase and timing budgets decrease, there is less time for switching between logic levels. Consequently, digital signal edges become faster, which results in the need for rigorous design requirements if signal integrity is to be maintained. High-performance digital systems are prone to two fundamental sources of signal degradation:

- Digital degradation that is timing-related, such as synchronization and setup and hold violations, which often generate metastability or race conditions that typically produce erratic system behavior
- Analog degradation, such as indeterminate signal amplitudes, power supply and ground variations, glitches, signal overshoot, crosstalk, and unwanted noise, that generates a diversity of system malfunctions

Both of these phenomena typically have their origins in printed circuit board (PCB) design or signal termination, but there are a myriad of other causes. Not surprisingly, there is a high degree of interaction and interdependence among digital and analog signal integrity requirements. The analog aspects of digital system design tend to cause the most concern. High-speed signals transmitted along PCB tracks tend to suffer from high-frequency attenuation, which makes it difficult for a receiver to interpret the information. The effect is similar to a low-pass filter, which decreases a signal's high-frequency content.

The main causes of high frequency attenuation are PCB dielectric loss, which is a capacitive effect, and the skin effect, which limits the signal to the surface of a conductor. As the data rate increases or the edge rate becomes faster, the signal frequency increases, and the dielectric loss becomes the dominant factor in high-frequency attenuation. The effect of dielectric loss is proportional to frequency, whereas the skin effect is proportional to the square root of the frequency. The skin effect describes how high-frequency currents tend to travel on the surface of a conductor, rather than on the whole cross section of the conductor. This is caused by the conductor's self-inductance, which increases the inductive reactance with frequency, forcing the current to travel on the surface of the material. This reduces the effective conductive area of a PCB trace, increasing the trace's impedance, which causes the signal to be attenuated.

While other PCB anomalies such as poor termination can cause crosstalk and reflections, the problem of high-frequency signal loss is aggravated as signal frequencies increase and PCB tracts lengthen. For example, Figure 1-2 shows the particularly demanding case of a 40-inch backplane where high-frequency signals are transmitted through a PCB stripline that is constructed with FR4-type PCB material.

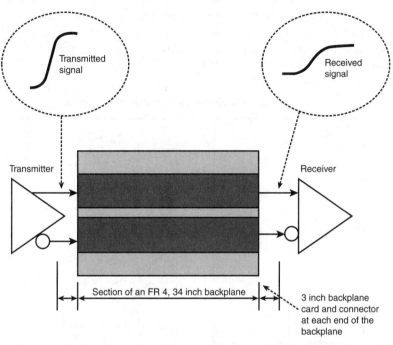

Figure 1-2 The particularly demanding case of a 40-inch backplane where the signals are transmitted through a PCB stripline that is constructed with FR4-type PCB material.

Figure 1-3 shows the dominant high-frequency dielectric loss effect in a 40-inch PCB stripline. The loss is caused by the capacitance formed by the trace and ground plane with a dielectric constructed with PCB-type FR4 materials. All PCB laminate materials have a specific dielectric constant, which will affect the impedance of the trace, especially at high frequencies, where the trace behaves as a transmission line. The value of a PCB dielectric constant is determined by comparing the capacitive effect of the PCB material to the capacitive effect of a conductive pair in a vacuum, where the vacuum has a dielectric constant of 1. In Figure 1-3, the FR4 material has a dielectric constant of about 4 to 4.7. A lower dielectric constant can allow a PCB to support a longer transmission line before the high-frequency losses become significant, but this is a simplification. Determining dielectric loss is a complex topic. Many materials are used as PCB laminates, and many have better propagation characteristics than FR4. However, the high-performance PCB becomes too expensive for large-volume, low-cost applications. Although the extensive length of the backplane used in this example has exaggerated the loss effects, Figure 1-3 clearly shows how the dielectric loss is directly proportional to an increase in signal frequency. The skin effect is somewhat constant at −10 dB throughout the frequency range 5 GHz to 10 GHz. Lossless transmission lines are a bit of a misnomer because they consider only impedance and timing. Attenuation is considered in a second-level approximation of the line.

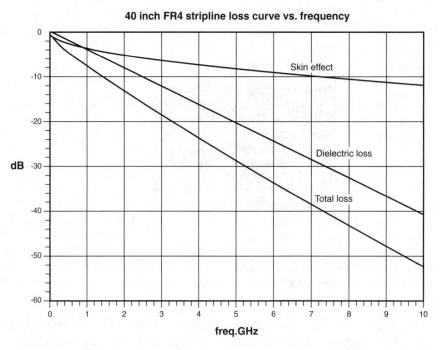

Figure 1-3   The dominant high-frequency dielectric loss effect in a 40-inch PCB stripline.

Both dielectric loss and skin effect can cause problems with intersymbol interference (ISI). The attenuation effectively prevents the signal from reaching its full amplitude within the required duration or its symbol time. As a result, the signal symbol, such as logic 1 or logic 0, spreads into the following symbol, mixing the symbols. The effect is pattern-dependent and is known as pattern-dependent jitter (PDJ) or data-dependent jitter (DDJ). If a string of data remains at the same level, such as a string of logic 0s, the energy in the signal has time to reach its peak, allowing the data to be transmitted and received correctly. However, for a quick switching signal, with alternating logic 1s and 0s, full signal strength is not reached within each symbol period. This causes the symbols to merge and the system to malfunction.

To maintain signal integrity in a modern digital system, differential signals and integrated signal processing functions within device drivers and receivers are becoming more common. Nonetheless, differential signals demand that designers pay special attention to PCB layout. Poorly designed differential traces and terminations can cause many of the signal integrity problems associated with conventional single-ended systems. Also, an intimate knowledge of signal pre-emphasis and equalization is necessary, because incorrectly applied pre-emphasis in effect generates unwanted overshoot, crosstalk, and noise.

1.2.1   The Effects of Increasing the Drive Signal

The simple solution to overcoming signal loss, or attenuation, is to increase the signal strength to overcome the attenuation. Unfortunately, increasing the signal strengths does not solve the problem of selective loss of high frequencies, or high-frequency roll-off. Also, the PDJ would deteriorate, because each symbol would be unable to achieve its full strength within its allotted time slot. Also, as a result of the increased signal level, each signal symbol will probably spread even further into the next symbol. Increasing the signal strength also affects the noise in the system, because noise increases proportionally with the increase in signal strength. What's more, the overall power consumption of the logic driver, or transceiver, also increases as the driver buffer increases the amount of current flowing into the PCB trace. Consequently, a simple increase in signal strength is not a solution to either the dielectric or skin-effect losses. In addition, increasing the signal strength may in fact make the situation worse.

1.3   PRE-EMPHASIS

Pre-emphasis is a way to boost only the signal's high-frequency components, while leaving the low-frequency components in their original state. Pre-emphasis operates by boosting the high-frequency energy every time a transition in the data occurs. The data edges contain the signal's high-frequency content. The signal edges deteriorate with the loss of the high-frequency signal components. A simple pre-emphasis circuit can be constructed from a two-tap finite impulse response (FIR) filter. The circuitry works by comparing the previously transmitted data bit to the current data bit, where the circuit block Z^{-1} provides the delay for a single data bit. If the two bits—the delayed bit and the current bit—are the same level, the current bit is transmitted at the normal level. If the two bits are different, the current bit is transmitted at a higher magnitude. Figure 1-4 shows the FIR filter block diagram and associated waveforms.

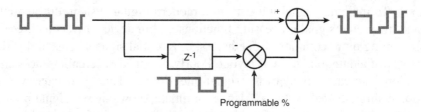

Figure 1-4   Block schematic of a FIR filter pre-emphasis circuit where the percentage of pre-emphasis is programmable.

The pre-emphasis circuit is primarily designed to overcome frequency-dependent attenuation.

1.3.1   Pre-emphasis Measurement

There are many different methods of measuring pre-emphasis. Although it is not important to follow a particular measurement method, it is important for the engineer to understand a particular definition when modeling his or her system. For example, Figure 1-5 shows the pre-emphasis measurement method used in an Altera programmable logic device. The waveform is part of a differential signal.

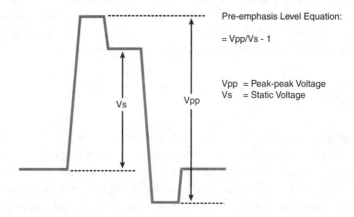

Figure 1-5   Waveform definition of pre-emphasis.

The pre-emphasis circuitry within a modern programmable device such as the Altera Stratix has an architecture that can be dynamically programmed to five different levels of pre-emphasis. The exact value of pre-emphasis cannot be pre-determined, because each device requires a percentage of pre-emphasis that is dependent on the output signal strength and transmission path characteristics. Table 1-2 shows five possible programmable pre-emphasis levels for a differential drive signal (V_{OD}) of 800 mV. In this case the internal termination is 100 ohms. The amount of pre-emphasis changes according to the transmission path parameters.

Table 1-2   Typical Percentage Pre-emphasis Levels for a Programmable Logic Device with an 800 mV Drive Signal (V_{OD})

Programmable Setting	Typical Pre-emphasis Level
1	11%
2	36%
3	65%
4	100%
5	140%

1.3.2   Receiver Equalization

An alternative to pre-emphasis is receiver equalization, which provides functionality in the receiver to help overcome the high-frequency signal losses of the transmission medium. Receiver equalization acts as a high-pass filter and amplifier to the data as it enters the receiver. In effect, equalization distorts the received data, correcting the distortion of the signal resulting from the high-frequency losses. This allows the receiver to rebuild the signal and interpret it successfully. External receiver equalization can be implemented with external filter networks. However, these filters require extra components with added PCB tracks and PCB stubs that require careful design if signal integrity problems are to be avoided. Furthermore, a fixed filter circuit is difficult to adapt for differing loss.

Some modern digital devices include equalization within the receiver. In a number of programmable logic devices, the equalization function is dynamically controllable. The equalization setting typically depends on the application and environment. For example, the receiver equalization would be up to 9 dB of gain for a 40-inch FR4 backplane. Moreover, data dispersion can be overcome when an equalizer is designed to cut off unwanted frequency components that spread symbols. The equalizer brings the symbols back into shape and time, thereby minimizing or eliminating PJD.

1.3.3   Maintaining Signal Integrity in Legacy Systems

The expansion of high-speed interfaces has led to some dilemmas because designers generally need to use legacy systems to support existing components or interfaces and reduce the cost of replacing an entire infrastructure. For example, this means that backplanes designed to operate at 1 Gbps are now required to run at 2.5 Gbps and faster to support existing and new components. In some circumstances this may be possible with the combined use of pre-emphasis and equalization. Equalization can compensate for many of the issues of the legacy backplane, such as narrow PCB tracking, which typically suffers from increased signal attenuation as transmission frequencies increase. However, equalization is relatively new as an integral part of logic receivers, so it is possible that legacy cards plugged into a system will not include equalization. This means that higher levels of pre-emphasis are required to ensure reliable communications. It is not uncommon for pre-emphasis levels in excess of 100% to be used in legacy applications. Normally each system and application require a unique setting for pre-emphasis, equalization, and drive strength. It is therefore important to model the entire data communication path using accurate model descriptions for the PCB interconnect and transceiver interfaces to ensure that the entire system is matched from within the driver to the internal receiver circuitry. In conclusion, it is clear that skin effect and dielectric loss can cause significant attenuation to the high-frequency content of signals. For example, this can impact the success of communicating high-speed data via a conventional FR4 PCB. The use of pre-emphasis and equalization can help the signal integrity of a transmission path, provided that you carefully select the parameters.

1.3.4   Simultaneous Switching Outputs

Although high-speed data rates correlate with high-frequency signals, it is the signal edge timing that has the most detrimental effect on signal integrity. This is particularly true as systems migrate to dense, highly integrated, high-speed switching systems where typically hundreds of pins are switching with edge rise and fall times that generally are faster than 500 picoseconds. A consequence of a large number of quickly switching device connections is the unstable power supply voltage, where greater power demanded in short time periods causes transient disturbances. Traditionally designers have decoupled power supplies to minimize transient charges and stabilize power sources. Nevertheless, new high-speed, high-density systems require very careful design to minimize power supply transients, or the result is a phenomenon called simultaneous switching noise (SSN).

Moreover, as digital circuitry increases in speed and output switching times decrease, higher transient currents occur within the device output circuits as the effective output load capacitors discharge. This sudden flow of current exits the device through internal inductances to a PCB ground plane. This causes a transient voltage to develop, which is a voltage difference between the device output and the board ground. This is known colloquially as ground bounce, but in actuality an unwanted signal return current causes transient variations in the ground voltage. The signal return currents or bounce effect can cause an output low ground signal to be seen as a high-output signal by other devices in the system. You can reduce unwanted signal return currents by following a number of classic design rules. Nonetheless, a number of programmable device manufacturers now provide pin slew rate control, which allows the designer to slow an edge rate and therefore reduce ground current transients or ground bounce effects. Additionally, most modern devices include multiple power and ground pins. This allows the designer to locate a high-speed input or output pin close to a ground pin to reduce the effects of simultaneous switching outputs (SSOs).

The challenges of high-speed design require some additional effort to ensure signal integrity. This can be achieved by following some simple analog design rules and by using careful PCB layout techniques. Nonetheless, contemporary integrated circuit manufacturers are providing many features to compensate for PCB anomalies and support high-speed design. Programmable slew rate control and programmable on-chip termination technology are helping make the designers' work somewhat easier. However, programmable pre-emphasis, equilisation, and slow rate control only work in stable systems, and they are not a substitute for good design practice.

1.4   THE NEED FOR REAL-TIME TEST AND MEASUREMENT

As data rates increase, it is ever more difficult to detect and debug noise and signal aberrations in a prototype or production model. A rigorous regime of signal integrity measurements can provide the means for the engineer to trace sources of noise or glitches and provide the wherewithal for him or her to eradicate the root causes of signal aberrations. Apart for the requirements of test and debug, there is often a need for a product to operate in a global market, and many designers are increasingly concerned with compliance measurements. Complying with industry standards ensures interoperability among system elements, where discrete system components from various manufacturers successfully interconnect and communicate. Compliance measurements usually entail a series of prescribed acquisition and analysis steps, which are carried out on a completed product. However, successful compliance design and testing often depend on eliminating signal integrity problems in the early phases of a design. Ideally the designer or test engineer locates signal integrity errors during the initial simulation of a product's development, as mentioned previously. However, a primary aim of the SI engineer must be to understand the role played by electronic bench instrumentation and portable test equipment in the pursuit of signal integrity. Today, the SI engineer needs to understand the latest methodologies that are used to achieve signal integrity. They are founded on the interrelationships between device simulation, circuit simulation, and real-time test and measurement. Apart from the fundamental theoretical concepts, or abstract mathematical models, the engineer must understand the practical issues that lie at the heart of signal integrity engineering. Remember that simulation is only a model, and a model is an imperfect replica of a real component or system. In many cases there is no substitute for a prototype and a real-time test phase, which at the very least will allow real-world data to be fed into the simulation.

One of the most demanding aspects of modern digital system design and debug is the successful measurement of the analog content of high-speed digital signals, where complex multilayer boards, the high density of interconnects, and highly integrated systems have made successful probing an exact science. Ideally the measurement system bandwidth, including that of the probe, should be at least three times the frequency of the signals to be observed. Above all, the edges of a digital signal must be considered when making a measurement. For instance, a standard 5 Gbps data rate serial bus signal requires at least a 15 GHz measurement bandwidth and a true differential probe that has a rise time of less than 35 picoseconds. Otherwise, the probe would at best fail to show any analog aberrations and at worse would simply stop the bus. Ultra-low loading and a diversity of attachment methods are needed to ensure fast, positive connection with minimal effect on signals.

Experienced engineers know that signal integrity is the result of constant vigilance during the design, simulation, and real-time test processes. It's all too easy for signal integrity problems to get compounded as a design evolves. An aberration that goes unnoticed in the early stages of a design can cause erratic behavior in a product, which can entail many hours of demanding test and debug to correct. An experienced engineer plans ahead and pays special attention to signal integrity issues in the early stages of a design. In particular, the engineer decides during the specification phase what signals need to be probed and how access is to be provided to signals of interest.

1.4.1   Timing Budgets and the Analog View

A signal integrity problem may first seem to be a misplaced digital pulse or timing error. However, the cause of the problem in a high-speed digital system can often be related to the analog characteristics of a digital signal. In many cases a digital problem can be fairly easy to pinpoint when an errant digital signal is successfully probed and the analog representation of the flawed digital signal is exposed. Analog characteristics can become digital faults when low-amplitude signals turn into incorrect logic states, or when slow rise times cause pulses to shift in time. An innovative test and measurement technique is to use a logic analyzer and oscilloscope in unison. The test technique is to show a digital pulse stream along with a simultaneous analog view of the same pulses, where both waveforms are shown on a single display, as shown in Figure 1-6. This technique is becoming a standard debug methodology, and the debug method is frequently the first step in tracking down an SI problem. Figure 1-6 is a real-time view of a digital data stream exhibiting a timing error. The analog view clearly shows the cause of the timing error as an amplitude aberration on the trailing edge of the digital pulse.

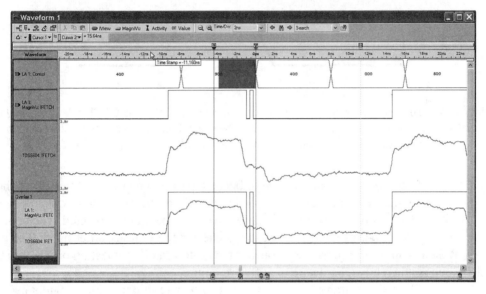

Figure 1-6   Digital pulse streams and a simultaneous analog view of the same pulses, where both waveforms are shown on a single display.

In any discussion of signal integrity, signal transitions deserve special attention. For example, the timing diagram shown in Figure 1-7 shows two digital inputs feeding an ordinary AND gate. The gray trace for Input A shows the correct pulse. Superimposed on it is an analog view, a distorted signal shown as a black trace, of the actual signal. Due to its slow rise time, the actual signal does not cross the required threshold value until much later than it should. Consequently, the output pulse from the AND gate is significantly narrower than it should be; the correct pulse width is shown in gray. The integrity of the signal on Input A is very poor, with serious consequences for the timing of digital elements elsewhere in the system. This type of SI problem typically causes timing errors in subsequent logic steps. Such dilemmas often require careful analysis for successful debugging.

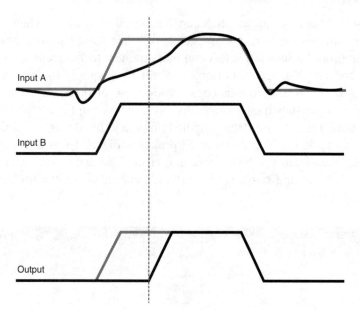

Figure 1-7 The gray traces have the required signal characteristics, and the black traces are the actual analog signals; this simple example is an AND gate. Clearly Input A has a slow rise time and crosses the logic threshold too late, which results in an output timing error.

Suppose the output went on to become part of a memory address. The short pulse might cause the memory to see logic 0 where logic 1 should exist and therefore select a different memory location than the one that is expected. The content of that location, of course, is inappropriate for the transaction at hand. The end result is an invalid transaction, raising the all-too-familiar question of whether the bug is inherently a hardware or software failure.

Slow signal transition edges can lead to intermittent system faults even if they are not causing repeatable errors. Timing budgets in high-speed digital systems allow very little time for signal rise and fall transitions. Setup and hold times have scientifically decreased in recent years. Modern electronic memory systems are a typical example of setup and hold times in the low hundreds of picoseconds. Slowly changing edges can leave too little margin in the timing budget for data transactions to be valid and stable, as implied in Figure 1-8. The relationships shown in Figure 1-8 are exaggerated to emphasize the concept. These two simple examples show some of the potential unwanted effects resulting from too slow an edge transition. The majority of SI problems in high-speed digital systems commonly are related to the effects of fast switching edges and their associated high-frequency signal content.

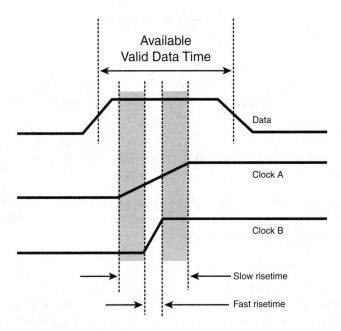

Figure 1-8   Although not to scale and exaggerated, Clock A is seen to clock the data with a slow rise time, leaving the system with greater susceptibility to noise, with the possibility of double clocking and metastability. The faster Clock B edge results in improved reliability, and the timing budget is improved.

One of the major timing concerns in SI engineering is the setup and hold timing values that are specified for clocked digital devices. Setup and hold timing are at the heart of application-specific integrated circuit (ASIC) functional verification measurements. Setup time is defined as how long the data must be in a stable and valid state before the clock edge occurs. Hold time is how long that data state must remain stable and valid after the clock edge. In the high-speed digital devices used for computing and communications, both setup and hold timing values may be as low as a few hundred picoseconds.

Transients, edge aberrations, glitches, and other intermittencies can cause setup and hold violations. Figure 1-9 is a typical setup and hold timing diagram. In this example, the data envelope is narrower than the clock. This emphasizes the fact that with today's high-speed logic, transition times and setup and hold values can be very brief, even when the cyclical rate or signal frequency is relatively slow.

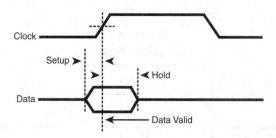

Figure 1-9   A typical setup and hold timing diagram.

There are three common approaches to evaluating a device's setup and hold performance. A number of other timing parameters can be verified using either a low-speed or high-speed functional test:

- Low-speed tests are a fairly coarse functional verification procedure, but they are often adequate. In some cases it is not necessary to take quantitative measurements of the actual setup and hold values that are specified on a device's data sheet. If a device can tolerate a broad clock placement range, the timing test may be as simple as running a low-speed functional data pattern, adjusting the position of the clock edge relative to the data, and observing the results on an oscilloscope. The oscilloscope trigger normally is set to a 50% level, which shows timing information before and after the clocking edge. A device tends to become metastable as it exceeds its setup and hold timing limitations, even at low speed. Metastability is an unpredictable state in which a device output may switch to either a logic 1 or logic 0 without any apparent regard for the logical input conditions. Similarly, excessive signal jitter may appear on the output when setup and hold tolerances are violated.

- High-speed tests typically require bursts of high-speed data, where the burst forms a functional test that exercises the device at rates approximating its intended operational frequency or higher. A signal source is used to deliver a block of data to the device under test (DUT) at data rates that are much higher than the basic low-speed functional test. However, this test process is still one of empirically finding a range of setup and hold values and specifying the system clock placement accordingly. Using a data generator with a repetitive data pattern, the recurring skew problems associated with high-speed setup and hold violations often can be isolated and rectified.

- Device self-test is an option on some contemporary source-synchronous and high-speed serial (HSS) components. The transmitter sends a known pattern to the receiver, which slides the location of the clock or strobe in time until data errors occur. The difference between the nominal clock or strobe location and the onset of data errors indicates how much margin exists in that interface.

1.4.2   Eye Diagrams

Eye diagrams have become one of the cornerstones of SI test and compliance measurements. The eye diagram is a regulatory measurement for validation and compliance testing of various industry-standard digitally transmitted signals. An eye diagram is a display that typically is viewed on an oscilloscope. It can efficiently reveal amplitude and timing errors in high-speed digital signals. The eye diagram shown in Figure 1-10 is built by overlaying digital signal waveform traces from successive logic signal cycle periods, or unit intervals. Waveform A in Figure 1-10 is a digital signal with exaggerated timing errors. Waveform B is the clock signal recovered from the signal shown as waveform A. Eye diagram A is generated by dividing waveform A into individual cycle periods and overlaying each cycle of waveform A. A number of measurements can be made on an eye diagram to quantify signal quality. One such measurement is the eye opening that relates to the signal timing, which is shown in Figure 1-10 as a gray area. Eye diagram B is produced from the recovered clock signal and shows the ideal eye opening, which indicates improved signal timing. Eye diagrams display serial data with respect to a clock that normally is recovered from the data signal using either hardware or software tools. In the eye diagram shown in Figure 1-11, the clock is recovered by a hardware-based reference or "golden" phase locked loop (PLL). The diagram displays all possible transition edges, both positive-going and negative-going, and both data states in a single window. The result is an image that somewhat resembles an eye, as shown in Figure 1-11.

In an ideal world, each new trace would line up perfectly on top of those that came before it. Also, the eye diagram would be composed of narrow lines representing the superimposed logic 1s and logic 0s. In the real world, signal integrity factors such as noise and jitter cause the composite trace to blur as it accumulates the logic 1s and logic 0s. The gray regions in Figure 1-11 have special significance; they are the violation zones used as mask boundaries during compliance testing. A compliance mask typically is produced by the instrument manufacturer in association with a standards body. In this case the gray polygon in the center defines the area in which the eye is widest. This encompasses the range of safe decision points for extracting the data content, the binary state, from a logic signal. The upper and lower gray bars define the signal's amplitude limits. If a signal peak penetrates the upper bar, for instance, it is considered a "mask hit" that will cause the compliance test to fail, although some standards may tolerate a small number of mask hits. More commonly, noise, distortion, transients, or jitter cause the trace lines to thicken. The eye opening shrinks, touching the inner gray polygon. This too is a compliance failure, because it reveals an intrusion into the area reserved for evaluating the logic state of the data bit. The compelling advantage of an eye diagram is that it enables a quick visual assessment of signal quality.

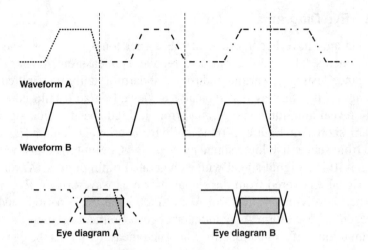

Waveform A a digital signal with exaggerated timing errors

Waveform B the recovered clock signal derived from waveform A

Eye diagram A is produced from wavefrom A and shows the eye opening with a grey masked area, which is one of the measures of the eye and indicates the quality of signal timing

Eye diagram B is produced from the recovered clock signal and shows the wider eye opening indicating improved signal timing over waveform A

Figure 1-10   Eye diagram formation.

Figure 1-11   An oscilloscope display showing a hardware-generated eye diagram. The gray areas are the mask violation zones.

1.4.2.1   Simulated Eye Diagrams

Although the ideal eye diagram measurement is carried out in real time, it is sometimes impossible to extract. For example, a programmable logic device receiver equalization circuit typically is embedded within the device. However, it is possible to simulate the results using SPICE modeling. The upper part of Figure 1-12 shows a simulated eye diagram into the programmable logic device receiver after traversing a 40-inch backplane, without pre-emphasis. The lower part of Figure 1-12 shows the same signal after it has passed through receiver equalization.

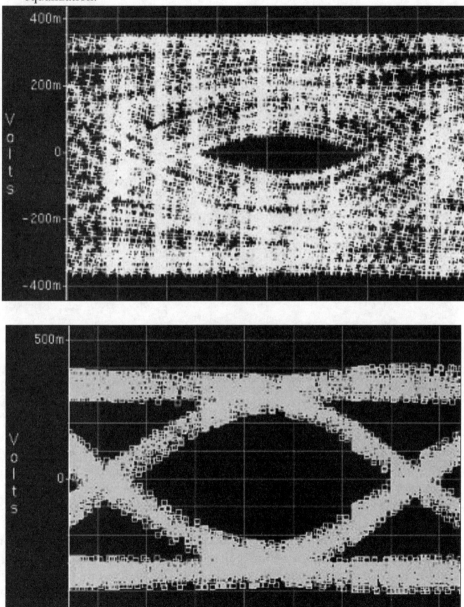

Figure 1-12   SPICE simulated eye diagrams.

The eye diagrams shown in Figure 1-12 are a clear example of how an equalization circuit significantly improves the quality of a received signal, allowing reliable detection of logic states. Another example, Figure 1-13, shows the effects of too much pre-emphasis. You must select the optimum settings for both pre-emphasis and equalization. Overcompensation can cause additional issues within the system. It adds extra jitter, which closes the eye, making it impossible for the receiver to interpret the information. Figure 1-13 clearly shows the effect of adding too much pre-emphasis and equalization.

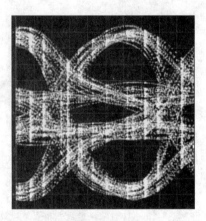

Figure 1-13   An eye diagram showing the exaggerated effects of too much pre-emphasis.

1.4.2.2   Real-Time Eye Diagrams

Real-time eye diagram debug methodologies often provide a shortcut that lets you quickly detect and correct SI problems. For example, some modern high-performance logic analyzers are combined with an oscilloscope and host troubleshooting tools that bring analog eye diagram analysis to the logic analyzer screen. The eye diagram is a real-time visualization tool that typically allows the designer to observe the data valid window, and general signal integrity, on clocked buses. This test methodology is a required compliance testing tool for many of today's buses, particularly the high-speed serial buses, but any signal line can be viewed as an eye diagram. Moreover, the logic analyzer eye diagram analysis can show wide parallel bus performance. It integrates hundreds of eye diagrams into one view that encompasses the leading and trailing edges of both positive-going and negative-going pulses that compose the bus signals. Figure 1-14 is an eye diagram where the contents of twelve address bus signals are superimposed. The benefit of observing bus lines simultaneously with an eye diagram is that it presents all possible logic transitions in a single view and allows fast assessment of the bus.

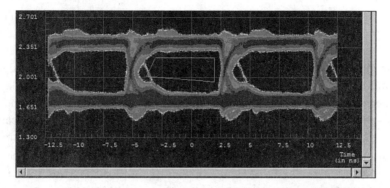

Figure 1-14   An eye diagram that simultaneously shows twelve bus signals.

An eye diagram can reveal analog problems in high-speed digital signals, such as slow rise times, transients, and incorrect logic levels. Figure 1-14 shows the performance of 12 parallel bus lines. The error encroaching into the mask is caused by an incorrect rise time in one of the bus signals.

The eye diagram shown in Figure 1-14 reveals an anomaly in the signals, which typically is shown in a distinctive color; the color indicates a relatively infrequent transition. In this example at least one of the signals has an edge that is outside the normal range. The mask feature built into the instrument helps locate the specific signal causing the problem. By drawing the mask in a particular way, such that the offending edge penetrates the mask area, the relevant signal can be isolated, highlighted, and brought to the front layer of the image. The result is shown in Figure 1-15, in which the flawed signal has been brought to the front of the display and highlighted in white. In this example the instrument identifies the aberrant edge and indicates a problem on the A3 (0) address bus signal. The origin of this particular problem is actually crosstalk, where the edge change is being induced by signals on an adjacent PCB trace.

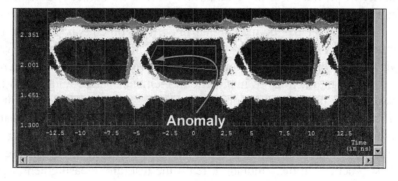

Figure 1-15   The flawed signal has been brought to the front of the display.

The real-time functional verification and troubleshooting phase of the design in this example has described how a common SI problem is detected and resolved. The logic analyzer is often the first line of defense when testing digital functionality. However, digital problems can stem from analog signal issues, including edge degradation due to improper termination or crosstalk, as demonstrated here. By teaming the logic analyzer with an oscilloscope and evaluating time-correlated digital and analog signals on the same screen, you can see problems affecting either domain using an eye diagram. Today, low-cost mixed signal instruments with real-time eye diagram capabilities are used to test and debug high-performance digital systems. Specialist oscilloscopes are used for the more demanding tasks of compliance and interoperability tests.

CONCLUSION

The design process has multiple steps, each with its own particular set of SI challenges in which the customer requirement is translated into specifications, simulation, and real-time measurement needs. This chapter has illustrated some of the more important points of SI engineering to promote the notion that today's SI engineers need innovative thinking if they are to keep pace with the digital bandwidth race. Increasing a system's operating rate is not simply a matter of designing a faster clock. As frequency increases, simulation issues become complex, and PCB traces on a circuit board become significantly more complicated. As frequencies increase, the trace begins to act like a capacitor. At the highest frequencies, trace inductance plays a larger role. All of these characteristics can adversely affect signal integrity. At today's clock frequencies, which are in the hundreds of megahertz and above, every design detail is important. Catching problems early and minimizing rework enables a new product to reach its market on time. History has shown that in the highly competitive embedded systems marketplace, the majority of profits go to the product that is first to market and that delivers premium performance.

This chapter has defined signal integrity problems as phenomena that can compromise a signal's ability to convey binary information. In real digital devices, these binary signals were shown to have analog attributes that result from the complex interactions of many circuit elements. These range from driver outputs to signal path transmission, terminations, and digital receivers. It is critical that today's designers follow a simulation, test, and debug strategy that is appropriate for the wide range of performance levels found in modern digital systems. The primary aim of this book is to advise you on and encourage the use of a range of best practices in the simulation, real-time test, and measurement aspects of SI engineering. However, we also hope that you will discover how to design digitally and think in analog. Moreover, the majority of simulation, test, and measurement tools show voltage and time. In a number of SI issues, the engineer must think in terms of signal currents, ground voltage, and electromagnetic effects, along with trace distance. Today the SI engineer has an ever greater dependence on automated real-time measurements and built-in signal analysis tools, such as eye diagram generation software. And what is most exciting is the interplay of simulation and real-time test. Device manufacturers are working with instrument providers to implant simulation models in benchtop instruments, allowing the real-time measurement and analysis of a partially populated system. In terms of digital design, simulation, test, and measurement, we live in exciting times. But as with any picture, the observers can see in the picture only what their experiences in life have equipped them with. Put another way, this book cannot teach experience. It can only point the way to good practice and, like a good companion, give sound advice.

Chip-to-Chip Timing and Simulation

As signal integrity engineers, we need to understand and quantify the dominant failure mechanisms present in each digital interface if we want to achieve the goal of adequate operating margins across future manufacturing and operating conditions. Ideally, we would have the time and resources to perform a quantitative analysis of the operating margins for each digital interface in the system, but most engineers do not live in an ideal world. Instead, they must rely on experience and sound judgment for the less-critical interfaces and spend their analysis resources wisely on the interfaces that they know will have narrow operating margins. Regardless of company size or engineering resources available, the goal of any successful signal integrity department is to perform the analysis and testing required to ensure that each interface has enough operating margin to avoid the associated failure mechanisms.

Most signal integrity engineers would agree about the fundamental causes of signaling problems in modern digital interfaces: reflections, crosstalk, attenuation, resonances, and power distribution noise. Although these physical phenomena may be part of a complex and unique failure mechanism process, the end result of each process is the same: A storage element on the receiving chip fails to capture the data bit sent by the transmitting chip. Because on-chip storage elements (flip-flops and registers) are typically beyond the scope of the models used in signal integrity simulations, it is easy to forget that the actual process of failure involves not just voltage waveforms at the input pin of a chip, but also a timing relationship between those waveforms and the clock that samples the waveform. In fact, the appearance of the waveform itself may be horrible so long as its voltage has the correct value when the clock is sampling it.

2.1   ROOT CAUSE

A virtual tour of an integrated circuit (IC) chip would uncover one or more flip-flops in close proximity to each signal IO pin. A driver or receiver circuit connects directly to the chip IO pin, and a flip-flop usually resides on the other side of the driver or receiver. The purpose of a digital interface is to transmit a data bit stored on the driving chip and reliably capture that same data bit some time later on the receiving chip. However, the process has one caveat: The data signal must be in a stable high or low state at the moment when the flip-flop samples it—that is, when the clock ticks. Because IC chips are subject to variations in manufacturing and operating conditions, the moment of stability translates to a window in time during which the data signal must remain stable. In essence, the job of the signal integrity engineer is to ensure that the data does not switch during this window.

The next few paragraphs offer a functional description of one type of complementary metal oxide semiconductor (CMOS) latch and the associated flip-flop. Not all CMOS flip-flops are identical, but this particular circuit lends itself well to description. It might seem that a discussion of latches and flip-flops is too fundamental for a book on timing for signal integrity. Careful consideration of these fundamentals, though, lays a foundation for understanding more complicated topics such as writing IO timing specifications for an ASIC or measuring bit error rate.

2.2   CMOS LATCH

At the heart of a latch lies a pair of pass gates. When the clock is low, the lower pass gate (P1 in Figure 2-1) allows data at the input, D, to flow through the latch to the output, Q. When the clock rises to a high state, the lower pass gate closes to block the data from the input, while the upper pass gate (P2 in Figure 2-2) opens to hold the output in the state that it was in just before the clock switched low. The mechanism for accomplishing data storage is positive feedback through inverters I4 and I5, which sample the outputs of both the lower and upper pass gates and feed them back to the input of the upper pass gate. When the lower pass gate is no longer open because the clock switched high, the upper pass gate turns on and activates the positive feedback loop. This allows the output of the upper pass gate to reinforce its input and hold the state of the data until the clock switches high again. The inverters I1 and I2 provide differential copies of the clock to the gates of the pass gates, as is necessary for their operation. The inverter I6 ensures that the sensitive internal data node is isolated from the next stage of logic; it also preserves the polarity of the output data, Q, with respect to the input data, D.

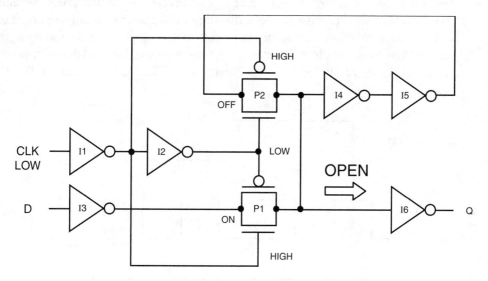

Figure 2-1   CMOS latch in open state.

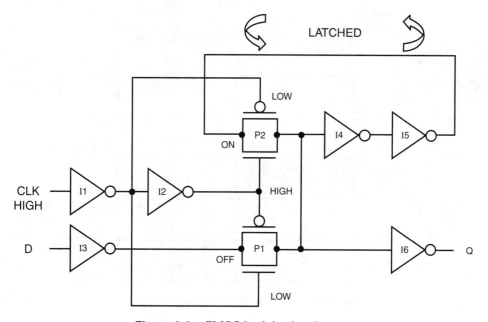

Figure 2-2   CMOS latch in closed state.

The problem with a latch is that it stores data only for half the clock cycle, which is not terribly efficient. During the other half, the output simply mimics the input, which is not useful. Figure 2-3 depicts this behavior. If one had a latch that stored data when the clock was high and another kind of latch that stored data when the clock was low, as described in Figure 2-4, this would cover the entire clock cycle. This is precisely what a flip-flop does. It is easy enough to arrive at a circuit that stores data during the high part of the clock cycle by swapping the connections from the clock inverters to the pass gates inside the latch. Connecting these two kinds of latches in series creates a flip-flop. But how does a flip-flop work?

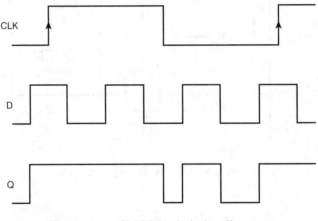

Figure 2-3   CMOS latch timing diagram.

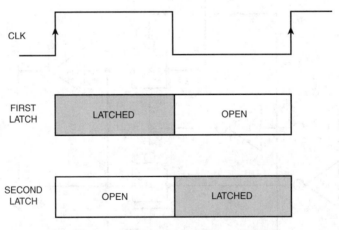

Figure 2-4   CMOS flip-flop timing diagram.

When the clock transitions from a low state to a high state, the lower pass gate in the first latch closes off, while the upper pass gate in the first latch opens to "capture" the data that was waiting at the input to the flip-flop when the clock switched high. During the time before the clock switched high, the data at the input of the flip-flop also showed up at the input of the second latch but did not make it any further because the second latch was in its closed state at the time. Now that the clock is high, any change in data at the input of the flip-flop goes no further than the first inverter. The flip-flop is essentially regulating the activity of all circuits that precede it.

Because the pass gates in the second latch have inverted copies of the clock, the second latch goes into its open state at the same time the first latch goes into its closed state. This means that the data stored by the first latch immediately passes through the second latch and shows up at the output of the flip-flop—after some brief propagation delay. When the clock switches high, it also "launches" the data that was present at the input of the flip-flop on to whatever circuitry is connected to the output of the flip-flop.

The first latch holds the data that was present at its input when the clock switched high for the remainder of the clock cycle. When the clock falls low, the second latch transitions back into its closed state, blocking the data path again and storing the data that the first latch was holding for it. Now the first latch goes back into its open state and allows the new data to enter it, ready to be latched again the next time the clock rises.

2.3   TIMING FAILURES

The previous discussion assumed that the data at the input of the flip-flop was in a stable state when the clock switched high. What would happen if the data and the clock were to switch at the same instant in time? CMOS logic is made from charge-coupled circuits. To change the state of the feedback pass gate in a flip-flop, it is necessary to move charge from one intrinsic capacitor to another. Some of this capacitance is a part of the transistor itself; a gate insulator sandwiched between a conductor and a semiconductor forms a capacitor. Some of the capacitance is associated with the metal that wires transistors together. It takes work to move electrons, and the alternating clock signals on either side of the feedback pass gate provide the energy to do the work.

If the data happens to change at the same time the clock is doing its work, it robs the circuit of the energy that it was using to change the state of the latch. This is not good. The result is metastability, an indeterminate state that is neither high nor low. This state may persist for several clock cycles, and it throws a wrench into the otherwise orderly digital state machine. Chip designers strive to avoid this troublesome situation by running timing analyzer programs during the process of placing and routing circuits. Signal integrity engineers strive to avoid the same situation, although their tools differ. Many of us are used to running voltage-time simulators, but few of us use timing analyzers.

The dual potential well diagram in Figure 2-5 is a useful analogy for visualizing the physics of metastability. Imagine a ball sitting at the bottom of a curved trough that is separated from an identical trough by a little hill. The first trough represents a latch that is in a zero state, and the second trough represents the same latch in a one state. Moving the ball from the first trough to the second requires the addition of some kinetic energy to get over the hump between the troughs. If the kinetic energy is just shy of the minimum threshold required, the ball may find itself "hung up" momentarily on the hill between the two troughs in a state of unstable equilibrium. A little bit of energy in the form of noise will push the ball into either of the two troughs.

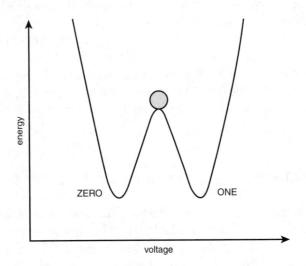

Figure 2-5   Dual potential well model for CMOS latch.

2.4   SETUP AND HOLD CONSTRAINTS

To prevent the condition of metastability from occurring in a flip-flop, the circuit designers run a series of simulations at various silicon manufacturing, temperature, and voltage conditions. They carefully watch the race between the data and clock signals on their way from their respective inputs to the four pass gates inside the flip-flop, as well as the time it takes for the upper pass gates to change state. Based on the sensitivity of these timing parameters to manufacturing and operating conditions, they establish a "safety zone" in time around the rising edge of the clock during which the data must not switch if the flip-flop is to settle in a predictable state when the clock is done switching. The beginning of this safety zone is called the *setup time* of the flip-flop, and the end is called the *hold time* (see Figure 2-6). These two fundamental circuit timing parameters form the foundation for the performance limitations of a digital interface.

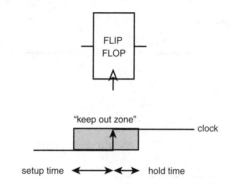

Figure 2-6   CMOS flip-flop setup and hold times.

The modern CMOS chip is a gigantic digital "state machine" that is governed by flip-flops, each of which gets its clock from a common source through a distribution network. In between the flip-flops lie various CMOS logic circuits that perform functions such as adding two binary numbers or selecting which of two data streams to pass along to the next stage of logic. Naturally, data does not pass through these logic circuits instantaneously. A circuit's characteristic propagation delay varies with the number and size of transistors and the metal connecting the transistors together.

Consider this simplest of state machines in Figure 2-7. When the left flip-flop launches a data bit into the first inverter in the chain at the tick of the clock, the data must arrive at the input of the right flip-flop before the tick of the next clock—less the setup time. If the data arrives after the setup time, the right flip-flop may go metastable or the data may actually be captured in the following clock cycle. In either case, the timing error destroys the fragile synchronization of the state machine, and the whole system comes to a grinding halt. This failure mechanism is called a *setup time failure*, and Figure 2-8 illustrates the timing relationship between data and clock. Depending on where you work, it may also be called a *late-mode failure* or *max-path failure*.

Now imagine that the clock signal does not arrive at both flip-flops at the same time. This is not an uncommon problem because so many flip-flops all get their clocks from the same source. Just like everything else in the world, it is impossible to design a perfectly symmetrical clock distribution scheme that delivers a clock to each flip-flop in a chip at the same time. The difference in arrival times of two copies of the same clock at two different flip-flops is called *clock skew*.

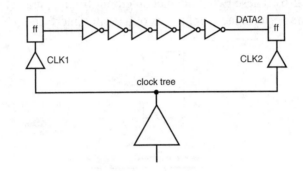

Figure 2-7   On-chip timing path.

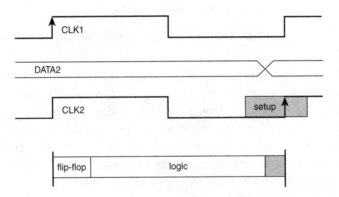

Figure 2-8   Setup time failure.

Imagine that the clock arrives at the capturing flip-flop later than the launching flip-flop. Now imagine that the propagation delay of the logic circuits, including the launching flip-flop, is smaller than the clock skew. In this case, the second flip-flop will capture the data during the same cycle it was launched—that is, one cycle early. Figure 2-9 depicts this condition. This failure mechanism is called a *hold time*, *early-mode*, or *min-path failure*, and it is the more insidious of the two failures because the only way to fix it is to improve the clock skew or add more logic circuit delay, both of which require a chip design turn. The setup time failure can be alleviated by slowing the clock. To prevent a hold time failure from occurring, it is necessary to ensure that the logic propagation delay is longer than the sum of the clock skew and the flip-flop hold time.

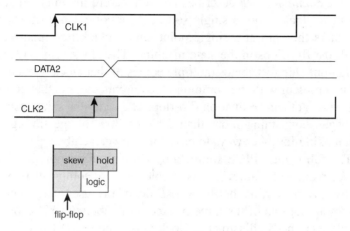

Figure 2-9 Hold time failure.

Perhaps this illustration will be helpful. Imagine a long hallway of doors spaced at regular intervals. The doors all open at the same instant and remain open for a fixed period of time. A man named Sven is allowed to run through exactly one door at a time, after which all doors slam shut again at the same instant and remain closed for two seconds before opening again and sending Sven on his merry way. Consider four cases, as follows:

1. If Sven is in good physical shape and times his steps, this process can continue indefinitely.
2. If Sven needs a membership to the health club, he may find himself in between the same two doors for two slams—or longer.
3. If Sven happens to be an overachiever, he may make it through two doors before hearing the last one slam behind him.
4. Sven runs into the door just as it is slamming in his face.

These four cases correspond, respectively, to reliable data exchange, setup time failure, hold time failure, and metastability.

2.5   COMMON-CLOCK ON-CHIP TIMING

The name *common clock* describes an architecture in which all storage elements derive their clocks from a single source. In a common-clock state machine, the clock enters the chip at a single point, and a clock tree repowers the signal to drive all the flip-flops in the state machine. The clock originates at an oscillator, and a fan-out chip distributes multiple copies of that clock to each chip in the system. The problem with the common-clock architecture is that copies of the clock do not arrive at their destination flip-flops at exactly the same instant. When clock skew began consuming more than 20% or 30% of the timing budget, people started looking for other ways to pass data between chips.

The following SPICE simulation "On-Chip Common Clock Timing" illustrates setup and hold failures. This simple network comprises two flip-flops separated by a noninverting buffer whose delay the user may alter by selectively commenting out two of the three instances of subcircuit x2. Clock sources vclk1 and vclk2 also need adjustment. Clock vclk1 (node 31) launches data from flip-flop x1, and clock vclk2 (node 32) samples data at flip-flop x3. In the default configuration of this deck, whose waveforms are shown in Figure 2-10, vclk1 launches data on the first clock cycle, vclk2 captures that data on the second clock cycle 5 ns later, and all is well.

```
On-Chip Common Clock Timing

*---------------------------------------------------------------------
* circuit and transistor models
*---------------------------------------------------------------------

.include ../models/circuit.inc
.include ../models/envt_nom.inc


*---------------------------------------------------------------------
* main circuit
*---------------------------------------------------------------------

vdat 10 0 pulse (0 3.3V 1.0ns 0.1ns 0.1ns 6.0ns 100ns)

* nominal clocks
vclk1 31 0 pulse (0 3.3V 2.0ns 0.1ns 0.1ns 2.4ns 5ns)
vclk2 32 0 pulse (0 3.3V 2.0ns 0.1ns 0.1ns 2.4ns 5ns)

* hold time clocks
*vclk1 31 0 pulse (0 3.3V 2.0ns 0.1ns 0.1ns 2.4ns 5ns)
*vclk2 32 0 pulse (0 3.3V 2.5ns 0.1ns 0.1ns 2.4ns 5ns)

* setup time clocks
*vclk1 31 0 pulse (0 3.3V 2.0ns 0.1ns 0.1ns 2.4ns 5ns)
*vclk2 32 0 pulse (0 3.3V 1.5ns 0.1ns 0.1ns 2.4ns 5ns)
```

```
*   d   q   clk vss vdd
x1 10  20  31   100 200 flipflop

* nominal delay buffer
*   a   z   vss vdd
x2 20 21 100 200 buf1ns

* hold time delay buffer
*   a   z   vss vdd
*x2 20 21 100 200 buf0ns

* setup time delay buffer
*   a   z   vss vdd
*x2 20 21 100 200 buf5ns

*   d   q   clk vss vdd
x3 21  22  32   100 200 flipflop

*-----------------------------------------------------------------------
* run controls
*-----------------------------------------------------------------------

.tran 0.05ns 20ns
.print tran v(31) v(32) v(21) v(22)
.option ingold=2 numdgt=4 post csdf
.nodeset v(x1.60)=3.3 v(x1.70)=0.0
.nodeset v(x3.60)=3.3 v(x3.70)=0.0

.end
```

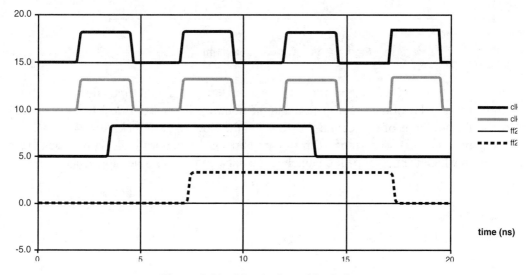

Figure 2-10   Nominal on-chip timing.

2.6   SETUP AND HOLD SPICE SIMULATIONS

In the setup time failure depicted in Figure 2-11, the first flip-flop launches its data during the first clock cycle, but the data does not arrive at the second flip-flop until after the second clock cycle, which means the second flip-flop does not capture the data until the third clock cycle. Clock skew exacerbates the problem by advancing the position of the second clock with respect to the first (clock skew = 2.0 ns – 1.5 ns = 0.5 ns). The sum of the flip-flop delay, noninverting buffer delay, and setup time is greater than the clock period, and a setup time failure occurs.

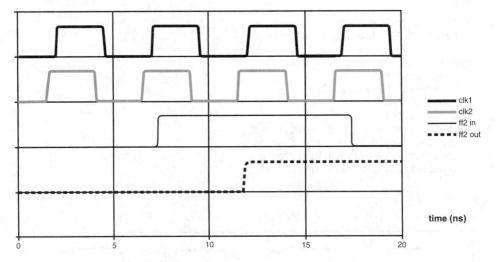

Figure 2-11   On-chip setup time failure.

Figure 2-12 demonstrates the other extreme: the hold time failure. The combined delay of the first flip-flop and the noninverting buffer is less than the sum of the hold time of the second flip-flop and the clock skew (clock skew = 2.0 ns - 2.5 ns = -0.5 ns). The second flip-flop captures the data during the same clock cycle in which the first flip-flop launched the data, and a hold time failure occurs.

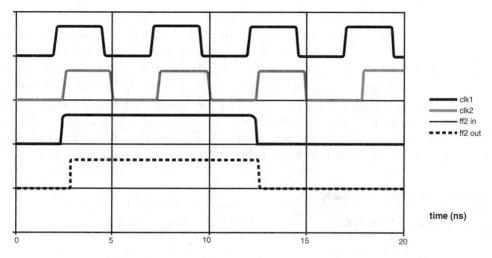

clk1
clk2
ff2 in
ff2 out

time (ns)

Figure 2-12   On-chip hold time failure.

2.7   TIMING BUDGET

The simple spreadsheet timing budget in Table 2-1 expresses the relationship between each of these delay elements and the corresponding timing constraints: one column for setup time and one column for hold time. Notice that the setup column contains an entry for the clock period while the hold column does not. That is because a hold time failure occurs between two rising clock edges separated by skew in the *same clock cycle*; it is entirely independent of clock period, which is why you cannot fix this devastating problem by simply lengthening the clock cycle. (Take it from one who knows: Devastating is no exaggeration.) The setup time failure, on the other hand, occurs when the sum of all timing parameters is longer than the clock period.

The second thing to notice is that flip-flop hold time appears only in the hold column, while flip-flop setup time appears only in the setup column. These two numbers represent the boundaries of the metastability window. The setup column numerically describes the physical constraint that data needs to switch to the left of the metastability window as marked by the setup time at the next clock cycle. The hold column conveys a different idea: Data cannot switch *during the same clock cycle it was launched* before the metastability window, as marked by the hold time.

Finally, the reason the hold column contains negative numbers is because the hold time failure is a race between delay elements (protagonists) on one side and the sum of clock skew and hold time (antagonists) on the other side. Therefore, the delay elements are positive, while the clock skew and hold time are negative. Summing all hold timing parameters in rows 1 through 5 results in a positive operating margin when the hold constraints are met and a negative margin when they are not. The numbers in rows 1 through 5 of the setup column are all positive, and setup operating margin is the difference between the clock period and the sum at row 6. Clock skew always detracts from operating margin.

Table 2-1   On-Chip Timing Budget

	Parameter	Description	Setup	Hold	Units
1	Tskew	Clock skew	0.50	-0.50	ns
2	Tclkq	Flip-flop clock-to-Q delay	0.35	0.35	ns
3	Tbuf	Noninverting buffer delay	5.00	0.00	ns
4	Tsu	Flip-flop setup time	0.10	n/a	ns
5	Th	Flip-flop hold time	n/a	-0.05	ns
6		TOTAL	5.95	-0.20	ns
7	Tclk	Clock period	5.00	n/a	ns
8		OPERATING MARGIN	-0.95	-0.20	ns

The following somewhat artificial example swaps three different delay buffers into the x2 subcircuit under nominal conditions that, of course, would never happen in a real chip. Setup and hold timing analysis on an actual design proceed by subjecting the logic circuits between two flip-flops to the full range of process, temperature, and voltage variations. To accomplish this in SPICE, the example decks in this chapter call an include file, which in turn calls one of three transistor model parameter files for fast, slow, and nominal silicon. It also sets the junction temperature and power supplies.

```
*-------------------------------------------------------------
* transistor model parameters
*-------------------------------------------------------------

.include proc_nom.inc

*-------------------------------------------------------------
* junction temperature (C)
*-------------------------------------------------------------

.temp 55
```

```
*-----------------------------------------------------------------
* power supplies
*-----------------------------------------------------------------

vss     100 0 0.0
vdd     200 0 3.3
vssio  1000 0 0.0
vddio  2000 0 3.3

rss     100 0 1meg
rdd     200 0 1meg
rssio  1000 0 1meg
rddio  2000 0 1meg
```

The fast and slow transistor model parameter files call the same set of parameters as the nominal file but pass different values representing the predicted extreme manufacturing conditions over the lifetime of the semiconductor process. A typical range of junction temperatures for commercial chips might be 25–85 C. Processor chips often run hotter. Chip designers commonly use +/- 10% for power supply voltage variation, but a well-designed power distribution network seldom sees more than +/- 5% from dc to 100 MHz. Transient IO voltages may droop more than 5% when many drivers switch at the same time, but that is the topic of another book.

2.8   COMMON-CLOCK IO TIMING

Several factors make timing a common-clock digital interface more complicated than timing an on-chip path:

- Delay from the chip clock input pin to the flip-flop (clock insertion delay)
- Skew between clock input pins on the driving and receiving chips
- Combination of driver and interconnect delay
- Receiver delay

A good part of this additional complexity is hidden beneath chip timing specifications. However, it is important to understand these underlying complexities, especially when writing timing specifications for a new ASIC.

```
*    a  oe  pad vss vdd vssio vddio
x2 20 200 21  100 200 100   200 drv

t1 21 0 22 0 td=1ns z0=50

*    a   z  vss vdd
x3 22 23 100 200 rcv
```

The differences between the on-chip and IO SPICE decks are not many. A driver, transmission line, and receiver replace the string of inverters. In addition to data input and output pins, the driver also has an output enable input pin (oe) that, when high, takes the driver out of its high-impedance state and allows the output transistors to drive the transmission line. An extra set of power supply rails (vssio and vddio) isolates the final stage of the driver from the core power and ground buses; the transient currents moving through the final stage are so large that the voltage noise they generate is best kept away from the core circuitry of the chip.

The ideal transmission line models a point-to-point connection between two chips on a printed circuit board. Typical common-clock interconnect simulations involve a multidrop net topology, lossy transmission lines, chip packages, and maybe a connector or two. The clock trees in Figure 2-13 appear in the timing budget but not in the simulation; they are simply too large to include in signal integrity simulations. The clock distribution chip does not appear in this simulation, either, although it is important to simulate these nets separately, as clocks are some of the most critical nets in any system.

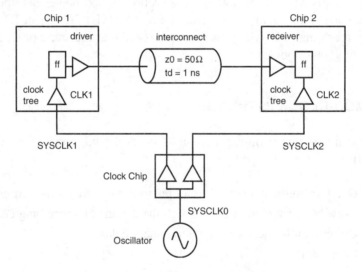

Figure 2-13   Common-clock interface.

When the input to the clock distribution chip switches, a clock signal travels down the wire from the output of the clock chip to the clock input of Chip 1, where the clock tree distributes it to every flip-flop on the chip. When the flip-flop in front of the driver sees the rising edge of the clock, it grabs the data at its input and launches it out toward the driver, which repowers the data and sends it off the chip and down the interconnect—a 1 ns 50 ohm transmission line, in this case. After a delay of 1 ns, the data signal shows up at the input of the receiver.

Meanwhile, a clock signal from the second tick of the clock travels from the clock chip to Chip 2, through its clock tree, and arrives at the clock input of the flip-flop on the other side of the receiver. If all goes well, the data signal will arrive at the data input of the flip-flop on Chip 2 slightly before the clock signal does—with a difference of at least the setup time. To ensure that this is always the case, you must construct a timing budget similar to Table 2-1 but with extra terms to account for the additional complexity. Table 2-2 adds three new lines.

Table 2-2   IO Timing Budget

	Parameter	Description	Setup	Hold	Units
1	Tskew	System clock skew	0.25	-0.25	ns
2	Tc1	Chip 1 clock insertion delay	0.40	0.20	ns
3	Tclkq	Flip-flop clock-to-Q delay	0.38	0.26	ns
4	Tpd	Driver and transmission line	2.65	2.00	ns
5	Tc2	Chip 2 clock insertion delay	-0.30	-0.30	ns
6	Trcv	Receiver delay	0.17	0.17	ns
7	Tsu	Flip-flop setup time	0.10	n/a	ns
8	Th	Flip-flop hold time	n/a	-0.05	ns
9		TOTAL	3.65	2.03	ns
10	Tclk	Clock period	5.00	n/a	ns
11		OPERATING MARGIN	1.35	2.03	ns

As was the case of the on-chip timing budget, clock skew comes right off the top and counts against the margin. The IO budget must account for system clock skew in addition to variations in two on-chip clock trees (rows 2 and 5 in Table 2-2). System clock skew has at least three possible origins. The output pins on the clock distribution chip never switch at exactly the same instant in time, and there is a specification in the chip datasheet for this. There are also differences in propagation delay down the segments L1 and L2. Even though these may be in the same printed circuit board, there will be layer-to-layer variation in dielectric material and trace geometry. Finally, the two chips will not have identical loading.

The timing budget in Table 2-1 assumed both flip-flops were on the same chip and neglected the clock insertion delay, defined as the difference in time between the switching of the chip's clock input and the arrival of that clock at the flip-flop. (This definition of clock insertion delay encompasses on-chip clock skew.) If clock insertion delay were identical for both chips, it would add to Chip 1 delay and subtract the same number from Chip 2 delay. However, silicon processing and operating conditions introduce variations that complicate things.

Assume the two chips depicted in Figure 2-13 are identical, and their nominal clock insertion delay is 0.3 ns. This delay is probably realistic for a 0.5 μm CMOS technology, which was state-of-the-art at the same time as 200 MHz common-clock systems. Also assume that the clock tree has a process-temperature-Voltage tolerance of 50%—a little worse than the transistor models predict for this imaginary CMOS process.

In the hold timing budget, worst-case timing occurs when the data path looking back from the capturing flip-flop is fast and the clock path looking back from the capturing flip-flop is slow. Tracing from the data input to the flip-flop on Chip 2 all the way to the source of the clock, which is the clock distribution chip, the clock tree on Chip 1 is actually in the *data* path. This implies that the clock insertion delay number in row 2 of the timing budget needs to be at its smallest possible value to generate worst-case hold conditions. Worst-case conditions for a setup time failure are the converse of those for a hold time failure. A large clock insertion delay on Chip 1 implies that the data arrives at the target flip-flop on Chip 2 later in time; this condition aggravates setup margin.

The flip-flop delay, setup, and hold times are present as they were in the on-chip budget. New to the budget is the combined delay of the driver and transmission line, considered here as one unit because the delay through the driver is very much a function of its loading. This delay is defined as the difference between the time when the driver input crosses its threshold, usually VDD/2, and the time when the receiver input exits its threshold window, usually some small tolerance around VDD/2 for a simple CMOS receiver. Note that the percent tolerance of this delay is less than it is for circuits on silicon because the transmission line delay is a large portion of the total delay, and its tolerance is zero for the purposes of this analysis.

Things get a little tricky on rows 5 and 6 because there is a race condition between the clock path and the data path on Chip 2. Using worst-case numbers for both these delays would result in an excessively conservative budget because process-temperature-Voltage variations are much less when clock and data circuits reside on the same chip. A crude assumption would be to simply take the difference between the nominal clock insertion delay of 0.3 ns and the nominal receiver delay of 0.17 ns. The clock would arrive at the flip-flop 0.13 ns later than the data if they both arrived at the pins of Chip 2 simultaneously. Clock insertion delay at Chip 2 is favorable for setup timing but detrimental for hold timing. To obtain a more accurate number, consult a friendly ASIC designer and ask for a static timing analysis run.

As in the on-chip budget, the hold timing margin is simply the sum of rows 1 through 7. Clock period does not come into play in a hold budget because all the action happens between two rising edges of the same clock cycle that are skewed in time. Setup timing margin is the clock period, 5 ns, less the sum of rows 1 through 8.

Early CMOS system architects realized that clock insertion delay was a pinch-point for CMOS interfaces, so they began using phase-locked loop circuits (PLLs) to "zero out" clock insertion delay at the expense of jitter and phase error. Contemporary digital interfaces rely heavily on PLLs, and signal integrity engineers need to pay close attention to PLL analog supplies to keep jitter in check.

2.9   COMMON-CLOCK IO TIMING USING A STANDARD LOAD

Unless they are directly involved in designing a new ASIC, signal integrity engineers do not typically have access to on-chip nodes in their simulations. If they succeed in convincing a chip manufacturer to part with transistor-level driver and receiver models, they can connect a voltage source to the input of a driver and monitor the output of a receiver. But that is where it ends. IO Buffer Information Specification (IBIS) models are the standard for interfaces running slower than 1 Gbps (they are capable of running faster), and IBIS only facilitates modeling that part of the IO circuit connected directly to the chip IO pad. This situation is not entirely unfavorable since chip vendors are forced to write their specifications to the package pin, a point accessible to both manufacturing final test and the customer. Having access to only the driver output and receiver input simplifies the chip-to-chip timing problem considerably—assuming the chip vendor did a thorough job writing the chip timing specifications.

How does a chip designer arrive at IO timing specifications that are valid at the package pins rather than the flip-flop? For input setup and hold specifications, they begin with the flip-flop setup and hold specifications and take into consideration the difference between the receiver delay and the clock insertion delay, including both the logic gate delays and the effects of wiring. Looking back from the flip-flip on the receive chip, the receiver circuit is in the data path, and the clock tree is in the clock path.

Therefore, to specify a setup time requirement at the package pins, they consider the longest receiver delay and the shortest clock insertion delay. Conversely, the hold time specification involves the shortest receiver delay and the longest clock insertion delay. Because both the receiver and the clock tree are on the same chip, their delays will track each other and never see the full chip-to-chip process variation. In addition to the on-chip delays, the setup and hold time specifications must also account for variations in package delay.

Chip output timing specifications are also relative to the package pins. The output data valid delay for a chip, sometimes called clock-to-output delay, is a combination of clock insertion delay, flip-flop clock-to-Q delay, and driver output delay with one caveat: The delay of the driver is a function of its load, and the driver may see a wide variety of loading conditions unknown to the chip designer.

The standard load or timing load resolves this ambiguity. The standard load usually comprises some combination of transmission line, capacitor, and termination resistor(s) intended to match a typical load as the chip designer perceives it. Along with the standard load, the vendor also specifies a threshold voltage that defines the crossing points for both the clock input and the data output waveforms. The output data valid specification quantifies the delay between the clock input switching and the data output switching, each through their corresponding thresholds, with the standard load connected to the package output pin. As a signal integrity engineer, your job is to understand and quantify the difference between the standard load and your actual load.

Extracting timing numbers from simulation using a standard load is straightforward. Simulate the driver with its standard load alongside the driver, interconnect, and receiver in the same input deck, as shown in Figure 2-14, using the same driver input signal for both copies of the driver. The interconnect delay timing interval begins when the output of the standard load driver crosses the threshold, VT, specified in the chip datasheet and ends when the input of the receiver crosses one of the two threshold values, VIH and VIL, that define its switching window.

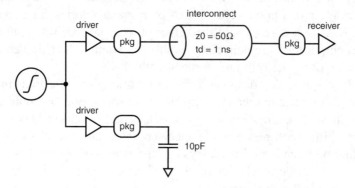

Figure 2-14   Actual load and standard load.

The interconnect timing interval in Figure 2-15 begins when the standard load crosses VT and ends when the actual load crosses VIL for the first time. Since the fast case corresponds to a hold time analysis, we use the earliest possible time the receiver could switch. In Figure 2-16, interconnect delay begins when the standard load crosses VT and ends when the actual load crosses VIH for the last time. The longest possible interconnect delay is consistent with a setup time analysis. Of course, some pathological combination of reflections could always result in the fast case producing the worst-case setup delay, so it pays to consider the possibilities.

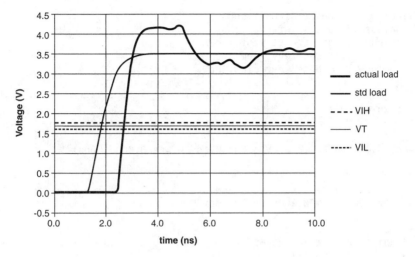

Figure 2-15   Fast interconnect delay with standard load.

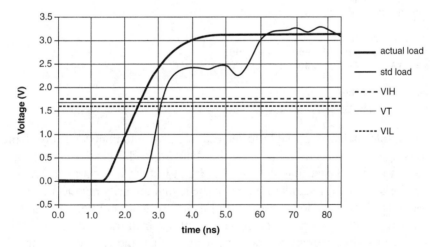

Figure 2-16   Slow interconnect delay with standard load.

The following analysis assumes no crosstalk-induced delay. Further discussion of the relationship between crosstalk and timing can be found in Chapter 4, "DDR2 Case Study," and Chapter 9, "PCI Express Case Study."

```
*-------------------------------------------------------------------
* circuit and transistor models
*-------------------------------------------------------------------

.include ../models/circuit.inc
.include ../models/intercon.inc
.include ../models/envt_fast.inc
*.include ../models/envt_nom.inc
*.include ../models/envt_slow.inc


*-------------------------------------------------------------------
* main circuit
*-------------------------------------------------------------------

vdat  10 0 pulse (0 3.3V 1.0ns 0.1ns 0.1ns 6.0ns 100ns)

* actual load
*    a   oe   pad vss vdd vssio vddio
x1   10 200 20   100 200 100 200 drv50

x2   20    21 0 bga_rlc
t1   21 0 22 0 td=1ns z0=50
x3   22    23 0 bga_rlc

*    a   z   vss vdd
x4   23 24 100 200 rcv

* standard load
*    a   oe   pad vss vdd vssio vddio
x5   10 200 40 100 200 100 200 drv50

x6   40 41 0 bga_rlc
c1       41 0 10pF
```

Note that it is possible to have a negative interconnect delay if the actual loading conditions are lighter than the standard load. This does not mean the signal is traveling faster than the speed of light! Rather, it is simply an artifact of the vendor's choice for standard load and the signal integrity engineer's choice of net topology. If the vendor chose a 50 pF standard load and you are timing a T net topology with two 5 pF loads at the end of a 4-inch transmission line, the actual load will switch before the standard load. In fact, the choice of standard load is not terribly significant so long as the simulations use the same load specified in the component datasheet.

It is also possible for the fast interconnect delay to be larger than the slow interconnect delay, as is the case for this configuration of driver and net. This looks strange in the budget, but rest assured that the sum of the fast driver and interconnect delays is actually smaller than the same sum in the slow case, as Table 2-3 proves. This is just another artifact of the choice of standard load.

Table 2-3   Driver and Interconnect Delay

Edge	Corner	Driver	Intercon	Sum	Units
Rise	Fast	0.74	0.90	1.64	ns
Fall	Fast	0.82	0.82	1.64	ns
Rise	Slow	1.41	0.61	2.02	ns
Fall	Slow	1.15	0.74	1.89	ns

The IO timing budget in Table 2-4 closely resembles the on-chip timing budget in Table 2-1 with Chip 1 taking the role of the first flip-flop, Chip 2 taking the role of the second flip-flop, and interconnect delay replacing the noninverting buffer delay. It is somewhat unusual to see a 0.41 ns setup and hold window in a datasheet for a chip running at 200 MHz. Although this is not difficult to achieve in silicon with a 0.5 μm CMOS process, chip vendors prosper when specifications are conservative and final test yields are high.

Table 2-4   IO Timing Budget Using Standard Load

	Parameter	Description	Setup	Hold	Units
1	Tskew	System clock skew	0.25	-0.25	ns
2	Tval	Chip 1 data valid	2.58	1.34	ns
3	Tpd	Interconnect propagation delay	0.74	0.82	ns
4	Tsu	Chip 2 setup time	0.23	n/a	ns
5	Th	Chip 2 hold time	n/a	-0.18	ns
6		TOTAL	3.80	1.73	ns
7	Tclk	Clock period	5.00	n/a	ns
8		OPERATING MARGIN	1.20	1.73	ns

Signal integrity engineers prosper when they successfully design and implement reliable digital interfaces. This fundamental conflict of interests often makes it difficult for signal integrity engineers to close timing budgets on paper. Fortunately, contemporary industry standard specifications that evolved with input from the user community have more realistic timing numbers.

Receiver thresholds are another area of creeping conservatism. Chapter 3 demonstrates that the actual receiver input window within which a receiver's output will transition from low to high is very narrow—on the order of 10 mV. Yet a typical VIH/VIL range for 3.3 V CMOS logic is 1.4 V, which is an artifact of transistor-transistor logic (TTL) legacy.

The two dashed lines in Figures 2-15 and 2-16 represent VIH and VIL at fast and slow process-temperature-Voltage corners. At least two philosophies exist for defining interconnect timing. In the first philosophy, setup timing uses the longest interconnect interval defined by the latest receiver threshold crossing; that is, VIH for the rising edge and VIL for the falling edge. Hold time timing uses the first VIL crossing for the rising edge and the first VIH crossing for the falling edge.

The second camp maintains the position that thresholds track with VDD by a factor of 50% for a typical CMOS receiver. Why time the receiver to fast thresholds when the driver is at low VDD? The two will never be off by more than 1% or 2% for a well-designed power delivery system. This example uses this second approach, but there is not much difference between the two because the receiver thresholds came from simulations of the receiver. When receiver thresholds come from a conservative component specification, the difference can be substantial.

2.10   LIMITS OF THE COMMON-CLOCK ARCHITECTURE

The common-clock IO timing budget demonstrates how large-scale distribution of a low-skew clock caps system performance somewhere between 200 and 300 MHz. Supercomputer makers were able to push this to 400 MHz, but not without a huge investment in custom technology. Multidrop net topologies impose their own set of limitations. In the past decade, point-to-point source synchronous and high-speed serial interfaces became the bridges that enabled the next level of system performance.

The common-clock architecture is still alive and well in digital systems everywhere. Although it may not be exciting to work on these nets, ignoring them is sure to bring more trouble than any signal integrity engineer would care to deal with. Any time the input to a flip-flop violates the setup and hold window, a timing failure will occur, whether the data is switching once every 10 ns or once every 200 ps.

2.11   INSIDE IO CIRCUITS

At first glance, it may seem that signal integrity engineering has more to do with packaging and interconnect technology than it has to do with silicon. After all, signal integrity engineers worry about things such as skin effect, dielectric losses, and coupling in three-dimensional structures; transporting electromagnetic energy from point A to point B is the primary goal. However, the output driver circuit is the source of the electromagnetic disturbances, and the receiver circuit must convert the electromagnetic energy incident on its input transistors into digital information. An elementary understanding of IO circuits helps us make decisions about what simulations are necessary to ensure the reliability of a digital interface.

For example, skin effect is a frequency-dependent phenomenon. The frequency content of a driver's output waveform determines whether a lossy transmission line model is required or whether an ideal transmission line is sufficient. Knowledge of a driver's rise time is necessary to make this decision. Unfortunately, the electrical characteristics of an IO circuit are often difficult to come by since many chip datasheets still contain only the most rudimentary dc specifications.

An IBIS model can fill in the gaps, but effective use of any component model implies a basic knowledge of the assumptions beneath the model. Some circuits make the transition between transistor-level modeling and behavioral modeling naturally, whereas others get trapped in between because the fundamental set of behavioral modeling assumptions is not consistent with the circuit function. This chapter provides an overview of the "workhorse" CMOS IO circuits and a platform from which to reach the more esoteric circuits.

2.12   CMOS RECEIVER

Although they often go overlooked, receiver thresholds are critical parameters for any digital IO interface. Perhaps the reason they go overlooked is that reliable and useful numbers remain so elusive. They don't need to be, though. Characterizing the input thresholds of a receiver is not difficult.

A receiver circuit interprets the signal on the net and translates it into a language the chip can understand—1s and 0s. The simple CMOS receiver in Figure 2-17 is a pair of inverters. If the n-channel and p-channel transistors have the right dimensions (that is, resistance), the inverter will change states when the input crosses VDD/2 and all four transistors are biased on. At this point, both the n- and p-channels have equal resistance, and the entire circuit functions as a pair of voltage dividers. The output will also be VDD/2.

The dc transfer characteristic in Figure 2-18 plots receiver output voltage against the input voltage, and a close look shows that the receiver does not switch exactly at VDD/2 in a vertical line; there is some region of uncertainty called the *threshold window*. Traditional circuit analysis defines the boundaries of the threshold window using the unity gain points.

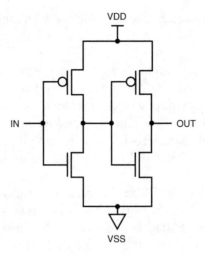

Figure 2-17 CMOS receiver.

Consider the receiver circuit as an amplifier. If biased near the threshold, the receiver will amplify a small signal superimposed on the dc bias level. By moving the dc bias level around, you will find two values at which the gain of the amplifier is equal to one—that is, output amplitude equals input amplitude. These are the unity gain points, and the receiver amplifies all signals that are biased between these points. An easier way to find these points is to plot a line whose slope is one on the dc transfer characteristic and find where this line is tangent to the curve. The input voltages that correspond to these two points of tangency are the unity gain points.

The distance in mV between unity gain points for the CMOS receiver in this fictional 0.5 μm circuit library is on the order of 10 mV—next to nothing. Variations in process and temperature affect the size of the threshold window very little, but the receiver threshold for this circuit does track differences between VDD and VSS by a factor of one half. This makes the thresholds sensitive to dc variation in supply voltage, as well as high-frequency ac noise from within the chip and mid- to low-frequency ac noise from without.

Chip designers must either account for high-frequency on-chip ac noise in the timing specifications or in the receiver thresholds. If they choose timing specifications, then typical receiver thresholds for a 0.5 μm CMOS process might vary ±165 mV around 1.67 V—that is, tracking a ±10% tolerance at the VDD pins by one half. If they choose to account for high-frequency on-chip ac noise in the dc input characteristics, the threshold window will probably be slightly larger. They had better not choose both ways!

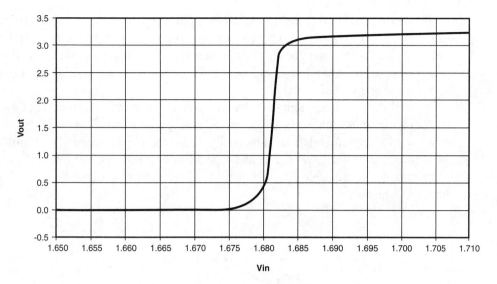

Figure 2-18   Receiver dc transfer characteristic.

In either case, the actual receiver threshold window is much narrower than the ancient 0.8 V and 2.0 V TTL specifications commonly seen in 3.3 V parts. If you use TTL specifications to define the right-hand side of an interconnect delay interval, as shown previously in Figures 2-15 and 2-16, this extra conservatism translates directly into lower system performance. We might expect IBIS models to contain more accurate threshold voltage numbers since the standard was intended to meet the needs of the signal integrity community, but that is not necessarily the case. It would appear that the authors of many IBIS models simply copy the receiver thresholds directly from the component datasheet rather than extracting them from simulations.

2.13   CMOS DIFFERENTIAL RECEIVER

As data rates climbed and supply voltages dropped, low-swing differential signaling technologies became attractive. The CMOS differential receiver resembles the emitter-coupled logic NOR gate (also known as the *bipolar differential amplifier*) found in Motorola's "MECL System Design Handbook." It is a current-mode circuit in which the two input transistors, M1 and M2, in Figure 2-19 are alternately biased into conduction. Beneath them sits a current source. The two resistors hanging from the VDD rail are really transistors biased into conduction and sized to give the desired resistance.

When M1 and M2 are driven differentially, the current from the source sloshes back and forth between the two parallel branches of the circuit, generating a voltage drop across whichever resistor happens to be carrying the current. An alternate implementation biases M2 with a dc reference voltage (Vref) and drives M1 with a single-ended signal.

The differential configuration has the advantage of an effective input edge rate that is twice as high as the single-ended configuration, making the circuit less sensitive to delay modulation induced by crosstalk or reflections. The single-ended implementation uses half as many package pins but requires distribution of a high-quality reference voltage, which is sensitive to noise and requires close attention to layout.

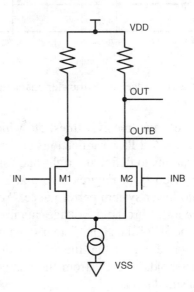

Figure 2-19   CMOS differential receiver.

2.14   PIN CAPACITANCE

Pin capacitance is yet another source of conservatism. A wide tolerance on a pin capacitance in a simulation model may not make or break a point-to-point net, but multiply that tolerance by eight for a memory address net and the simulation results can indicate that the interface will never function at the desired performance level when the actual system may never fail. This leaves the signal integrity engineer with the uncomfortable decision of accepting the risk of a design that is inconsistent with the component specification (the vendor *could* ship hardware to that spec) or lowering the performance target. A third solution exists: Write a purchase specification that calls for hardware-verified models and design to those models.

Pin capacitance originates in several places. Looking into a chip output pin, we see the sources and drains of the parallel output transistors that make up the driver's final stage. Both drivers and receivers require electrostatic discharge (ESD) protection devices, which are often just large transistors connected in such a way as to expose their intrinsic source-to-well or drain-to-well junction capacitances to the IO pad. In a wire bond package, the bond pad and the underlying substrate form a parallel plate capacitor. In a flip-chip package, the solder ball contributes capacitance, as does the metal between the IO circuit and the solder ball.

Finally, there is the package itself, which may take on many different forms. Because the chip datasheet specifies parameters at the package pins, the pin capacitance specification also includes the package. However, the package capacitance and the silicon-related capacitance are separate in most chip models used by signal integrity engineers. At rise times below 500 ps, it doesn't make much sense to model a 0.5-inch stripline in a ball grid array (BGA) package as a lumped capacitance.

With the exception of the package, these various sources of capacitance lie buried inside the model of an IO circuit. Even if the SPICE code for the circuit model were unencrypted, it would be difficult to calculate the pin capacitance from the transistor sizes, device parameters, and transistor model equations. Fortunately, there is a relatively easy way to extract IO pin capacitance from a SPICE simulation if the equivalent model for this circuit is a simple capacitor and a high impedance. For a driver, this implies that there is a tri-state enable function. For a receiver, this implies that there are no on-die termination devices. Agilent Technologies has published an excellent application note on this technique titled, "Measuring Parasitic Capacitance and Inductance Using TDR."

The reflection of a rising step from an ideal capacitor is a negative-going pulse whose area is proportional to the value of the capacitor. The SPICE simulation network that produced the pulse shown in Figure 2-20 is simple: an ideal 50 ohm source driving an ideal 50 ohm transmission line with an ideal 50 ohm termination. Place the device-under-test (driver or receiver) in the middle of the transmission line, which must be long enough relative to the pulse width and the rise time of the source to allow reflection to occur well after the source is done switching. First, integrate the area between dc level and the negative pulse. Second, multiply by two and divide by the transmission line impedance. This calculation yields the equivalent capacitance of the IO circuit. Test this technique using a known capacitor prior to using it to extract circuit capacitance.

$$C_{DUT} = \frac{2}{Z_o} \int 1 - V_{reflected}(t) \cdot dt$$

Equation 2-1

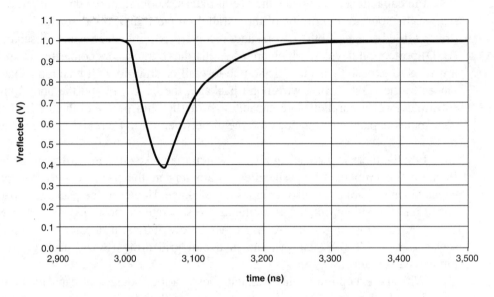

Figure 2-20   TDR extraction of IO pin capacitance.

2.15   RECEIVER CURRENT-VOLTAGE CHARACTERISTICS

If a receiver has some form of on-die termination (ODT), its current-voltage (IV) curves indicate how well it can absorb excess energy present on a net. There is strong motivation to study the current-voltage characteristics of a new IO circuit before putting it to use.

On-die termination is a useful solution to two common problems: 1) a transmission line stub between the package pin and the termination resistor, and 2) scarce printed circuit board real estate for large buses. Common configurations for on-die termination are parallel 100 ohm resistors to VSS and VDDIO for push-pull drivers, 50 ohm to VTT for open-drain drivers, or 100 ohm differential for low-voltage differential signaling (LVDS). Active termination is a more exotic circuit that turns on when the input passes some set of thresholds but draws no dc power when the input is below the thresholds.

The aggregate IV curve for a receiver with on-die termination is the super-position of the ESD diode curve and the termination curve. One word of caution: Resistors on silicon typically have a much wider tolerance (~30%) than what you might expect from a resistor pack or discrete component available to a PC board designer, so it is essential to ensure that the interface will still function with these wide tolerances when using on-die termination. See Chapter 4, "DDR2 Case Study," for a practical example of DDR2 ODT.

2.16   CMOS PUSH-PULL DRIVER

Driver IV characteristics naturally vary with IO circuit design, the most common of which is the CMOS push-pull driver. A pair of inverters accomplishes the physical function of a push-pull driver, which is to move large quantities of charge on and off the die. A few extra logic gates give this circuit tri-state capability.

The output stage of a push-pull driver is just a large number of parallel n-channel devices and a larger number of p-channel devices connected to the IO pad. In Figure 2-21, the multipliers indicate how many 0.5 μm x 20 μm transistors are in parallel. The current through the channel of a MOSFET transistor is a function of its drain-source voltage, Vds (the voltage across the channel resistance), *and* its gate-source voltage, Vgs. Once Vgs reaches a critical threshold, the channel comes alive and begins to conduct. Raising Vgs beyond the threshold increases the conductivity of the channel. As the output of the driver switches from a low to a high voltage, the Vgs voltages for the n-channel and p-channel output transistors transition through an entire continuous family of curves!

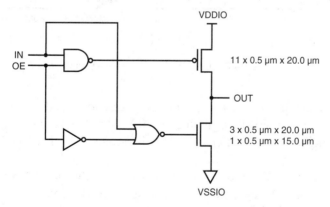

Figure 2-21   CMOS push-pull driver.

This family of curves is not visible from the output pin of the chip because the user cannot directly control the internal nodes within the driver circuit. The output of a chip under dc test conditions is either in a high or a low state, and the user can only see one curve from the family of curves—the curve that represents the driver's final state after the transient event of switching has passed (dashed lines in Figure 2-22). It is not possible to vary the voltage at the gate of the output transistors and produce the classical family of curves. This important fact influences the generation and simulation of behavioral models.

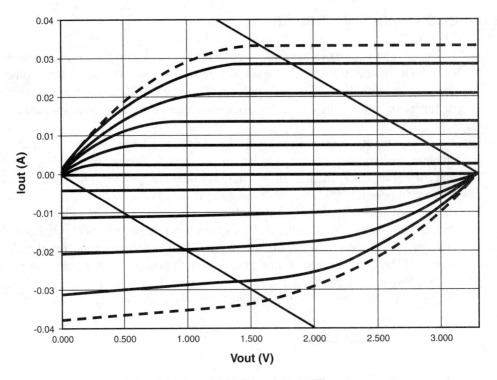

Figure 2-22 CMOS driver output IV curves.

Note that the solid curves in Figure 2-22 correspond to Vgs voltages spaced evenly at 0.5 V intervals beginning at 1.0 V. The threshold of the p-channel transistor is higher than the n-channel transistor, and one of the p-channel curves is flat at zero.

The output IV curves for a driver obey a nonlinear relationship that is difficult to quantify. In fact, device physicists have divided the IV space into two regions in which different sets of equations apply. These equations and their derivatives must remain continuous across the boundary between the two regions; if they are not, nonconvergence of SPICE simulations will occur, and this never made anybody's day.

2.17   OUTPUT IMPEDANCE

Is it possible to extract any useful information from these curves without running a simulation? Imagine holding the input of the driver in a low state and connecting a resistor between the output and VDDIO. What will the output voltage and current be? An easy way to answer this question graphically is to draw the linear IV characteristic of the resistor on top of the "final" IV curve of the n-channel pull-down transistor—that is, the curve that represents the largest value of Vgs achieved when switching is complete. The intersection of the line and the curve is the operating point of the driver with that particular load. Dividing the voltage by the current gives the output impedance of the driver—at that operating point.

What does output impedance mean, and what is its relevance? If we load the driver with an ideal resistor whose value is the same as the output impedance of the driver, the output voltage will be VDDIO/2 by virtue of the voltage divider equation. If the value of the resistor is same as the impedance of the transmission line connected to the driver in a system application, the output voltage will be the same as the input to the transmission line when the driver is in the middle of the switching event—that is, before the reflected wave has returned to it. This is a useful thing to know because it is important that the driver be strong enough to deliver the energy it takes to cause a full-swing waveform at the receiver input.

The driver will deliver its maximum power to the load when its impedance is equal to that of the load. Keep in mind that the output impedance of the driver is a function of its load. The IV curve of another resistor will intersect the IV curve of the driver at a different point. Another important point is that there are really *two* output impedances for a push-pull driver: one for the p-channel devices and one for the n-channel devices. In a well-designed driver, these two impedances will be nearly equal, but this is not always the case. An unbalanced driver can be difficult to work with because optimizing the net topology for the rising edge will cause the falling edge to be too strong or too weak.

The dc operating point simulations in Figures 2-23 and 2-24 define another method to extract driver output impedance using the voltage divider equation to calculate the equivalent resistance of the driver given a particular resistive load. This technique is also useful in a transient simulation or in the lab where sweeping the output current of a driver requires special test equipment.

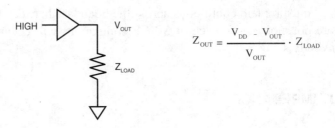

$$Z_{OUT} = \frac{V_{DD} - V_{OUT}}{V_{OUT}} \cdot Z_{LOAD}$$

Figure 2-23   Calculation of output high impedance.

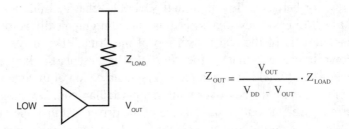

$$Z_{OUT} = \frac{V_{OUT}}{V_{DD} - V_{OUT}} \cdot Z_{LOAD}$$

Figure 2-24   Calculation of output low impedance.

2.18   OUTPUT RISE AND FALL TIMES

Output rise and fall time are, in a sense, a combination of output impedance and capacitance. When a driver switches, it must move charge between its own intrinsic capacitance and the power supply rails. The size of these capacitors and the channel resistance of the transistors govern the rate at which the driver is able to move charge. We can define the intrinsic rise and fall times of the driver as the fastest possible time that a driver can change states when no load is present. The switching time of the pre-drive stage—that is, the transistors immediately preceding the final output stage—also influences the output rise and fall times. As is the case with output impedance, the rise and fall times may be significantly different if the circuit designer did not intentionally make them the same.

Output rise and fall times have a domino effect on the system design. At a bare minimum, the knee frequency or corner frequency associated with a given rise or fall time determines the bandwidth requirement for all interconnect between the driving and receiving chips. Likewise, it also determines the bandwidth requirements for models used in simulation, and this bandwidth is typically higher than that required for the actual interconnect since simulators must accurately reproduce the corners of waveforms and noise pulses of short duration. The minimum rise or fall time determines the length at which a piece of wire begins behaving like a transmission line.

On the negative side, the instantaneous changes in current that accompany sharp edges produce disturbances on the power distribution system that may ultimately result in the violation of flip-flop setup and hold times. Rapidly changing electric and magnetic fields surrounding a discontinuity such as a connector cause coupling between adjacent signal conductors—another detractor from system performance.

As signal integrity engineers, one of the first tasks we face at the beginning of a new project is cataloging rise and fall times for each bus and each chip, for they will dictate the interconnect technology and models requirements (refer to Table 1-2). The intrinsic rise and fall times will be misleading, however. The actual rise and fall times measured at a package pin will most certainly be lower as the wave encounters extra capacitance, attenuation from conductor and dielectric losses, and perhaps the inductance of a bond wire array. After passing through the first layers of interconnect, the corners of the waveform will no longer be as sharp.

It is no coincidence that the derivative of a typical digital waveform has a Gaussian-like shape—the same shape as a crosstalk pulse. Coupled noise is a direct function of dV/dt. Cataloging this metric in mV/ns will help you understand where crosstalk hotspots are and how to control them. Be aware that instantaneous dV/dt can vary as much as 30% from a linear estimate using the 20% and 80% crossing points. Lower speed interfaces are not sensitive to this subtlety, but it can make a difference worth paying attention to when you're counting mV. The dV/dt in Figure 2-25 is 9.1 V/ns using the linear approximation; the actual instantaneous dV/dt in Figure 2-26 is 12 V/ns.

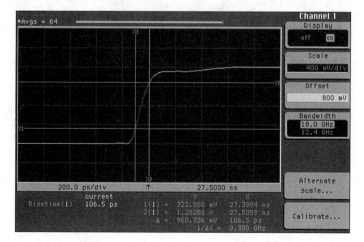

Figure 2-25   Edge rate calculated at 20% and 80% points.

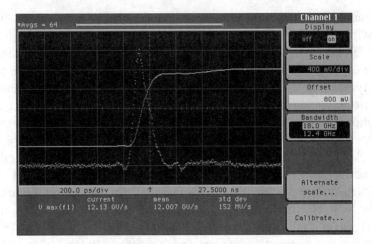

Figure 2-26   Instantaneous dV/dt.

2.19 CMOS CURRENT MODE DRIVER

Most high-speed serial interfaces do not use CMOS push-pull drivers; they use current mode drivers that bear some resemblance to the schematic in Figure 2-27. This is really just the same circuit as the differential receiver shown previously in Figure 2-19, except the transistors are much larger and the resistors are in the neighborhood of 50 ohm.

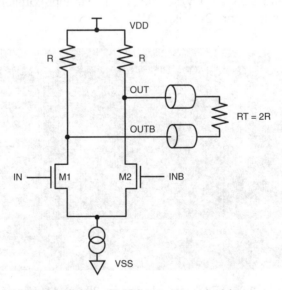

Figure 2-27   CMOS current mode driver.

Let's do a quick dc analysis of this circuit. The predrive stage drives M1 and M2 differentially, so when M1's channel is conducting, M2's channel is off. The equivalent circuit is shown in Figure 2-28. To get the voltage at the positive output pin, calculate the parallel resistance of R1 combined with R2 + RT. It's 37.5 ohm, which means the voltage at node P is 750 mV. Because the right branch has three times the resistance as the left branch, it carries ¼ the current or 5 mA. The drop across RT is 500 mV. The mirror analysis applies when M2 is on, and each output swings between 750 mV and 1250 mV for a net differential swing of 1000 mV at the receiver input.

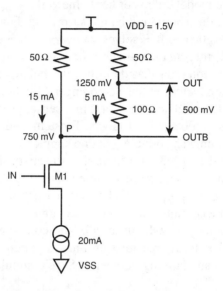

Figure 2-28   DC analysis of CMOS current mode driver.

This circuit has some favorable qualities. Ideally, there is no net change in current through the source when the output switches states. If the legs are balanced properly and driven "exactly" out of phase, the net change in current through the VDDIO supply will be zero as well. This means no simultaneous switching noise between VDDIO and VSS on the die and no noise-induced jitter. Well, almost none. If any circuit were ideal, we would all be out of jobs.

The CMOS push-pull driver shown previously in Figure 2-21 uses both p-channel and n-channel FETs. In an ideal circuit, both channels would have the same resistance. In real life, the p-channel and n-channel resistances mistrack each other, leading to asymmetrical rise and fall times. The current mode driver solves this problem by using only n-channel transistors.

2.20 BEHAVIORAL MODELING OF IO CIRCUITS

Behavioral modeling has not always enjoyed a good reputation. In the early days of the signal integrity boom, many die-hard proponents of SPICE simulation (myself included) could not understand how such a simple model could accurately reproduce the behavior of dozens of transistors in a complex IO circuit. The complexity of the two models seemed to be different by one or two orders of magnitude. The remarkable fact, however, is how closely a behavioral model *does* mimic the characteristics of an IO circuit—*if* certain assumptions are satisfied and *if* the author of the model did his or her homework.

In hindsight, this should not be too surprising. For all its complexity, the set of device equations that SPICE solves for every instance of a transistor is really a behavioral model, too, albeit a much more complicated behavioral model. The more fundamental equation that describes the behavior of quantum mechanical systems in semiconductors is the Schrödinger Equation, and no one would ever consider using it to simulate transistors. In a sense, every mathematical model of a physical system is a behavioral model. Einstein said we should strive to make these models as simple as possible—but no simpler!

IBIS emerged in 1993 with a mission to satisfy the modeling needs of the growing signal integrity community while preserving IO circuit and semiconductor process intellectual property. Today, another urgent need presses the community toward advanced time- and frequency-domain simulation techniques: the need for simulations that will run at all. The combination of advanced CMOS device technologies, lossy transmission lines, s-parameters, and large-scale coupling has rendered an alarming number of SPICE simulations a quivering blob of pudding.

The primary job responsibility of a signal integrity engineer is to ensure the reliability of chip-to-chip interfaces, not to debug models. Time spent resolving nonconvergence problems and waiting for multiday simulations to finish is time wasted. In a more efficient paradigm, we would use behavioral simulation to define the boundaries of the design space, which probably accounts for 80% of all simulations. Then, for the most demanding interfaces, SPICE simulation would provide back-end verification that the behavioral modeling was indeed correct.

IBIS version 1.1 is a natural platform on which to build an understanding of behavioral modeling and simulation of IO circuits. A variety of industry-standard bus technologies is available to signal integrity engineers today: Gunning Transceiver Logic (GTL), High-Speed Transceiver Logic (HSTL), and Stub Series Terminated Logic (SSTL), to name a few of the JEDEC standards. There is an even wider variety of circuit design techniques, some of which are company jewels and may never be known to the world at large. Among the well-known techniques are the simple push-pull driver, open-drain driver, multiple phased output stages, dynamic termination, de-emphasis, equalization, and so on. HSTL and SSTL both use a push-pull output, while classic GTL is an open-drain output with a feedback network.

Any behavioral model assumes one of these known circuit design techniques as a *template*. The information contained in an IBIS model is really not a model at all; the model resides under the hood of the simulator. An IBIS model is merely a database of circuit parameters and tables that a simulator will load into its own unique behavioral model before initiating the simulation. If the circuit template in the simulator does not match the information in the IBIS model, the results of the simulation may not be worth much at all.

2.21   BEHAVIORAL MODEL FOR CMOS PUSH-PULL DRIVER

The 50 ohm CMOS push-pull driver from the previous SPICE simulations will serve as a demonstration vehicle for understanding how behavioral simulation works, how it is different from SPICE simulation of the same circuit, and what its accuracy limitations are. How do the transistor model and the behavioral model compare to one another?

The p-channel and n-channel transistor symbols in Figure 2-29 represent many parallel transistors that share a common gate, source, and drain. Beneath each of those transistors lies one instance of the device equations that govern the current-voltage behavior of that transistor. The SPICE model statement personalizes the device equations for a particular semiconductor process. In the behavioral model, a single time-dependent voltage-controlled current source (VCCS), whose current-voltage characteristics are defined by a look-up table, replaces all the p-channel transistors in the output stage.

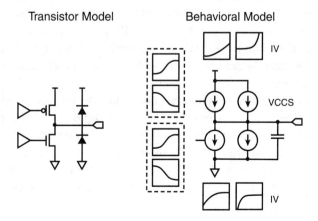

Figure 2-29   Behavioral model for CMOS push-pull driver.

The behavioral simulator ensures that whenever a given current is flowing through the source, the voltage across it will be the same value found in the lookup table for that current. The lower VCCS represents all of the n-channel transistors. Herein lies one of the *fundamental assumptions* of behavioral modeling: Although SPICE has access to the continuous family of IV curves for any transistor, the behavioral model has only one curve and must make assumptions about what all the other curves in the family look like. This is true for both the n-channel and p-channel devices.

Another fundamental assumption involves the predrive stage of the IO circuit, represented in this diagram by the two triangular buffers that independently drive the pull-up and pull-down transistors. (Recall that in the CMOS driver, these were actually NAND and NOR gates rather than simple buffers.) The switching rates of these predrive buffers determine the rise and fall times of the output stage, together with the IV curves and loading.

Programs that generate IBIS models do not have access to node voltages that are internal to the driver SPICE subcircuit. This is particularly true if the SPICE subcircuit is encrypted or if the IBIS model was extracted from lab measurements. Therefore, the behavioral simulator must make further assumptions about the rise and fall times at the input of the driver final stage based on rise and fall times of the output.

In the case of an IBIS 1.1 model, the simulator also assumes that the pull-up and pull-down transistors switch at the same time, which is not necessarily the case if the predrive stage has two independent buffers. This also is a fundamental assumption. In the diagram of the behavioral model shown previously in Figure 2-29, the third terminal on the VCCS represents the time-domain waveform that turns the source on and off, which is analogous to the Vgs waveform at the input of the final stage in a SPICE driver subcircuit.

Two sets of voltage-time waveforms are to the left of the VCCS enclosed by dashed boxes. One set of waveforms controls the pull-up transistors, and the other controls the pull-down transistors. In IBIS 1.1, both sets of waveforms begin switching at the same time. IBIS 2.1 allows independent control over the pull-up and pull-down transistors via a set of voltage-time waveforms rather than a simple rise and fall time.

The second set of voltage-controlled current sources corresponds to the two ESD protection devices. Compared to the complexity of the pull-up and pull-down elements, these seem trivial as they have no time-dependent waveform turning them on and off. Each VCCS draws current from its corresponding supply rail. If the circuit were a receiver with some form of on-die termination, the IV curves for the termination would be superimposed on top of the ESD curves.

Finally, the lumped capacitor, called C_comp in IBIS, represents all capacitances that are connected to the output node of the circuit: transistors, ESD protection diodes, metal, and bond pad or solder ball. The value of this capacitor is not a function of time or voltage, which is a good assumption if the semiconductor junctions are reverse biased. As soon as they begin to conduct this assumption breaks down.

The physical location of C_comp may also be relevant. Drivers with on-chip series termination will have their capacitance distributed on either side of the resistor, and this can become an accuracy issue. In simultaneous switching simulations, the assumption that the other node of the capacitor is connected to ground breaks down. In the case of the CMOS driver, however, a constant lumped C_comp remains a valid assumption.

2.22 BEHAVIORAL MODELING ASSUMPTIONS

In summary, the fundamental assumptions for an IBIS 1.1-compatible behavioral model of the CMOS driver are as follows:

1. The behavioral model topology that the simulator employs is consistent with the IO circuit design (in this case, a simple CMOS push-pull final output stage).
2. The behavioral simulator interpolates between a single pull-up (p-channel) IV curve and a single pull-down (n-channel) IV curve.
3. The behavioral simulator infers the Vgs waveform at the input to the final stage from the rise and fall times measured at the output of the final stage with a 50 ohm load.
4. The pull-up and pull-down transistors begin switching at the same time.
5. A single constant-value capacitor represents all capacitances seen looking into the pad node.

2.23 TOUR OF AN IBIS MODEL

Armed with an understanding of the IO circuit models used in signal integrity simulations, you can confidently assess the quality and accuracy of the models needed to perform your own analysis of the timing and voltage margins for a given interface. Model assessment is a significant step toward engineering a reliable digital interface.

A close look at an example IBIS model points out the differences and similarities between a behavioral model and a SPICE transistor model. The data structures within this IBIS model file illustrate the assumptions a behavioral simulator makes about the electrical characteristics of an IO circuit and how those assumptions compare to a SPICE model of the same circuit. The IBIS Committee has published a good deal of educational material, including instructions on how to create IBIS models.

The following simple IBIS model has three main sections: the header, the pin table, and the model data. IBIS is fundamentally a chip-centric specification, so there is a one-to-one correspondence between an IBIS model and the chip it represents. In some cases a chip manufacturer may publish a library of individual models from which the user can assemble a complete IBIS model—for example, ASICs.

The header contains various bits of background information about the file. The pin table lists each pin in the chip and, if the pin is a signal, associates a model with it. In the language of IBIS, *keyword*s are enclosed in brackets, [ ], and may be followed by *subparameters* that are associated with the preceding keyword. For example, the [Model] keyword defines a new set of IO circuit model parameters, including pin capacitance, rise and fall times, and IV curves. The "Model Type" subparameter is hierarchically below the [Model] keyword, and it can take on the values Input, Output, I/O, 3-state, or Open_drain. The [Model] keyword covers the first assumption of behavioral modeling: It defines the circuit topology that the simulator will use.

```
[IBIS Ver]      2.1

[File Name]     74lvc125.ibs

[File Rev]      1.0

[Date]          October 5, 2005

[Source]        SPICE simulation of 50 Ohm push-pull driver
                MOSIS 0.5 um CMOS process SCN05H
                Designed for demonstration purposes by Greg Edlund

[Notes]         Correlated:
                HSPICE behavioral simulations using IBIS model with
                HSPICE transistor simulations using SPICE model

[Disclaimer]    none

[Copyright]     none

[Component]     74LVC125
```

```
[Manufacturer]   none

[Package]
|
| variable              typ               min               max
|
R_pkg                   1.0m              NA                NA
L_pkg                   0.01nH            NA                NA
C_pkg                   0.01pF            NA                NA

|
|
[Pin]    signal_name    model_name    R_pin   L_pin   C_pin
|
|
   1       oe1           rcv           0.1     5.0nH   1.2pF
   2       in1           drv50         0.1     4.0nH   1.0pF
   3       out2          rcv           0.1     4.0nH   1.0pF
   4       gnd           gnd           0.1     5.0nH   1.2pF
   5       in2           rcv           0.1     5.0nH   1.2pF
   6       out1          drv50         0.1     4.0nH   1.0pF
   7       oe2           rcv           0.1     4.0nH   1.0pF
   8       vdd           power         0.1     5.0nH   1.2pF

|-------------------------------------------------------------------
| rcv model definition
|-------------------------------------------------------------------

[Model]      rcv
Model_type   Input
Polarity     Non-Inverting
Vinl =       1.51
Vinh =       1.84


|-------------------------------------------------------------------
|
|                       typ               min               max
|
C_comp                  3.88pF            3.82pF            3.92pF
[Temperature Range]     55.0              85.0              25.0
[Voltage Range]         3.300V            3.135V            3.456V

|-------------------------------------------------------------------

[GND_clamp]
|
| voltage               I(typ)          I(min)          I(max)
|
-1.0000e+00             -5.0000e-01      -5.0000e-01      -5.0000e-01
-9.5000e-01             -5.0000e-01      -5.0000e-01      -5.0000e-01
-9.0000e-01             -5.0000e-01      -5.0000e-01      -5.0000e-01
-8.5000e-01             -5.0000e-01      -5.0000e-01      -5.0000e-01
-8.0000e-01             -5.0000e-01      -5.0000e-01      -5.0000e-01
-7.5000e-01             -5.0000e-01      -5.0000e-01      -1.4170e-01
-7.0000e-01             -1.1840e-01      -5.0000e-01      -2.1350e-02
```

-6.5000e-01	-2.0660e-02	-1.0370e-01	-3.8280e-03
-6.0000e-01	-3.7910e-03	-2.0660e-02	-1.0450e-03
-5.5000e-01	-7.6490e-04	-4.1380e-03	-4.1960e-04
-5.0000e-01	-1.5430e-04	-8.2430e-04	-1.6330e-04
-4.5000e-01	-2.9830e-05	-1.6670e-04	-3.1700e-05
-4.0000e-01	-6.6550e-06	-3.4160e-05	-7.3930e-06
-3.5000e-01	-1.5620e-06	-7.1260e-06	-1.7670e-06
-3.0000e-01	-3.7850e-07	-1.5200e-06	-4.1260e-07
-2.5000e-01	-9.2930e-08	-3.3230e-07	-9.3940e-08
-2.0000e-01	-2.2780e-08	-7.4750e-08	-2.0850e-08
-1.5000e-01	-5.5490e-09	-1.7490e-08	-4.5410e-09
-1.0000e-01	-1.3730e-09	-4.5040e-09	-1.0100e-09
-5.0000e-02	-3.8400e-10	-1.5210e-09	-2.6790e-10
0.0000e+00	-9.4850e-11	-7.0640e-10	-8.0010e-11
3.3000e+00	0.0000e+00	0.0000e+00	0.0000e+00

[POWER_clamp]

voltage	I(typ)	I(min)	I(max)
-1.3000e+00	5.0000e-01	5.0000e-01	5.0000e-01
-1.2500e+00	5.0000e-01	5.0000e-01	5.0000e-01
-1.2000e+00	5.0000e-01	5.0000e-01	5.0000e-01
-1.1500e+00	5.0000e-01	5.0000e-01	5.0000e-01
-1.1000e+00	5.0000e-01	5.0000e-01	4.1438e-01
-1.0500e+00	5.0000e-01	5.0000e-01	1.5942e-01
-1.0000e+00	3.2280e-01	5.0000e-01	2.2776e-02
-9.5000e-01	5.5080e-02	5.0000e-01	3.2546e-03
-9.0000e-01	9.3990e-03	5.0000e-01	4.6537e-04
-8.5000e-01	1.6040e-03	5.0000e-01	6.6593e-05
-8.0000e-01	2.7370e-04	5.0000e-01	9.5409e-06
-7.5000e-01	4.6720e-05	5.0000e-01	1.3697e-06
-7.0000e-01	7.9740e-06	5.0000e-01	1.9724e-07
-6.5000e-01	1.3610e-06	5.0000e-01	2.8590e-08
-6.0000e-01	2.3250e-07	5.0000e-01	4.2249e-09
-5.5000e-01	3.9780e-08	5.0000e-01	6.8324e-10
-5.0000e-01	6.8760e-09	5.0000e-01	1.6312e-10
-4.5000e-01	1.2550e-09	5.0000e-01	8.4801e-11
-4.0000e-01	2.9330e-10	5.0000e-01	7.1698e-11
-3.5000e-01	1.2750e-10	4.7291e-01	6.8343e-11
-3.0000e-01	9.7720e-11	2.0228e-01	6.6467e-11
-2.5000e-01	9.5670e-11	4.0026e-02	6.4857e-11
-2.0000e-01	9.3680e-11	7.9205e-03	6.3250e-11
-1.5000e-01	9.1690e-11	1.5677e-03	6.1647e-11
-1.0000e-01	8.9700e-11	3.1022e-04	6.0040e-11
-5.0000e-02	8.7710e-11	6.1390e-05	5.8437e-11
0.0000e+00	8.5720e-11	1.2147e-05	5.6827e-11
3.3000e+00	0.0000e+00	0.0000e+00	0.0000e+00

```
|----------------------------------------------------------------
| drv50 model definition
|----------------------------------------------------------------
```

```
[Model]        drv50
Model_type     3-state
Polarity       Non-Inverting
Enable         Active-High
Vmeas =        1.5V
Cref  =        10pF
Vref  =        0


|-----------------------------------------------------------------
|
|                          typ          min          max
|
|C_comp                    3.81pF       3.76pF       3.86pF
|[Temperature Range]       55.0         85.0         25.0
|[Voltage Range]           3.300V       3.135V       3.456V


|-----------------------------------------------------------------
|

[Pulldown]
|
| voltage                  I(typ)       I(min)       I(max)
|
-1.0000e+00                -5.0000e-01   -5.0000e-01   -5.0000e-01
-9.5000e-01                -5.0000e-01   -5.0000e-01   -5.0000e-01
-9.0000e-01                -5.0000e-01   -5.0000e-01   -5.0000e-01

[Pullup]
|
| voltage                  I(typ)       I(min)       I(max)
|
-1.3000e+00                5.0000e-01    5.0000e-01    5.0000e-01
-1.2500e+00                5.0000e-01    5.0000e-01    5.0000e-01
-1.2000e+00                5.0000e-01    5.0000e-01    5.0000e-01

[GND_clamp]
|
| voltage                  I(typ)       I(min)       I(max)
|
-1.0000e+00                -5.0000e-01   -5.0000e-01   -5.0000e-01
-9.5000e-01                -5.0000e-01   -5.0000e-01   -5.0000e-01
-9.0000e-01                -5.0000e-01   -5.0000e-01   -5.0000e-01

[POWER_clamp]
|
| voltage                  I(typ)       I(min)       I(max)
|
-1.3000e+00                5.0000e-01    5.0000e-01    5.0000e-01
-1.2500e+00                5.0000e-01    5.0000e-01    5.0000e-01
-1.2000e+00                5.0000e-01    5.0000e-01    5.0000e-01


|-----------------------------------------------------------------
|

[Ramp]
```

```
!
!
!                               typ            min            max
!
dV/dt_r                         1.022/0.345n   0.756/0.406n   1.268/0.281n
dV/dt_f                         1.039/0.340n   0.822/0.367n   1.252/0.292n
R_load = 50


[End]
```

2.24   IBIS HEADER

The header stores some valuable information. The version keyword allows the
syntax checker to decide which features are legal. The filename keyword and the
actual name of the file must match. This file happens to represent a fictional low-
voltage CMOS (LVC) dual tri-state buffer in a standard 125 form factor that uses
the CMOS push-pull driver circuit, drv50.

The source, notes, and disclaimer keywords are of special interest. A consci-
entious author will give some indication of the conditions under which he or she
created the IBIS model. Did the information come from production-level SPICE
models that were extracted from an integrated circuit layout? Or perhaps the
source was lab measurements of an actual component, which only apply to one
process-temperature-Voltage point. If the author's name does not appear under the
source keyword, the user might wonder whether anyone is willing to support the
IBIS model.

Many signal integrity engineers find this common disclaimer to be particularly disturbing: "…for modeling purposes only." This suggests that the vendor
does not stand behind the IBIS model in the same way it stands behind the component datasheet. This is disconcerting because model parameters are in some
ways even more influential in determining the success or failure of a design than
the simpler parameters found in a component datasheet. However, if the quality of
an IBIS model is not consistent with the needs of the design, signal integrity engineers have a tool for affecting change that is more potent than complaining:
Include model quality and accuracy in the purchase order.

2.25   IBIS PIN TABLE

The pin table, designated by the [Pin] keyword, associates a [Model] keyword
with a chip pin number. Remember, the information found in an IBIS model is not
really the model itself but rather the *input parameters* to the model; the model

resides inside the simulator. The language can be a bit confusing on this point. The first two columns in the table store the pin number and name as found in the component datasheet. If the user is running automated post-route simulation, the software will match the pin number in the logical netlist with a pin number from this table. Every model name found in the third column must have a matching [Model] keyword under the model section of the datasheet. The remaining three columns contain lumped RLC parameters for the purpose of modeling the package pin.

There are three ways to code IC package model data in IBIS. If the optional per-pin parameters do not exist under the pin table, the simulator assigns the default values found under the [Package] keyword to every pin. IBIS 2.1 offers a more sophisticated means of coding package models using the [Define Package Model] keyword. In the following example, the model for pin 10 of a BGA package comprises three sections: a 2 nH lumped inductance for the bond wire, a 0.5 in length of 50 ohm transmission line for the copper wire, and a 1 pF lumped capacitance for the solder ball and pad. In this case, the information truly does define a circuit-based model for the package; there is no ambiguity left for the simulator to resolve.

```
|
10 Len=0 L=2nH / Len=0.5 C=3pF L=9nH R=1m / Len=0 C=1pF /
|
```

2.26 IBIS RECEIVER MODEL

The first set of model data following the pin table corresponds to the receiver circuit, rcv. The previous section on IO circuit characteristics discussed three primary attributes of a simple CMOS receiver, and each has its place in the IBIS receiver model: threshold voltages, pin capacitance, and IV curves. Most IBIS models for low-voltage CMOS (3.3 V) use the default TTL thresholds commonly found in component datasheets—that is, 0.8V and 2.0V. I extracted the thresholds for this IBIS model from SPICE simulations of the receiver circuit, which means they are more accurate and result in less conservatism in the system design.

The [Temperature Range] and [Voltage Range] keywords document the conditions under which the model author derived the other keywords and subparameters. Some degree of confusion has surrounded these keywords because the values found under the min column are not always the smallest numerical values, and the values found under the max column are not always the largest. The values found under the min column for the [Temperature Range] and [Voltage Range] keywords correspond to the conditions that gave the highest channel resistance, also called weak, slow, or worst, depending on convention at your place of employment. For a CMOS process, high temperature and low voltage go along with weak silicon, and these conditions are the ones that produced the IV curves found in the min column of the [GND_clamp] and [POWER_clamp] keywords. Conversely, the values found under the max column for the [Temperature Range] and [Voltage Range] keywords represent the simulation conditions that gave low channel resistance, also called strong, fast, or best.

The confusion arises in two places. First, the min value of the [Temperature Range] keyword is the highest temperature for CMOS chips. Min really refers to current. Second, the C_comp subparameter does not obey the rules described in the previous paragraph! The min value for C_comp should be the smallest numerical value, and the max value should be the largest numerical value—regardless of the process, temperature, and voltage conditions.

The [GND_clamp] keyword stores the IV curves for the current that flows in through the pad node and out through the VSS node. In the CMOS receiver circuit, the current flows only through the ESD protection devices; the input to the buffer is two high-impedance MOSFET gates that draw very little current. In more sophisticated receivers, current may also flow through passive or active termination devices, usually transistors.

The [POWER_clamp] keyword stores the IV curves for the current that flows in through the VDDIO node and out through the pad node. According to the current convention in IBIS, negative currents flow into the pad node and positive currents flow out. Another important IBIS convention: The voltages in the [POWER_clamp] tables are referenced to VDDIO rather than VSS. Because this is a CMOS circuit whose VDDIO rail is 3.3V, this means that the first column in the table reads -3.3V when the actual input is at ground and 0V when the input is at VDDIO! This feature allows the simulator to change VDDIO voltages without having to recalculate the IV curves.

In this IBIS receiver model, the voltage-controlled current sources in the corresponding behavioral model are always in an active state. Some receiver circuits activate their termination devices dynamically; special syntactical constructs in IBIS 3.2 facilitate modeling these circuits.

2.27   IBIS DRIVER MODEL

Compare the first few lines of code immediately following the drv50 and rcv [Model] keywords; the only difference is that the subparameters Vmeas, Cref, Vref have replaced the subparameters Vinl and Vinh. The model for a tri-state driver does not need receiver switching thresholds, but it does require a description of the standard load found in the component datasheet. During automated post-route simulation, the simulator will load drv50 with Cref, measure the time at which the package pin node crosses Vmeas, and store this value for later use as time zero in all timing calculations. If you are extracting interconnect timing from post-route simulations, it is critical that these values exactly match those in the component datasheet.

The code under the drv50 [Model] keyword contains the same [GND_clamp] and [POWER_clamp] tables found under the rcv [Model] keyword because these IV curves represent the ESD protection devices, which are the same for both IO circuits. The two keywords [Pulldown] and [Pullup] are unique to drivers, and they store the IV curves for the n-channel and p-channel transistors in the output stage of this simple CMOS push-pull driver. Like the [POWER_clamp] table, the [Pullup] table is also referenced to VDDIO.

Extracting the [Pulldown] and [Pullup] tables for a driver is not as simple as extracting the [GND_clamp] and [POWER_clamp] tables for a receiver because the transistor IV curves and the ESD device IV curves are embedded in the same set of curves—unless the author manually edits the SPICE subckt for the driver and comments out the ESD devices. The common solution to this problem is to tri-state the driver, measure the ESD device IV curves, and subtract them from the composite IV curves to get the [Pulldown] and [Pullup] tables. This procedure is the source of the nonmonotonic warnings IBIS users often see when they run ibis-chk. The SPICE model for a diode contains an exponential equation, which can generate huge differences in current for very small changes in voltage:

$$I = I_{SAT} \left[e^{\frac{Vq}{nkT}} - 1 \right]$$

Equation 2-2

Subtracting one huge number from another huge number causes round-off error in the less significant digits. These errors are usually innocuous and do not have a significant effect on simulation results; simulators and users usually ignore them. However, they do have the potential to generate a large number of warning messages that need to be sorted.

The one remaining piece of information needed to make a behavioral driver model work is how fast the [Pulldown] and [Pullup] tables transition from an off state to an active state. In an IBIS 1.1 datasheet, the [Ramp] keyword stores this information, which is simply Δt and ΔV measured at the 20% and 80% points when the driver is loaded with R_load, which is usually 50 ohm to VSSIO for a rising edge and 50 ohm to VDDIO for a falling edge. The 20% and 80% points are a fraction of the output swing rather than VDDIO.

From this information, the behavioral simulator constructs an "internal stimulus function" that activates the [Pullup] table while deactivating the [Pulldown] table in the case of a rising edge. It also has to account for the fact that the [Ramp] values were extracted in the presence of C_comp, the pin capacitance. The behavioral simulator must be able to correctly replicate the values in the [Ramp] table when loaded with R_load.

Not as simple as it appears at first glance!

2.28   BEHAVIORAL MODELING ASSUMPTIONS (REPRISE)

It is now possible to restate the fundamental assumptions of behavioral modeling cast in the language of IBIS 1.1, as follows:

1. The behavioral simulator assumes the IO circuit topology to be one of the following forms: Input, Output, I/O, 3-state, or Open_drain.
2. The behavioral simulator interpolates between the [Pullup] table and the [Pulldown] table to find the trajectory of the driver output through IV space during the switching event.
3. The behavioral simulator constructs an internal driving function from the rise and fall times measured with a 50 ohm load and uses this function to turn the [Pullup] and [Pulldown] tables on and off.
4. The [Pullup] and [Pulldown] tables begin switching at the same time.
5. C_comp represents all capacitances seen looking into the pad node.

It is extremely important to understand that the behavioral simulator has no information about anything that happens prior to the driver output stage or after the receiver input. This implies that the concept of IO circuit delay has no meaning in an IBIS-based behavioral model. *It also implies that all timing constraints associated with storage elements inside the chip reside elsewhere!*

In a behavioral model based on IBIS 1.1 assumptions, the [Pulldown] and [Pullup] tables begin switching at the same point in time even though this may not be the case in a real-life IO circuit. The circuit designer may have intentionally skewed the turn-on of the p-channel devices from the turn-off of the n-channel devices to prevent a temporary short circuit between VDDIO and VSSIO. This current does nothing useful and generates di/dt noise. To enable modeling of non-linear skewed internal driving functions, IBIS 2.1 implemented a feature known as VT tables, which look like this:

```
[Rising Waveform]
R_fixture = 50
V_fixture = 0.0
|
|
|  Time                        V(typ)          V(min)          V(max)
|
|

[Rising Waveform]
R_fixture = 50
V_fixture = 3.3
|
|
|  Time                        V(typ)          V(min)          V(max)
|
|

[Falling Waveform]
R_fixture = 50
V_fixture = 0.0
|
|
|  Time                        V(typ)          V(min)          V(max)
|
|

[Falling Waveform]
R_fixture = 50
V_fixture = 3.3
|
|
|  Time                        V(typ)          V(min)          V(max)
|
|
```

Like the [Ramp] keyword, there is a subparameter, R_fixture, in the VT table that specifies the load resistance. Unlike the [Ramp] keyword, the termination voltage of the R_fixture is not assumed; the V_fixture subparameter defines the termination voltage. Another difference is that two rising waveforms replace the single dV/dt quotient: one for 50 ohm to VSS, and one for 50 ohm to VDDIO. The same is true for the falling edge. With more complicated information at its disposal, the behavioral simulator can construct a more realistic internal stimulus function, which results in a better match between behavioral and transistor simulation results—at least for loads close to R_fixture.

2.29   COMPARISON OF SPICE AND IBIS MODELS

The most obvious difference between these two kinds of models is that many fewer lines of code are in the drv50 SPICE subckt than are under the corresponding IBIS [Model] keyword. This is not always the case, since some SPICE models have hundreds of transistor model calls and a plethora of resistors and capacitors that model the wires between transistors. The reason for the apparent simplicity of the SPICE model for this demonstration circuit is that most of the complexity is hidden in the transistor model equations, one instance of which is called every time a transistor appears in the subckt. SPICE calculates the terminal currents and voltages of every transistor rather than relying on tables to calculate only the pad node voltage and the currents flowing in and out of the pad node.

```
*-------------------------------------------------------------------*
*             a   en  pad  vss  vdd  vssio  vddio
.subckt drv50 10  20  30   100  200  1000   2000
*-------------------------------------------------------------------*

xnand  10 20 40 100 200   drvnand
xnor   10 60 50 100 200   drvnor
xinv      20 60 100 200   inv

mpout01 30 40 2000 2000 cmosp l=0.5u w=20.0u
mpout02 30 40 2000 2000 cmosp l=0.5u w=20.0u
mpout03 30 40 2000 2000 cmosp l=0.5u w=20.0u
mpout04 30 40 2000 2000 cmosp l=0.5u w=20.0u
mpout05 30 40 2000 2000 cmosp l=0.5u w=20.0u
mpout06 30 40 2000 2000 cmosp l=0.5u w=20.0u
mpout07 30 40 2000 2000 cmosp l=0.5u w=20.0u
mpout08 30 40 2000 2000 cmosp l=0.5u w=20.0u
mpout09 30 40 2000 2000 cmosp l=0.5u w=20.0u
mpout10 30 40 2000 2000 cmosp l=0.5u w=20.0u
mpout11 30 40 2000 2000 cmosp l=0.5u w=20.0u

mnout01 30 50 1000 1000 cmosn l=0.5u w=20.0u
mnout02 30 50 1000 1000 cmosn l=0.5u w=20.0u
mnout03 30 50 1000 1000 cmosn l=0.5u w=20.0u
mnout04 30 50 1000 1000 cmosn l=0.5u w=15.0u

desd1  1000   30  esd area=200
desd2    30 2000  esd area=200

cpad 30 0 1pF

.ends
```

The SPICE subckt contains circuit topology and connectivity information, while the IBIS model only makes reference to a predefined IO circuit type. IBIS depends on the behavioral simulator to define the circuit topology and load the corresponding model with the information from the IBIS model. Although IBIS covers the majority of IO circuits, some circuits employ more complex topologies and functions that do not fall under the umbrella of predefined IBIS model types.

One example is a driver that employs de-emphasis, a technique used to combat attenuation and intersymbol interference. In this type of driver, the output impedance is a function of the data pattern history. When driving two 0s and then a 1, the output impedance is stronger than it would be when driving one 0 and then a 1 (see Chapter 9, "PCI Express Case Study"). Although IBIS 4.1 does not include a predefined model type for de-emphasis, it does allow the user to call external models written in other languages that may facilitate modeling of this and other circuits. The user must be aware that a behavioral simulation may not faithfully reproduce the waveforms you would expect to see in a SPICE simulation of the same circuit. So how is the user to know whether to trust the output of a behavioral simulator? This question leads into the multifaceted topic of simulation accuracy.

2.30   ACCURACY AND QUALITY OF IO CIRCUIT MODELS

Many factors muddy the waters around the issue of accuracy; a simple working definition for engineers adds some clarity. The accuracy of a simulation is a measure of how close the simulation results come to lab measurements. Lab data are not easy to come by. Although it may be possible to characterize an IO circuit on a simple buffer chip in the lab, getting an output to switch on a microprocessor or a DRAM requires a live system, software, and a diagnostics console. Hopefully the circuit designer spent the time to correlate lab measurements on prototype silicon with SPICE simulations during the design verification process. Figure 2-30 is a clear graphical representation of model-to-hardware correlation. For complex IO circuits found in high-performance interfaces, it is entirely appropriate for a signal integrity engineer to request proof of accuracy from the chip vendor.

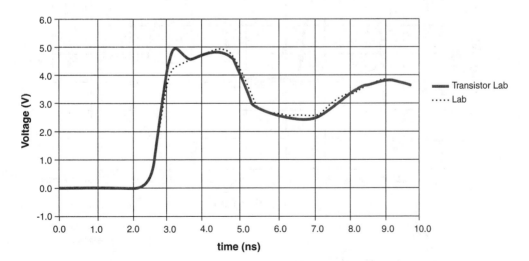

Figure 2-30   Correlation of transistor-level simulation to lab data.

What about the other IO circuits that get used in midrange and low-end applications and account for a majority of all silicon in production today? Is it reasonable to request rigorous proof of accuracy to a 10% tolerance on all these circuits? Probably not, since operating margins are not tight enough to justify the expense associated with the extra effort. In the end, we must rely on the process the semiconductor manufacturer implements to model transistors and the process the chip designer implements to extract transistor-level models from IO circuit layout data.

If the model data take the form of an IBIS model or some other behavioral format, a secondary question persists: How well does the behavioral simulation agree with the transistor-level simulation? This question also touches on the topic of accuracy, though it is one level removed from lab measurements. If the behavioral simulation agrees with the transistor-level simulation, which in turn agrees with the lab, then the behavioral simulation is accurate by proxy.

Much debate has surrounded the topic of behavioral model accuracy, the perspectives ranging all the way from behavioral models being useless to behavioral models being sufficient to design multigigabit interfaces. Reliable studies have demonstrated close agreement between behavioral simulation results and transistor-level simulation results (see Figure 2-31) across a wide range of operating conditions subject to two caveats.

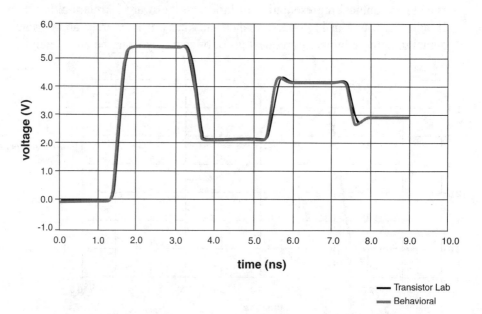

Figure 2-31   Correlation of transistor-level and behavioral simulation.

First, the topology of the behavioral model must mirror the topology of the actual IO circuit. If the driver has three parallel multiphased output stages but the behavioral model has only one, the accuracy may be less than acceptable. Second, the engineer who builds the behavioral model must understand the IO circuit. If that engineer merely runs a translator without bothering to find out how the circuit works or without checking the results, then simulations that use that behavioral model are unlikely to be accurate. Both these caveats imply that the end user is entirely dependent on the chip designer since it is unlikely that the end user will ever see a circuit schematic.

How can someone with little or no access to information about an IO circuit feel confident in the results of a simulation that uses a behavioral model derived from an IBIS model? Fortunately, IBIS version 4.1 has a little-known mechanism for bundling a set of "Golden Waveforms" together with the normal model parameters and tables found in an IBIS model. The model builder can include waveforms for a combination of realistic loads such as a transmission line, termination network, and pi network. The user can simulate the behavioral model under the same loading conditions and then overlay results on top of the Golden Waveforms. In an ideal world, the simulator would do this automatically and report the results to the user.

The [Test Data] keyword marks the beginning of the Golden Waveforms, and the [Test Load] keyword describes the network corresponding to the measurement or simulation conditions that produced the waveforms. The [Test Load] keyword allows the model builder to select any combination of the elements from the template in Figure 2-32.

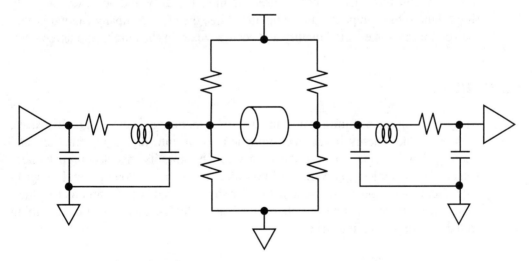

Figure 2-32   Golden Waveform test load.

When comparing transistor-level and behavioral simulation results, be aware that different behavioral simulation engines may produce different results for the same IBIS model and the same test load. Even though the input data is the same, the algorithms for translating the data, constructing an equivalent circuit, and solving that circuit vary from simulator to simulator. The package model is another significant factor that may cause more correlation trouble than the IO circuit itself. Correlating with and without the package model can shed light on the source of the discrepancy.

Of course, the preceding discussion assumes that the models will run in the first place. The entire issue of accuracy becomes a moot point if fundamental syntactical errors prevent the simulator from ever running. Both SPICE and IBIS models are susceptible to general quality issues. SPICE models are particularly susceptible to numerical issues embedded deep within the transistor model parameters. They may also be missing fast or slow transistor model parameters. IBIS models may likewise be missing min or max tables and subparameters. Sometimes an IBIS file doesn't even pass the syntax checker, which leads the user to question everything else in the file! Lack of documentation is a generalized problem that is endemic to both types of models. The IBIS Committee has published a quality checklist of errors commonly seen in IBIS models, as well as a few errors pertaining to SPICE models.

In the face of all that can go wrong with the IO circuit modeling process, signal integrity engineers understandably feel vulnerable because the success of our designs depends so strongly on models developed by people who don't share the same motivations. However, consumers of electrical component models hold one powerful tool that is seldom used. If model quality and accuracy are truly important to the company purchasing the chips, then that company can work with its vendors to make model quality and accuracy part of the purchasing agreement.

CONCLUSION

Integrated circuits require some means for generating electromagnetic waves, converting the energy in an electromagnetic wave into digital information, and storing that information near the chip pin. The circuits that accomplish these functions are subject to a variety of non-ideal behavior. Although signal integrity engineers rarely get to see schematics for the IO circuits they simulate in their daily work, understanding their behavior is nevertheless an element of designing reliable chip-to-chip interfaces.

3

Signal Path Analysis as an Aid to Signal Integrity

Signal integrity is a growing priority as digital system designers pursue ever-higher clock and data rates in computer, communications, video, and network systems. At today's high operating frequencies, anything that affects a signal's rise time, pulse width, timing, jitter, or noise content can influence reliability at the system level. To ensure signal integrity, it is essential to understand and control impedance in the transmission environment through which signals travel.

The development of true analog CMOS processes has led to the use of high-speed analog devices in the digital arena. High clock speeds have become common for digital logic. Systems considered high-end and high-speed a few years ago are now easy and inexpensive to implement in the new CMOS technology. However, this integration of fast edge rates and system speeds brings the challenges of analog system design to a digital world. "High-speed" does not just mean faster communication rates—say, faster than 1 gigabit per second (Gbps). More importantly, a signal with a 600-picosecond rise time is also a high-speed signal. This opens the entire printed circuit board (PCB) to careful and targeted board simulation and design. The designer must consider any discontinuities on the board. Some common sources of discontinuities are vias, right-angled bends, and passive connectors.

A relatively low-frequency digital signal or, more importantly, a signal with slow edge rates, may see a connecting wire or PCB trace as an ideal connection without resistance, capacitance, or inductance. Moreover, a relatively high frequency, or fast edge, signal could similarly see a well-designed short section of PCB track as an ideal connection. Consequently, the signal integrity engineer carefully needs to determine specific signal and connection parameters to establish whether a connection will be transparent to a signal or, more importantly, if the connection will exhibit transmission line properties. Digital signals with fast edge rates are our main concern, because they create connection path conditions where analog circuit characteristics dominate, causing impedances, inductances, and capacitances to become principal characteristics in the signal transmission path. These SI issues are a secondary consequence of Moore's Law, which has been the driving force in modern microprocessor and digital design for a number of decades. The twofold increase in system speed every eighteen months has constantly reduced signal timing budgets. Today Moore's Law manifests itself as a significant decease in signal rise times that currently results in a corresponding increase in signal integrity issues. Edge rates are now a pivotal concern in the SI engineering of data transmission paths.

The traditional method of assessing transmission line parameters or signal path discontinuities is to apply sinusoidal waves to the line and measure the resultant analog signals. Frequency response techniques allow engineers to determine a line's standing wave ratio, which is the ratio of the incident signal to the reflected signal, and other frequency domain characteristics. Likewise, a number of modern test instruments, such as vector network analyzers (VNAs), accurately measure the frequency domain scatter parameters (S-parameters) of two-port networks, such as a transmission line. However, digital design engineers essentially live in the time domain, where they instinctively design and interpret time-related signals and intuitively perform time domain measurements with oscilloscopes and logic analyzers. Therefore, this chapter introduces the concept of transmission line impedance and other signal path parameters by intuitive time domain reflectometry (TDR) measurement methodology. Although this chapter focuses on TDR, it is important to note that the frequency domain and S-parameters are significant measurements in the characterization of a modern high-speed communication channel. Today many TDR systems feature built-in mathematical functions and specialist integrated software that provide both time and frequency information. Even so, there is often a need for both TDR and VNA measurements. Consequently, this chapter considers the merits of both TDR and VNA but focuses on time domain measurements as a vehicle to intuitively understand modern high-speed digital signal path design.

3.1   THE TRANSMISSION LINE ENVIRONMENT

A common definition of a PCB transmission line is the signal path or conductive connection between a digital transmitter and receiver. Traditionally, transmission lines were telecommunication cables operating over long distances, and reflected signals appeared as annoying echoes that often made a line inoperable. However, with high-speed digital signal transmission, even the shortest passive PCB trace can exhibit transmission line effects. Transmission line theory encompasses electromagnetic field concepts and generally attracts complex mathematical analysis. With TDR, the engineer can intuitively understand what is happening with his device. It is important for the signal integrity engineer to acquire equally intuitive insight into transmission line theory, even though some of the basic concepts require a few simple calculations. In transmission line theory, we need to consider the propagation characteristics of a connection from the viewpoint of a digital signal as it traverses the connection system. The signal path in this case is a linear passive connection system that is simply called the interconnection. It often consists of a complex signal path, such as a printed circuit board trace, cable, IC package, flexible PCB, and connectors.

As an example, consider Figure 3-1, which shows two commonly used PCB transmission line topologies. If an electromagnetic signal, such as a wireless signal, travels in free space, typically it would have the velocity of light, which is 3.10^8 meters per second. As a signal enters a denser transmission medium, such as a visible ray of light passing through glass, the signal would slow to a lower velocity. The microstrip PCB topology has one half of its track in free space and the other half bounded by a dielectric, compared to a stripline PCB track, which is surrounded by a dielectric. This makes the signal velocity calculation for the microstrip somewhat more complicated. The simplified calculation for the signal velocity for stripline is the ratio of the speed of an electromagnetic signal in free space to the square root of the dielectric constant of the material surrounding the PCB trace.

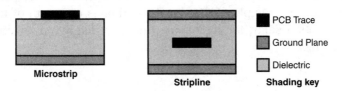

Figure 3-1   Cross sections of microstrip and stripline PCB transmission line topologies.

Therefore, the velocity (v) of a signal in stripline is

$$v = \frac{c}{\sqrt{\varepsilon_r}}$$

Equation 3-1

where c is the speed of light in free space (3.10^8 meters per second) and ε_r is the dielectric constant of the material surrounding the PCB trace.

A common PCB dielectric material is FR4, which has a dielectric constant in the region of 4.

We can put some values into Equation 3-1 and determine the approximate velocity of a signal in stripline, assuming a dielectric constant of 4:

$$v = \frac{300.10^8 \text{ centimeters per second}}{\sqrt{4}} = 15.10^9 \text{ centimeters per second}$$

Given that 1 inch equals 2.54 cm:

$$v = \frac{15.10^9 \text{ centimeters per second}}{2.54} = 5.9.10^9 \text{ inches per second}$$

Therefore, the velocity of the signal in stripline is on the order of $5.9.10^9$ inches per second, which is approximately 6 inches per nanosecond. The velocity of microstrip is somewhat faster at about 7 inches, or more, per nanosecond.

The implication of this important calculation is that a digital signal with a 1 ns rise time would take about 6 inches of stripline to reach its full amplitude. In other words, the spatial boundary of the 1 ns edge would be approximately 6 inches as the digital signal propagates along the stripline, as shown in Figure 3-2.

When does a PCB track become a transmission line? It depends on the signal rise time and the electrical length of the interconnection that the signal is propagating. For example, if the signal rise time is one tenth of the interconnection propagation delay, the interconnection most likely will exhibit transmission line properties.

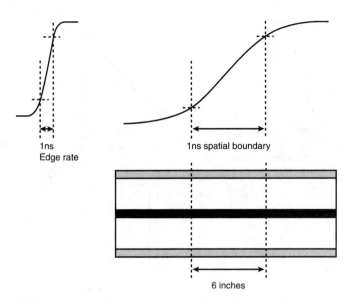

1ns
Edge rate

1ns spatial boundary

6 inches

Figure 3-2   A 1 nanosecond leading edge applied to a length of stripline.

The key concept shown in Figure 3-3 is that the leading edge of the digital signal has a spatial boundary that is long enough to be unaffected by the electrical characteristics of the stripline. In essence, the 1 ns edge becomes a 6-inch leading edge traversing the short length of stripline. In this case, the stripline is more or less transparent to the signal. From a signal integrity point of view, this is a good thing. Of course, if you wanted to troubleshoot or determine the characteristics of a short length of interconnect by way of a fast edge, such as the case when using TDR, you would have to use an instrument that generated a step pulse with a rise time that is sufficiently faster than the interconnect's propagation delay. To use a digital edge to expose multiple discontinuities in a transmission line, as is the case with TDR, we have to apply a digital edge with a rise time that is at least twice as fast as the propagation delay between discontinuities.

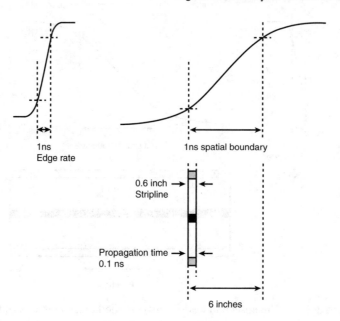

Figure 3-3   A 1 nanosecond leading edge applied to a 0.6-inch length of stripline.

3.2 CHARACTERISTIC IMPEDANCE, REFLECTIONS, AND SIGNAL INTEGRITY

Today engineers develop digital systems such as computers and telecommunication networks that depend on reliable high-speed digital data transmission and reception. An important part of developing such a system is the complete communication channel analysis to confirm the system's signal integrity. The linear passive interconnect between a transmitter and receiver is often analyzed as a transmission line. In addition, the effects of the digital transmitter and receiver interface circuitry have to be considered, because they can have a significant effect on the data that propagates throughout the whole channel. As we have seen at low frequencies, or slow edge rates, the frequency response of the transmission line has little influence on the signal, unless the interconnection path is particularly long. However, as signal frequencies increase and edge rates decrease, the frequency response of the transmission environment as a whole, including transmitter and receiver, takes effect. Even the shortest lines can suffer from signal integrity problems. A first step in understanding transmission line problems is having a firm grasp of the basic concepts of characteristic impedance and impedance matching.

Transmission line theory is fundamentally about the transient response of a data path. As such, the theory attracts complex mathematical equations. A common problem in applied engineering is that practicalities become submerged in abstract mathematical models, which is often the case in transmission line theory. Sometimes it is worthwhile to stand back from the complex theory and consider the basic concepts of transmission line behavior that are in essence intuitive and straightforward. The first point to note is that the analysis of the signal traveling along the stripline, shown in Figure 3-4, concerns AC impedance. This is a first-order model of a transmission line, and we essentially ignore the DC characteristics of the stripline. Imagine a fast edge applied to the stripline shown in Figure 3-4. The time taken for the signal to travel along the line is the time taken for the signal to charge each capacitive section distributed along the line. Moreover, the signal attenuation depends on the line's AC impedance, which is proportional to the inductance and indirectly proportional to the capacitance. In other words, if the line's shunt capacitance increases, the stripline's impedance decreases. But if the series inductance increases, the line's impedance increases. What this means in practice is that if the line's capacitance increases, the signal travels more slowly down the line. If the inductance increases, the signal suffers a greater amount of attenuation.

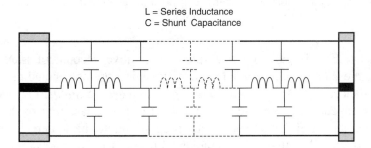

Figure 3-4   Looking into a section of stripline to view the LC model seen by a transient signal.

A greater worry in SI engineering is the continuity of impedance in a transmission path and the effect that discontinuity has on digital signals. Consider a worse case, in which the signal-carrying conductor in the stripline shown in Figure 3-5 is unterminated. You can see that the incident signal climbs to twice its amplitude at the open end before reflecting along the line. This phenomenon of signal reflection at an open-ended line is obvious when you consider that the end

of the line is an open circuit. Removing the line from the signal transmitter terminals would be the same as applying an open circuit to the transmitter. Therefore, it
should come as no surprise that a line with no termination produces a reflected
signal of twice the incident signal amplitude, similar to removing the load from
the transmitter terminals. Likewise, a line with a short circuit produces a negative
reflected voltage. This voltage subsequently cancels the incident signal and ultimately produces zero voltage at the transmitter terminals, as if they had a short
circuit applied.

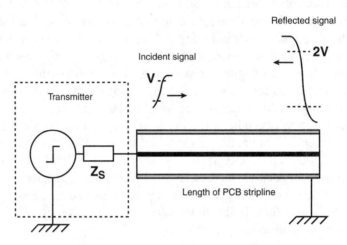

Figure 3-5   An intuitive image of a signal edge as it traverses and reflects at an open-ended
stripline.

Dr. Eric Bogatin gives a splendid intuitive account of transmission line
properties, along with detailed mathematical descriptions, including characteristic
impedance, in his renowned book *Signal Integrity Simplified*. He shows that a
line's impedance is inversely proportional to the square root of the capacitance
per unit length of line and is directly proportional to the square root of the inductance per unit length of line. Figure 3-6 illustrates the relationship between transmission line impedance and capacitance. As it implies, the characteristic
impedance of a transmission line is the line's instantaneous impedance as seen by
the signal traversing the line, given as Z_0 and measured in ohms. The characteristic impedance is dependent on a line's physical properties, such as the dielectric
material that determines the inductance and capacitance per unit length of the
line. But the characteristic impedance is unrelated to the actual length of the
transmission line. As the dimensions of the transmission line change from
the standard section A to the narrower section B, the line's capacitance increases,

and therefore the line's characteristic impedance decreases. Likewise, as the transmission line section B in Figure 3-6 widens in section C, the capacitance decreases. Consequently, the line's characteristic impedance increases.

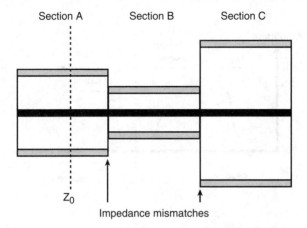

Figure 3-6 A length of stripline with three distinct characteristic impedances.

An intuitive way to think about transmission line theory is to remember that if excess inductance exists at one point in the transmission line, this causes a corresponding increase in the line's impedance at that point. Likewise, any excess in capacitance in the transmission line results in a corresponding decrease in the line's impedance. For example, capacitive loading is common in many via structures found in high-speed backplanes. Typically this causes the characteristic impedance of the transmission line to drop dramatically at the feed-through structure. Consequently, the possibility of an impedance discontinuity at a via and its effect on signal integrity is a significant concern in the successful design of high-speed systems. Therefore, we need to investigate further the effects of impedance mismatches in high-speed systems.

A digital signal traverses a transmission line as an electromagnetic wave. Any discontinuities in a line result in unwanted signal reflections. Reflected signals appear at transmission line discontinuities because there cannot be instantaneous changes in voltage or current at the discontinuities. The significance of a reflected signal is the effect it has on the incident signal, whereby the resultant signal is the phase and amplitude addition of the incident and reflected signals, as shown in Figure 3-7. Reflected signals can cause a range of unwanted SI manifestations, such as ringing. Ringing can lead to high-frequency signal components, resulting in unwanted electromagnetic radiation, crosstalk, and noisy return or ground currents. Ideally, the transmission line will have continuous characteristic impedance without discontinuities throughout its entire length, from transmission source to receiver termination.

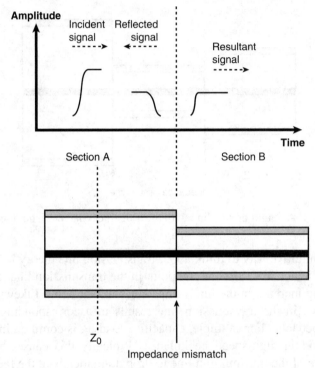

Figure 3-7   The general effect that impedance discontinuities have on a digital signal traversing a line.

To summarize, we noted that a high-speed digital signal would see the transmitter, the PCB track with its interconnections, and the receiver as an entire transmission system. The communication system for a high-speed signal is often a complex arrangement, although many tools and techniques are available to make the design and verification of transmission line systems manageable. Moreover, you have seen that it is critical for the digital signal designer to have an intuitive understanding of basic transmission line theory to avoid signal integrity problems.

3.3 THE REFLECTION COEFFICIENT, IMPEDANCE, AND TDR CONCEPTS

TDR involves measuring and visualizing signal paths, interconnects, and terminations. TDR is a diverse measurement methodology with applications in optical, electronic, and mechanical systems. The significant feature of TDR is its ability to produce an image of a signal route, impedance, or communication path. The electric version of TDR enables the SI engineer to translate abstract models and signal propagation hypotheses into real-world observations of the physical layer structures within the interconnect path.

Visualizing the characteristics of a transmission line can be tricky. This is unlike in Roman times, when an engineer had little trouble visualizing the dimensions of his foot and the inch, which was more or less the thickness of his thumb. A key skill in designing or debugging a high-speed digital system is the SI engineer's perception of signal path parameters. TDR gives the SI engineer accurate knowledge of critical signal path characteristics. Moreover, TDR provides illustrative data that allows the engineer to determine if transmission environment parameters are within allowable limits.

Figure 3-8 shows that the impedance of a signal source output (Z_S), along with an associated transmission line impedance (Z_0) and receiver impedance or load (Z_L), which should correspond or match Z_0. Signal reflections generated by unequal impedances are the enemy of good signal integrity. An important transmission line parameter measured by TDR is the ratio of the reflected signal to the incident signal, which is the reflection coefficient, given the symbol ρ (rho). The reason for the importance of the reflection coefficient in TDR is the direct relationship between reflection coefficients and impedance discontinuities. The principal purpose of a TDR instrument is to measure incident and reflected signals, calculate the reflection coefficients, and display the mathematically translated reflection coefficients as impedance or other relevant units.

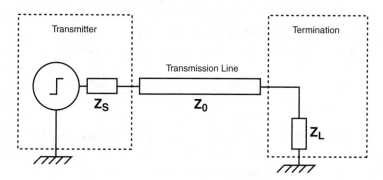

For the ideal data transmission path $Z_S = Z_0 = Z_L$

Figure 3-8 The ideal transmission path.

The reflection coefficient ρ (rho) = $\dfrac{\text{Reflected signal voltage}}{\text{Incident signal voltage}}$

Equation 3-2

If we plug numbers into Equation 3-2 to represent three classic conditions—a matched load, a short circuit, and an open load—we can see that ρ has a range of values from plus one (+1) to minus one (−1), with 0 representing a matched load:

- A matched load occurs when Z_L is equal to Z_0, and the load impedance and transmission line characteristic impedances match. This is the ideal case. The reflected wave, Vreflected, is equal to 0, and ρ is 0. There are no reflections.
  Where there is zero reflected signal voltage:

The reflection coefficient (ρ) = $\dfrac{\text{Zero reflected signal voltage}}{\text{Incident signal voltage}}$ = zero

- A short circuit occurs when the load impedance Z_L is 0, which implies a short circuit. The reflected wave is equal to the incident wave, but opposite in polarity. The reflected wave negates part of the incident wave, and the reflection coefficient ρ is minus 1 (−1).
  Where the reflected signal voltage equals the minus incident signal voltage:

The reflection coefficient (ρ) = $\dfrac{\text{Minus incident signal voltage}}{\text{Incident signal voltage}}$ = -1

- An open circuit load occurs when the load impedance Z_L is infinite; an open circuit is implied. The reflected wave is equal to the incident wave and is of the same polarity. The reflected wave reinforces part of the incident wave, and the reflection coefficient ρ has a value of 1 (+1).
  Where the reflected signal voltage equals the incident signal voltage:

The reflection coefficient (ρ) = $\dfrac{\text{Incident signal voltage}}{\text{Incident signal voltage}}$ = +1

TDR is in essence a simple measurement concept based on the determination of the reflection coefficient ρ. TDR instruments typically generate a very fast digital edge as an incident signal and measure reflected signals produced by the unit under test, as shown in Figure 3-9. Mathematical functions built into a TDR instrument compute ρ and fundamentally apply Equation 3-3 to determine the characteristic impedance (Z_0) of the unit under test, as shown in Figure 3-9. It is important to note in Equation 3-3 that the accuracy of any TDR calculation of Z_0 depends not only on measured signals but on the precision and accuracy of the known source impedance Z_S.

$$\text{Characteristic impedance } (Z_0) = Z_S \cdot \frac{1+\rho}{1-\rho}$$

Equation 3-3

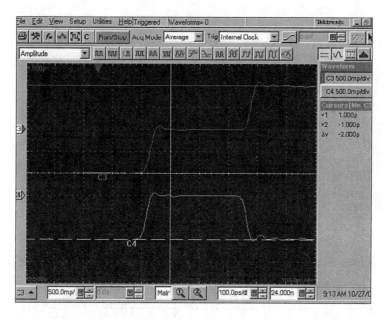

Figure 3-9   The oscillogram shows the time domain measurement of a step waveform passing through a short transmission line. The top waveform illustrates an open circuit line where the reflection coefficient ρ is +1. The bottom waveform shows a short circuit line where the reflection coefficient ρ is –1.

You will see that a key skill in designing or debugging high-speed digital systems is the ability to use TDR to visualize signal path parameters. So far, you have seen the need for the signal integrity engineer to have accurate knowledge of critical signal path characteristics. More importantly, the engineer often needs to know whether the parameters are within allowable limits.

3.3.1   TDR Concepts

TDR measures the reflections that result from a signal traveling through a transmission environment of some kind—a circuit board trace, a cable, a connector, and so on. The instrument generates a known step with known source impedance (Z_S) and rise time (Δ_r). Therefore, the initial signal launched into the device under test (DUT) is a known incident signal (V_S), as shown in Figure 3-10. Normally the generated step signal has a programmable edge rate, where an edge rate of 35 picoseconds would provide excellent resolution of DUT impedances. To be able to resolve the very fast edge rates of the incident and reflected signals, the equivalent time sampler module typically requires a bandwidth on the order of 50 GHz.

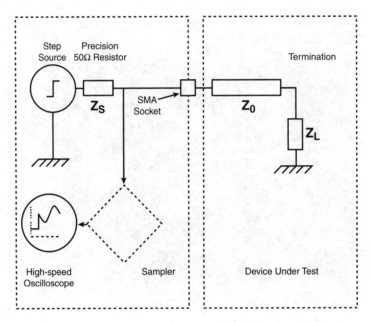

Figure 3-10   A simplified block diagram of a TDR measurement system.

If the DUT has a characteristic impedance (Z_0) and termination impedance (Z_L) that match the source impedance (Z_S), as shown in Figure 3-11, the signal seen by the sampler is half the amplitude of the generated pulse, given the step generator output impedance is also equal to Z_S. The measurement circuit forms a simple potential divider network. In addition, the time taken for the reflected signal to return to the sampler is twice the propagation delay of the device under test. Fortunately, the engineer is unconcerned with the mechanics of TDR, because the instruments incorporate mathematical functions to allow the instrument to display actual device impedances and distance, or voltage against time. However, we must consider a few classic cases in which the DUT has common impedance discontinuities and the reflected signal can take a number of standard forms. An experienced TDR engineer generally recognizes a dip in a reflected signal as a shunt capacitance and a spike as a series inductance in the transmission path. The ability to read and decipher the TDR trace, relating trace figures to circuit elements, is valuable, as shown in Figures 3-12 and 3-13.

A greater skill, however, is the ability to relate the effect that particular circuit discontinuities or features will have on transmitted data.

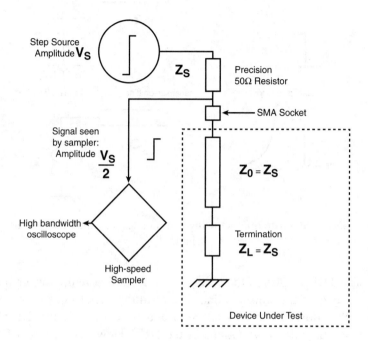

Figure 3-11 An ideal DUT with no reflected signals.

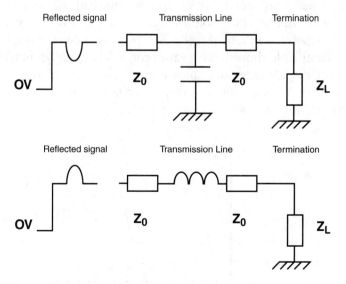

Figure 3-12 TDR trace illustrations showing common circuit discontinuities and their associated trace shape.

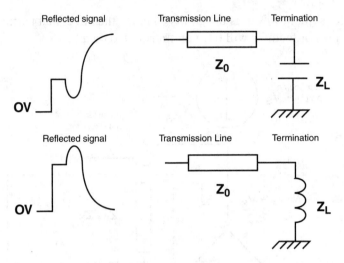

Figure 3-13   TDR trace illustrations showing common terminations and their associated trace shape.

The TDR display is the voltage waveform that returns when a fast step signal propagates a transmission line. The resulting waveform seen by the sampler is the combination of the incident step and reflections generated when the step encounters impedance variations in the DUT. However, in a real DUT, multiple reflections are commonplace. Although TDR measurements are simple in concept, their real-word interpretation has complicated mathematical foundations. Most TDR instruments incorporate dedicated software to compute complex algorithms to resolve multiple images and produce numeric results displayed as clearly identifiable impedance waveforms. Additionally, most modern TDR instruments display magnitude measurements in units of volts, ohms, or ρ per division on the vertical scale and display units of time (seconds per division) on the horizontal axis, as shown in Figure 3-14.

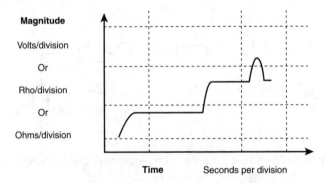

Figure 3-14   A simplified illustration of a TDR waveform.

It is worth noting that to fully appreciate the TDR display, you need to keep in mind the characteristic reflection waveforms reproduced by idealized transmission line impedances. Moreover, as you have seen, you should be familiar with the effects of diverse transmission line impedances and terminations, such as capacitive and inductive loading.

3.4   LOOKING AT REAL-WORLD CIRCUIT CHARACTERISTICS

In general, etched circuit boards have impedance-controlled microstrip and stripline transmission lines. Over the span of these transmission lines, components, vias, connectors, and other "interruptions" create impedance discontinuities. Typically, a simulation model of these discontinuities centers on lumped inductors, capacitors, and transmission line losses. A real-world TDR trace of a prototyped DUT shows these discontinuities as a road map of the impedance variations. The waveform shown in Figure 3-15 is the result of a TDR step launched into a complex transmission medium.

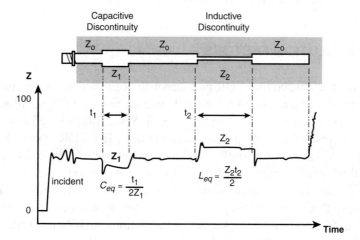

Figure 3-15   The TDR waveform shows the effect of all the reflections created by all the impedance discontinuities in the DUT. The waveform is like a road map of the impedance variations in the DUT.

A TDR sampling module requires a very accurate, controlled pulse with a fast rise time and minimal aberrations. Imagine sending a typical data waveform down the transmission path shown in Figure 3-15. The data typically would have its own aberrations that would interact with the discontinuities in unpredictable ways. Such situations lend themselves to erratic, intermittent SI problems. Characterizing a transmission environment with TDR measurements and correcting impedance discontinuities can significantly improve signal integrity. However, to achieve accurate TDR measurements, it is important that the SI engineer appreciate the necessary requirements for an accurate incident pulse, precise reference impedance, and meticulous instrument calibration. Some TDR instruments have an architectural design that allows the automatic minimization of step aberrations to less than 4%. This function maintains high accuracy of the reflection coefficient and the resultant extracted impedance profile.

A high-performance TDR instrument provides attributes that allow intuitive test and measurement and a high level of confidence in the measurement accuracy. Some high-performance TDR instruments have a PC or workstation platform and host both the simulation tools and real-world measurements. A useful facet of high-performance TDR is the ability to compare real-world TDR measurements and simulated waveforms to determine how much impedance deviates from its required value and quickly rework the simulation. Reducing the development time is a prime aim of the SI engineer, who typically is under pressure to minimize a product's time to market.

3.5 TDR RESOLUTION FACTORS

We have established that TDR measurements can produce useful insights into circuit impedance and signal integrity. However, not all TDR solutions and, in particular, TDR instruments are created equal. Several factors affect a TDR system's ability to resolve closely paced discontinuities. If a TDR system has insufficient resolution, small or closely spaced discontinuities typically merge or are smoothed together into a single aberration in the waveform. This effect may not only obscure some discontinuities, but it also may lead to inaccurate impedance readings. Rise time, settling time, and pulse aberrations can also significantly affect a TDR system's resolution. Many factors contribute to the accuracy of a TDR measurement. These include the TDR system's step response, interconnect reflections and DUT losses, step amplitude accuracy, baseline correction, and the accuracy of the reference or source impedance (Z_S) used in the measurements.

A TDR system can only resolve two closely spaced discontinuities by half the rise time of the incident pulse, often quoted as the TDR resolution. However, realize that for a single discontinuity, the quoted TDR resolution is unimportant. A single discontinuity, such as a via, typically would be seen at a fifth of the TDR incident pulse rise time. Therefore, a TDR instrument with 35 picoseconds of rise time would have sufficient resolution to identify two discontinues spaced 0.1 inch apart on an FR4 PCB and show a single aberration where the feature size is in the region of 0.025 inch.

3.5.1 Rise Time

A reflection from an impedance discontinuity in a DUT has a rise time equal to or longer than (slower) that of the incident step. The physical spacing of any two discontinuities in the circuit determines how closely positioned their reflections will be relative to one another on the TDR waveform. Two neighboring discontinuities may be indistinguishable to the measurement instrument if the distance between them amounts to less than half the system rise time of the incident signal. Equation 3-4 summarizes this concept.

$$\text{TDR resolution} = \frac{\text{incident signal (TDR pulse) rise time}}{2}$$

Equation 3-4

3.5.2 Pre-aberrations

Aberrations that occur before the main incident step can be particularly troublesome because they arrive at a discontinuity and begin generating reflections before the main step arrives. These early reflections reduce the resolution of the TDR measurement by obscuring closely spaced discontinuities.

3.5.3 Settling Aberrations

Aberrations, such as ringing, that occur after the incident step, as shown in Figure 3-16, cause corresponding aberrations in the reflections. These aberrations are difficult to distinguish from the reflections caused by discontinuities in the DUT. Note that aberrations in the TDR instrument's step generator and aberrations in the step response of its sampler have virtually the same effect.

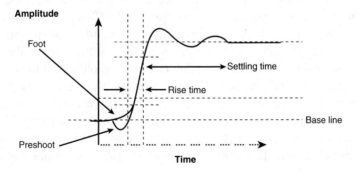

Figure 3-16 A simplified diagram of a TDR generated pulse that contains signal aberrations.

3.5.4   Reference Impedance

All TDR measurements are relative. In essence they are direct comparisons between incident and reflected signal amplitudes. Modern TDR instruments perform all the associated calculations and show results directly in rho or ohms. However, apart from a comparison of incident and reflected voltages, Equation 3-3 confirms that the TDR process is notably dependent on the accuracy of the reference or source impedance (Z_S). High-performance TDR instruments use high-precision 3.5 mm air line connectors as stable impedance references to calculate rho and ohms. Alternatively, proprietary 50-ohm standard connectors are specifically designed for the standard calibration of a TDR instrument.

3.5.5   Step Amplitude and Baseline Correction

In general, modern TDR instruments provide built-in calibration of the incident step amplitude and baseline level for accurate computation of test impedances. Moreover, high-performance TDR instruments place a known air line, or standard 50-ohm component, in the sampling module specifically for calibration purposes. This lets the instrument automatically and periodically monitor the baseline and incident step amplitude. Consequently, the system automatically compensates step amplitude and offset drifts, ensuring precise, repeatable measurements.

3.5.6   Incident Step Aberrations

The most obvious problem caused by incident step aberrations, such as a long setting time, is the inability to measure accurately reflected step amplitudes. When test impedances are significantly different from 50 ohms, reflections typically are of relatively low amplitude. TDR measurement accuracy is highly dependent on the precise measurement of the reflected signal amplitude. If test impedances notably differ from 50 ohms and the incident pulse fails to settle in a relatively short time, in comparison to the propagation delay of the line under test, serious measurement errors will result. The closer the DUT impedance is to 50 ohms, the greater the accuracy of the measurement system and, in particular, the higher the measurement system's tolerance to incident step (settling time) aberrations.

A second type of aberration that can cause problems is the foot or pre-shoot that precedes the incident step. If the transmission line under test has an open circuit termination, the step's foot or pre-shoot generates reflections at the open circuit. These reflections appear before the step's reflected rising edge comes into view. This second type of step aberration causes obvious, unwanted end-of-line measurement errors. However, low-frequency step aberrations cause a more subtle effect.

Low-frequency incident step aberrations can appear as a slope in the TDR trace, which is an instrument-generated artifact. Such an artifact can even appear when an external calibration 50-ohm termination replaces the DUT. In effect, a low-frequency aberration in an incident step function can cause the TDR instrument to display an erroneous offset measurement.

3.5.7 Noise

Random noise can be a significant source of error when making TDR measurements, particularly when measuring small impedance variations. Fortunately, modern instruments can perform signal averaging to reduce the effects of random noise. The drawback in many instruments is that averaging can dramatically slow the processing speed, particularly when displaying automatic measurements. Modern high-performance oscilloscopes, which form part of a TDR measurement system, addresses this problem with built-in multiple processors that can share the processing workload and produce real-time denoised measurements.

3.5.8 Interconnect Accuracy and Reflections

TDR measurements that require a long probe cable are prone to measurement inaccuracies. To reduce the effect of the cable loss in a long probe cable, it is wise to measure the DUT relative to the end of the probe cable. The rho of the probe cable (ρ cable) causes a shift in the reference level of the TDR measurement. The incident step amplitude presented to the DUT will be 1 minus the rho of the cable ($1 - [\rho$ cable]). Consequently, for maximum accuracy, the SI engineer has to measure the probe cable rho and use a measurement technique that compensates for losses in the probe cable. Moreover, reflections from poorly designed interconnect components and unsatisfactory probe-to-DUT interfaces cause TDR measurement problems.

Therefore, it is worth noting that an unsatisfactory TDR probe interface can create a large inductive reflection, and the reflected signal must settle before you can make an accurate measurement. To minimize measurement errors, it is extremely important to keep the TDR instrument's probe tip and ground leads as short as possible. This reduces the ground loop inductance and creates a good launch point for the TDR step.

3.5.9 Cable Losses

Cable losses in the test setup can cause several problems. Both conductor loss and dielectric loss can occur, but conductor loss usually dominates. Conductor loss results from the finite resistance of the metal conductors in the cable. Due to the skin effect, the resistance increases with frequency. The result of this incremental series resistance is an apparent increase in impedance, as you look further into the cable. Therefore, with long test cables, the DUT impedance looks higher than it actually is.

The second problem is that by the time an incident pulse reaches the end of a long probe cable, both the rise and settling times are degraded. Degradation in the incident pulse can severely affect TDR measurement resolution and accuracy, because the effective amplitude of the incident step differs from the expected calibrated value. Fortunately, incident pulse amplitude inaccuracy does not cause a serious error when the DUT impedance is close to 50 ohms. But for DUT impedances that are noticeably larger or smaller than 50 ohms, the measurement error can be significant.

3.5.10   Controlling Rise Time

Although in many cases the fastest available incident step rise time is desirable, in other cases very fast rise times can give misleading results on a TDR measurement. In essence, it is unnecessary to overdesign a board by finding and correcting impedance discontinuities that will be invisible to the real-world or applied signal. For example, testing the impedance of a microstrip on a circuit board with a 35 picosecond rise time system provides excellent resolution. Reflections from small discontinuities such as stubs or sharp corners in the microstrip are quite visible when excited by the TDR step. However, if the same transmission line were to convey a 10 Gbps signal, a higher-resolution TDR signal would be appropriate for the board's test and characterization.

This difference between measurement and reality could be the beginning of a signal integrity problem. In attempting to correct the misleading impedance readings, you might compromise the environment that the real operational signals need. It is often preferable to see the transmission line's TDR response to rise times, similar to the actual circuit operation. Some TDR instruments provide a means of increasing the apparent rise time of the incident step. Moreover, some high-performance sampling oscilloscopes implement filtering using a double-boxcar averaging technique that is equivalent to convolving the waveform with a triangular pulse. Fast filtering using live waveform mathematics shows the filtered response in near-real time. This technique provides live filtered waveforms that quickly respond to changes, and it requires no additional calibration steps. The alternative measurement method is to use an external pulse generator, although the high-performance TDR instrument filtered waveform result corresponds very well with similar measurements made with an equivalent pulse from an external generator.

High-bandwidth TDR sampling heads are extremely static-sensitive. To ensure their continued performance in impedance and signal integrity applications, they require a high level of static protection. This is beyond the simple protection that suffices for conventional oscilloscopes. The following guidelines are essential when handling or using any high-bandwidth TDR sampling module:

- Cap or terminate probes and sampling inputs when not in use.
- Use grounded wrist straps of less than 10 megohms when in contact with the TDR probes and sampling module.
- To prevent the possibility of probing a device that may have built up a charge, always discharge the device before probing or connecting it to the module. You can do this by touching the device with the probe ground lead or your finger—assuming you are using a wrist strap—to the point you want to TDR test. Some high-performance instrument manufacturers, such as Tektronix, provide a static isolation unit that uses a microwave relay to isolate and automatically discharge the DUT. With this unit, the DUT cable remains connected to ground until a foot switch is depressed and the relay switches, connecting the DUT to the TDR input.
- Remember that long cables stored in a draw can build up a static charge over time. They must be grounded before being connected to the TDR instrument.

3.6   DIFFERENTIAL TDR MEASUREMENTS

Differential transmission lines offer many advantages over conventional transmission channels. Many high-performance designs, such as high-speed serial buses, use differential impedance or balanced lines to minimize SI issues. Although differential signaling offers many advantages, there remains a need to make differential TDR measurements to support signal integrity. All the single-ended TDR measurement concepts discussed so far also apply to differential transmission lines. However, the basic principles of TDR require additional measurement concepts and instrumentation techniques to take into account the unique requirements of measuring differential impedance. The additional instrumentation hardware requirement is a four-channel TDR or four-port VNA.

A differential transmission line has two unique modes of propagation, each with its own characteristic impedance and propagation velocity. Much of the literature calls these the odd mode impedance and the even mode impedance:

- A definition of odd mode impedance is the impedance measured by observing one line referenced to ground while driving the other line by a complementary signal.

- The differential impedance is the impedance measured across the two lines with the pair driven differentially. Differential impedance is twice the odd mode impedance.
- A definition of even mode impedance is the impedance measured by observing one line referenced to ground while driving the other line with an equivalent signal.
- The common mode impedance is the impedance of the lines connected in parallel, which is half the even mode impedance.

Tying together the two conductors in a differential line and driving them with a traditional single-ended TDR system yields a good measure of common mode impedance. However, common mode impedance is often less important than the differential impedance. To provide true differential impedance measurements, the TDR sampling module needs to provide a polarity-selectable TDR step for each of two channels. This approach drives the differential system under test in true differential mode, just as it does when the DUT is functioning in its intended application. Moreover, separate responses, acquisitions, and evaluations of the differential line provide true TDR measurements of the differential line. Alternatively, a mathematical calculation based on linear supposition is used to correct the resultant reflections produced by exciting the differential circuit with staggered step inputs. This mathematical function built into some TDR instruments provides an accurate analysis of most differential circuits.

However, to achieve a real-time differential TDR measurement, both sampler channels must have a matched step response. Also, the two incident steps must be matched but complementary, and the incident steps must be time-aligned at the DUT. To meet these requirements, the TDR instrument typically incorporates two acquisition and polarity-selectable TDR channels in a single sampling module containing a common clock source. Moreover, the relative timing of the two-step generators should be deskewed to precisely align the two incident steps at the DUT and remove any mismatch due to cabling.

With the TDR system set up with complementary incident steps, and using ohm units, adding the two channels yields the differential impedance. The individual traces give the odd mode impedance. If the line is not balanced, the two traces do not match exactly.

3.6.1 Deskewing the Step Generators

Differential TDR measurements require precise matching of both the stimulus and the acquisition systems in terms of timing and step response. The TDR instrument fixes the timing of each incident step response; however, the relative timing of the two channels is usually adjustable. Both the acquisition timing and the

TDR step timing must match to yield a valid TDR measurement. Notice, however, that the matching requirement lies at the DUT rather than the front panel of the instrument. Poor matching between the cables or interconnect devices and the DUT skews the TDR incident steps in time when they arrive at the DUT, even if they are aligned at the instrument's front panel. This causes significant measurement error.

The most efficient way to eliminate or reduce measurement skewing is to minimize the DUT interconnect lead length and to use carefully matched probe cables. Extending the sampling head 2 meters from the mainframe is possible in some high-performance TDR instruments and incurs no loss in performance. This helps minimize probe cable lengths with associated mismatch, loss, dispersion, and other interface errors. Deskewing the acquisition paths is very important. In some instruments the timing of the step generators and acquisition can be difficult to detect and correct. Individually adjusting the incident steps and setting the acquisition is not always obvious. However, a number of high-performance instruments provide a specific trace that shows the effect of a timing mismatch in real time. Displaying the effects of a mismatch in real time makes the adjustment of the instrument simple and intuitive.

3.7   Frequency Domain Measurements for SI Applications

The two principal measurement techniques for passive signal integrity characterization of data path transmission and modern digital system gigabit interconnects are time domain TDR and frequency domain VNA. Differential transmission lines coupled with the microwave effects of high-speed data have created the need for the VNA and characterization of physical layer components. There is often a need to test and characterize complex microwave behavior in bound conductors with a frequency measurement, such as a high-speed digital interconnect. In fact, many digital standards groups recognize the importance of specifying frequency domain physical layer measurements as a compliance requirement.

Nonetheless, as you have seen, the TDR instrument is a very wide bandwidth equivalent sampling oscilloscope with an internal step generator. The TDR sends a step stimulus to the DUT. Based on reflections from the DUT, the designer can deduce a lot of information about the device properties, such as the location of failures, impedance, and time delay information. TDR allows the engineer to gain insight into the DUT's topology. With the aid of built-in mathematical functions and an additional sampling channel, or module, the TDA instrument makes possible Time Domain Transmission (TDT) measurement, as shown in the top part of Figure 3-17. TDT measures crosstalk or characterizes lossy transmission line parameters, such as rise time degradation, return loss, and skin effect and dielectric loss. However, a TDT measurement system requires a high-performance TDR with an additional sampling channel, as well as comprehensive dedicated software to translate time domain measurements into DUT frequency-dependent behavior. VNA instruments work in the frequency domain and are altogether different in concept from time domain TDR tools. VNA instruments apply sinusoidal waves as a signal source to stimulate the DUT. They use a very narrow band filter at the receiver to determine the device's response. Moreover, because of the narrow bandwidth receiver inside the VNA, the VNA's resultant noise floor is inherently lower than a TDR instrument. By sweeping the signal source and the receiver's filter in a synchronized fashion, the VNA instrument makes a swept-frequency, steady-state measurement, as shown in the bottom part of Figure 3-17.

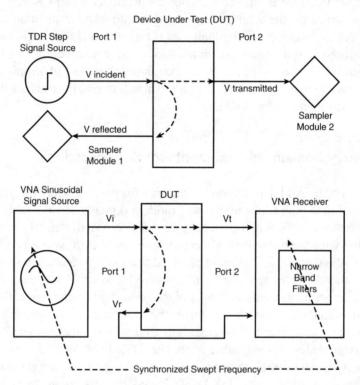

Figure 3-17   Simplified block diagrams of two-port networks where the signals at the input and output terminals characterize the frequency domain scatter measurements or S-parameters.

A frequency domain VNA measurement separates and measures scattered signal amplitudes and phase produced at a two-port network. A VNA instrument uses built-in mathematical functions to translate scatter measurements into S-parameters, which are ratios of incident wave voltages to reflected or transmitted wave voltages for a two-port measurement. The VNA instrument can display S-parameters as signal power against frequency, frequency magnitude and phase, or graphically on a Smith chart. The DUT is effectively a black box where the applied signals, reflected signals, and resultant output signals determine the behavior and frequency domain S-parameters of the complete component or DUT. The ratio of the Vincident and Vreflected signals gives the S11 and S21 parameters, called the reflection and transmission coefficients. To some extent they are similar to reflection and transmission coefficients in a time domain measurement.

The fundamental difference between the TDR and VNA instruments is that the TDR measures voltage against time, and the VNA measures power against frequency. The displayed measurements are quite different. Moreover, there is a fundamental difference in that VNA measurements relate to the frequency characteristics of a DUT in total, whereas TDR measurements give information about each discrete discontinuity in a transmission path. Nevertheless, we know from basic signal transformation theory that the Fourier transform relates the time and frequency domains. The time and frequency domains show the same information in different forms. Each domain contains the same information, but in a different representation. Moreover, there are clear relationships between TDR reflection and transmission coefficients and VNA S-parameters. These relationships allow dedicated software in a TDR system or VNA instrument to show both frequency and time domain information. The preferred representation is the prerogative of the engineer. In general, the RF or microwave engineer prefers the frequency domain. Digital design engineers live in the time domain, where they are most at ease with their time-related measurement environment. However, the advent of special signal integrity software to control both TDR and VNA instruments has created a new generation of engineers who use both time and frequency domain analysis.

TDR is visual and intuitive due to the transient nature of the measurement technique. The incident step propagates through the discontinuities in the DUT, and the reflections indicate the exact location and size of discontinuities. The fast TDR rise time of 35 ps ensures that the time domain is a broadband measurement that captures a wide range of frequencies. This allows specialist software to translate TDR measurements into frequency domain S-parameters.

However, a dedicated VNA instrument has a better frequency domain performance than an equivalent TDR instrument and therefore greater accuracy in characterizing some high-frequency signal paths. However, it comes at a price, because the dedicated VNA instrument is more expensive compared to a TDR-based system. When choosing an instrument, it is important to consider the cost-benefit ratio of a VNA instrument. An equivalent TDR instrument does not provide the same frequency domain performance as a VNA. It is up to the digital design engineer to decide whether the parameters within their system justify the extra performance and cost of a dedicated VNA instrument.

The dynamic range of VNA measurements tends to be substantially higher than that for TDR measurements. Typically a VNA instrument has a 110 dB dynamic measurement range, compared with 50 to 60 dB for an equivalent TDR instrument. Originally, the development of VNA instruments centered on microwave design. Engineers were concerned primarily with narrowband and resonant systems, such as mixers, filters, resonators, power splitters, and combiners. Microwave engineers required exact data about the frequency band over which a circuit could operate, its center frequency, and its Q-factor. These requirements caused instrument manufacturers to continuously improve VNA accuracy and achieve very high dynamic range, wide signal-to-noise ratio, in the frequency domain. Today a high-performance VNA instrument easily exceeds 100 dB when careful measurement techniques are used. To achieve this high dynamic range and accuracy, a narrowband filter analyzes data at each frequency point. The alternative would be to use a TDR and, where necessary, use dedicated software to translate the time domain measurements into frequency domain S-parameters. However, the high dynamic range in a VNA makes a significant difference to the measurement of transmission lines that exhibit microwave properties.

A high-performance sampling oscilloscope TDR system with an incident pulse of 35 ps rise time typically allows the digital design engineer to measure up to 12 GHz of S-parameters, with up to –60 dB of dynamic range. With the addition of dedicated software, the fast rise time TDR enables the engineer to perform S-parameter measurements up to 65 GHz, with up to –70 dB of dynamic range. Nonetheless, a VNA offers a significantly improved measurement performance in the microwave frequency bands. Moreover, the engineer often prefers to work in the time domain and to translate results into the frequency or to work purely in the frequency domain.

The similarities and differences between TDR and VNA allow the two instruments be applied to different aspects of gigabit interconnect modeling. With the addition of specialist software, the designer can use either instrument to generate gigabit interconnect SPICE or IBIS models and to analyze eye diagrams, jitter, losses, crosstalk, reflections, and ringing in PCBs, device packages, sockets, connectors, and cable assemblies. Time domain-based analysis provides the designer with topological interconnect models, which have one-to-one correlation between the model components and the physical interconnect structure. Such models can include frequency-dependent losses and resonances. They are most convenient for troubleshooting where they enable the SI engineer to identify causes of signal integrity problems, within required time scales and budget.

CONCLUSION

High data rates of 10 Gbps and above are particularly sensitive to mismatches and variations in transmission path impedance—especially resultant signal reflections that significantly decrease signal quality. The integrated oscilloscope and logic analyzer display shown in Figure 3-18 shows the classic effect that reflected signals have on the edges of a digital signal. The signal edges appear to have a midpoint step or a gnawed edge, as if bitten. These particular effects were produced by an incorrectly terminated PCB track.

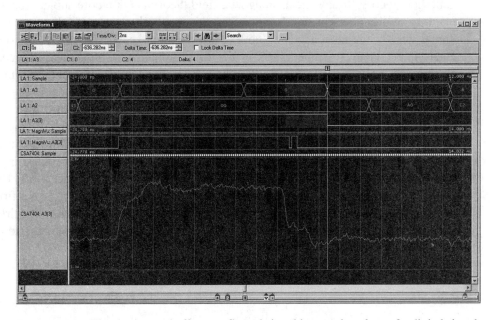

Figure 3-18 The detrimental effect a reflected signal has on the edges of a digital signal.

Impedance tolerances are part of the electrical specifications for many of today's high-speed buses and digital system components. While it is standard practice to model high-speed circuits where modeling hastens the design cycle and minimizes errors, there remains a need to prototype a modeled design to verify its parameters with hardware measurements. Moreover, calculating or modeling the time domain response of a section of printed circuit board track from first principles typically involves an intimate knowledge of electromagnetic field theory, such as the solution of Maxwell's equations. Electromagnetic theory is all-inclusive and typically takes into account all the electrical parameters and device geometry of a particular electronic component or connection system. Although microwave engineers and device manufactures have traditionally used specialist electromagnetic toolsets to model connectors and transmission lines, digital engineers generally have only working knowledge or an appreciation of electromagnetic field theory. Today, electromagnetic field theory is not a core subject for the majority of digital engineers. Nonetheless, this is a high-speed digital world. The digital design engineer needs to understand the high-speed signal transmission environment, analyze impedances in the time domain, and characterize signal channels in the frequency domain. You have seen that TDR is inherently an intuitive measurement system. Used wisely, it should enable the SI engineer to design and produce reliable high-performance digital systems. Understandably, there is a limitation to the depth and detail that we can give to signal path analysis. One of the victims is the scope given to S-parameters and frequency domain measurements. In essence, this chapter has attempted to describe the fundamental principles of TDR and make you aware of VNA. Even so, many other facets of TDR and VNA remained unexplored in this chapter. Our hope is that you have a firm understanding of TDA and an awareness of VNA. We also hope you're inspired enough to investigate further the applications of TDA and VNA in the myriad of white papers and trade articles.

4

DDR2 Case Study

Those who have experienced the joy of debugging a memory interface realize that it's not enough to simply make the waveforms look pretty. They also have to switch at the right time. A DDR2 interface is a complex state machine that involves a lot more than just distributing clock to a couple of flip-flops. The memory interface controller (MIC) must initialize and refresh the DRAM memory cells. It also has the job of enabling and disabling on-die termination. A common clock samples address and command, but a source-synchronous strobe samples the data. And every MIC has its own technique for synchronizing incoming read data with the core clock domain. The signal integrity engineer has to be well-versed in all these subtleties.

This chapter does not delve into state machine protocol. For a thorough treatment of this topic, read Granberg. Nor does it discuss the complications of strobe-to-strobe and clock-to-strobe skew so dependent on the MIC. You can expect an in-depth treatment of data-to-strobe timing for both read and write transactions that examines the contribution in ps of each relevant signaling effect. If you apply this approach to your own design, you will understand how close it is to failure and what areas to concentrate on if you want to improve margins.

The DDR2 case study features the SigXplorer simulator and Model Integrity tools from Cadence Design Systems.

4.1 EVOLUTION FROM A COMMON ANCESTOR

For nearly three decades, bipolar transistor technology dominated integrated circuit and system design, the most common circuit families being emitter-coupled logic (ECL) and transistor-transistor logic (TTL). The rise of high-performance business and technical computing platforms in the 1960s and 1970s carried these circuit technologies into the commercial marketplace and set the stage for the personal computer boom of the 1980s. These early computing systems employed the common-clock architecture discussed in Chapter 2, "Chip-to-Chip Timing and Simulation," and illustrated in Figure 4-1.

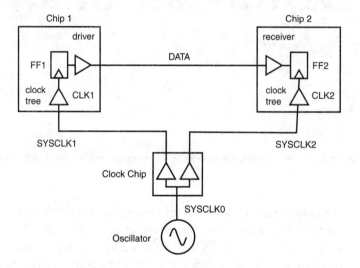

Figure 4-1   Common-clock interface.

Something disturbing began happening in the mid-1990s: The common-clock architecture began to run out of gas in high-performance systems. Maxwell showed that electromagnetic disturbances travel at the velocity of light in that medium, and Einstein postulated that there is not much point in trying to go any faster. Furthermore, the time it took for the clock to propagate through its on-chip distribution network and launch data into a transmission medium was too large compared to the clock cycle. Chip designers turned to phase-locked loops (PLLs) to combat unacceptable CMOS manufacturing variations, but these analog circuits had their own set of problems.

Around the same time that the client-server train began gathering steam, source-synchronous interfaces emerged as a means to satisfy the ever-increasing demand for IO bandwidth. The genius of the source-synchronous interface is that clock and data originate on the same chip and traverse the interconnect medium together. In an ideal world, both clock and data experience identical conditions on the silicon and in the interconnect, and the only trick that remains is how to center the clock in the middle of the data eye, which is the purpose of the phase-shift circuit labeled "φ" in Figure 4-2.

Naturally, the ideal world exists only in the realm of imagination. If it really did exist, we would be out of a job much more often than we usually are. The physical effects of the real world nibble away at the operating margins of a source-synchronous interface. Even though the schematics for each IO circuit on a given chip may be the same, the etching and doping conditions that form the transistors are not identical; nor is the on-chip metal identical between each instance of the IO circuit. The wires in the package are not identical, either. Chip timing specifications cover these variations between clock and data seen at the package pins.

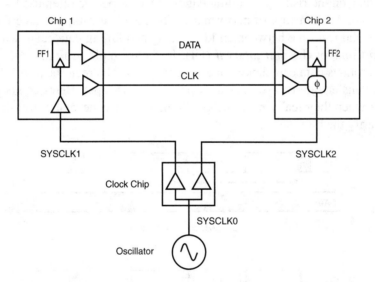

Figure 4-2   Single data rate source-synchronous interface.

Manufacturing variation of interconnect components, once thought to be negligible, now rival silicon process variations. Etching and weave variations impact the electrical properties of printed circuit board traces, as do copper and dielectric material properties. When a net jumps from one layer to another, we must account for the fact that both layers may be at opposite ends of their manufacturing extremes. Even when the net stays on the same layer, alignment of the line to fibers surrounding it can cause unexpected impedance disturbances. Long nets and narrow line widths conspire to make attenuation a dominant effect.

Crosstalk is a force to be reckoned with in every source-synchronous interface; connectors and tightly spaced microstrip lines are two common problem areas. Reference voltages are susceptible to noise, which effectively shifts the switching point of a receiver away from its ideal position in time.

The signal integrity engineer must account for how each source of variation affects clock-to-data skew and understand its relative contribution to the timing budget. This may be a tedious task the first time through, but it will pay off in future designs as the initial budget becomes a reference point that will help you decide when to rely on previous work and when to perform new analysis. The exercise also identifies the "big hitters" that warrant more attention than the smaller-budget items; they become opportunities for performance enhancement.

Before continuing, a few words of clarification are in order. Common-clock interfaces typically use the frequency of the clock signal as the primary performance metric. Because double data rate (DDR) source synchronous interfaces capture data on the rising and falling edges of clock, people often use bits transferred per second as a metric of performance. To calculate data rate, take the reciprocal of the shortest time between an ideal rising and falling data edge, which is called the *bit time* or the *unit interval* (UI). In the example in Figure 4-3, the source-synchronous data can change once every 1.5 ns. The reciprocal of 1.5 ns yields a data transfer rate of 667 megabits per second (Mbps). Beware: People still say MHz when they really mean Mbps. Sometimes it is necessary to ask the speaker to clarify the bit time.

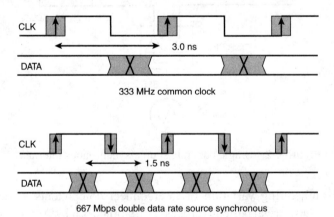

Figure 4-3   Clock frequency versus data rate.

4.2   DDR2 SIGNALING

To design a successful source-synchronous interface, you need to balance skew between data and strobe, both on-chip and off-chip, with jitter from power supply noise, crosstalk, reflections, attenuation, and PLLs. DDR2 memory serves as an excellent example of this balancing process since its position as an industry standard has made it familiar to a wide audience. Because it shares many characteristics with other source-synchronous interfaces, studying DDR2 can serve as a springboard to understanding other buses. This design example focuses on the DDR2 data bus and omits the address and command bus, which is more like a common-clock interface.

A typical DDR2 interface comprises a memory interface controller, a cluster of DRAMs, and buffer chips that redistribute clock and address to the DRAMs. Each group of single-ended bidirectional data bits (DQ) shares the same differential strobe signal (DQS). During a write cycle, the controller sources DQS. During a read cycle, the DRAM sources DQS. The clock signal (CK) typically originates on the controller and samples the address and control bus at the DRAM. Both CK and DQS run at the same frequency, but DQS samples data on both edges while CK samples address and control only on the rising edge. These first two categories of timing constraints are obvious because they involve capturing data at the primary flip-flops—that is, those connected directly to an IO circuit:

1. DQ to DQS (both read and write)
2. Address and control to CK
3. DQS to CK
4. DQS to DQS

Categories three and four are more subtle. When writing to the DRAM, the controller must satisfy the DDR2 specification for DQS to CK timing. While DQS captures data at the first layer of flip-flops within the DRAM, an internal copy of CK moves data from the DQS domain to the next layer of flip-flops within the DRAM. Therefore, a well-defined relationship between DQS and CK is necessary to facilitate reliable transfer of data across this second boundary. The DRAM also converts a copy of CK into a read strobe.

During a read transaction, the DRAM sends multiple bytes of DQ and DQS back to the controller, whose job it is to assemble the bytes into a word. Like the DRAM, the controller must also move data between the primary and secondary layers of flip-flops, as indicated by the gray flip-flops in Figure 4-4. If a large amount of skew is between DQS signals from different byte lanes (bits that share the same strobe), the controller may fail to resynchronize each byte with all the other bytes.

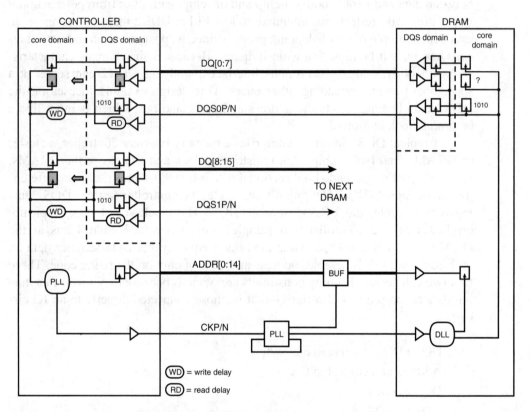

Figure 4-4   DDR2 interface functional diagram.

Figure 4-4 depicts the function of a DDR2 interface; it is not intended to be logically or electrically correct. Note that during a write cycle, the DRAM expects to see DQS aligned in the middle of the data bit. During a read cycle, the DRAM sends DQ and DQS at the same time and expects the controller to shift DQS into the middle of the data bit. The difficult work of phase alignment occurs in the controller rather than the DRAM. The delay blocks in Figure 4-4 represent the phase shift function, which may be accomplished differently by different controllers. If the controller and interconnect are under concurrent development, then signal integrity engineers and chip designers can make trade-offs to optimize system performance, but this is usually not the case.

The arrows in the controller represent a cloud of logic whose job it is to transfer data from the DQS domain to the core domain—a critical and nontrivial function that varies from core to core.

Imagine the DQ unit interval as having three distinct regions used by the controller, the DRAM, and the interconnect. Figure 4-5 depicts a well-balanced and "equitable" interface in which each of these three shares an equal slice of the unit interval. Alternatively, one could allocate more ps to the party who has the most difficult job, if representatives from each party had an equal voice in the decision.

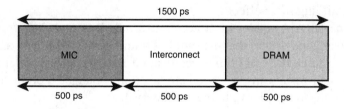

Figure 4-5 Equitable allocation of unit interval.

4.3 WRITE TIMING

In a world of perfect silicon with zero setup and hold times, interconnect could consume the entire unit interval, and we could all go home in time for dinner. Unfortunately, before a signal ever begins its journey across the interconnect, the controller has already consumed part of the unit interval in the form of skew between DQ and DQS. Compared to other source-synchronous interfaces, DDR2 has a small number of data bits per strobe, and this decreases the skew between each data bit and its corresponding strobe. For example, when the power distribution system droops during a noise event, all IO circuits slow down together because they are close enough to each other on the chip to see the same power disturbance. However, differences in on-chip wiring between DQS and each DQ bit cause skew. Furthermore, the DQS driver is differential, but the DQ driver is single-ended, which means the two IO circuits are not identical. The controller timing specification must account for these and other sources of skew and jitter between DQ and DQS. After the controller write timing is accounted for, the DRAM and interconnect can use the remainder of the unit interval.

On the way from the controller to the DRAM, the signal accumulates more skew and jitter from the interconnect. Meanwhile, the DRAM waits on the other end of the net, ready to capture data using the strobe sent by the controller. The DRAM expects the strobe to be nicely centered inside the DQ window; it has no automated strobe centering circuitry of its own. Like the controller, the DRAM introduces its own sources of skew, both in the package and on-chip. Again, IO circuits play a role in generating jitter. In a good DRAM design, both DQ and DQS use the same differential receiver circuit. The DQ inputs derive their switching thresholds from a common reference voltage (Vref), but you have a choice whether to use the same reference voltage for the DQS signal or to use differential signaling. If you choose differential signaling, a noise event on the reference voltage will create jitter between DQ and DQS. This implies a trade-off between the benefits of differential signaling and the benefits of sharing a common switching threshold between DQ and DQS, which is why it is important to develop a means for quantifying these effects.

The DRAM vendor must account for all skew and jitter from the package pins inward and ensure that the sum lies within the DDR2 setup and hold specifications. The DDR2 hold time specification is nearly twice as large as the setup time, which results in an asymmetrical timing budget. Fortunately, both these numbers are small relative to the unit interval.

The dark gray regions in Figure 4-6 represent the portion of the unit interval consumed by the controller, and the DRAM write timing specifications define the boundaries of the light gray regions. The white regions are available for the interconnect. The DDR2 specification, JEDEC JESD79-2, defines the DRAM data setup (tDS) and hold (tDH) times in the typical fashion with respect to DQS. In this example, the controller also specifies data valid (tDV) with respect to the strobe; subtraction is necessary to find the number of ps consumed by the controller:

$$tMIC = tCK/2 - [tDV(max) - tDV(min)] = 3000ps/2 - [550ps + 550ps] = 400ps$$

Equation 4-1

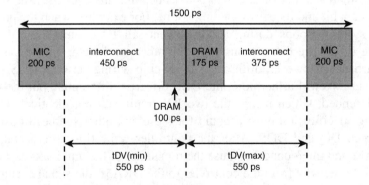

Figure 4-6   Write unit interval allocation.

These are the makings of a rudimentary eye mask without the voltage dimension. Keep in mind that simulations do not model either of the gray regions, so an eye mask must lump the light and dark gray regions together in the middle. This particular write cycle is favorable for the signal integrity engineer. The DRAM only uses 18% of the UI, and the controller uses 27%, leaving a generous 55% for the interconnect.

The write timing budget in Table 4-1 expresses the allocation of the unit interval in tabular format. Row five is the sum of the controller, DRAM, and interconnect contributions, and operating margin is the difference between this sum and half the unit interval. Assuming the signal integrity engineers and chip designers have accounted for all relevant effects, the operating margin must be zero or positive for the interface to function reliably over all manufacturing and operating conditions for the working lifetime of the product. If the interconnect uses 825 ps (450 ps + 375 ps), the operating margin will be exactly zero.

Table 4-1   Preliminary Write Timing Budget

	Parameter	Symbol	Setup (ps)	Hold (ps)
1	Memory interface controller	tMIC	200	200
2	DRAM input setup time	tDS	100	
3	DRAM input hold time	tDH		175
4	Interconnect skew and jitter		450	375
5	TOTAL		750	750
6	Unit interval	tCK/2	750	750
7	OPERATING MARGIN		0	0

4.4   READ TIMING

Timing a read transaction is slightly more complicated than timing a write transaction. The controller contributes a setup and hold time to the budget in the typical fashion. Finding the DRAM's share of the unit interval involves calculating the "read data valid window" using four DRAM timing specifications, as illustrated in Figure 4-7.

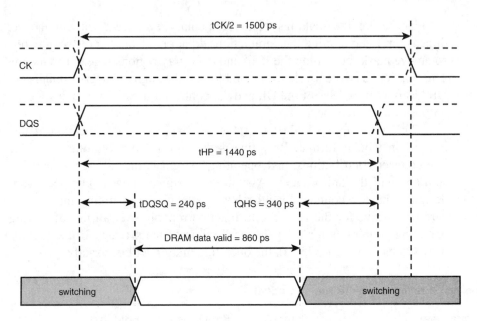

Figure 4-7   Read data valid window for a byte lane.

The first calculation applies the clock low level width parameter, tCL, to the ideal clock period, tCK, to arrive at the shortest possible DQS pulse width, tHP. This parameter is the equivalent of the clock duty cycle, and it turns out to be 48% of 3000 ps for DDR2 667 or 1440 ps. Two other timing parameters define the size of the data valid window, which is the region between the gray zones where data is guaranteed not to switch. First, the worst-case skew between the strobe and the last DQ bit to become valid (tDQSQ) erodes the leading edge of the data valid window by 240 ps. Second, the DQ hold skew factor (tQHS) erodes the trailing edge by 340 ps. The combination of all these parameters leaves a data valid window of 860 ps. The difference between 1500 ps and 860 ps is the amount of the unit interval the DRAM consumes during a read transaction, which is row 3 in Table 4-2.

Table 4-2   Preliminary Read Timing Budget

	Parameter	Symbol	Setup (ps)	Hold (ps)
1	MIC input setup time	tDS	250	
2	MIC input hold time	tDH		250
3	DRAM		320	320
4	Interconnect skew and jitter		180	180
5	TOTAL		750	750
6	Unit interval	tCK/2	750	750
7	OPERATING MARGIN		0	0

The controller setup and hold times (tDS and tDH) complete the picture. Figure 4-8 illustrates how much of the unit interval remains for the interconnect. In this example, the DRAM consumes 43% of the unit interval, and the controller uses 20%, leaving 37% for the interconnect, which is close to the equitable 33% slice depicted previously in Figure 4-6. However, this situation stresses the controller, whose task it is to center the strobe in the middle of the read data eye and align eight or nine bytes of incoming data for consumption by logic in the core clock domain. The logic required to implement this function is considerably more complicated than simply using a strobe to capture data in a flip-flop. Complexity translates to skew and jitter.

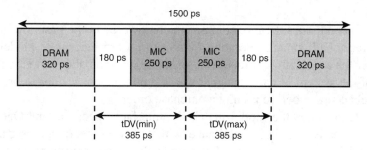

Figure 4-8   Read unit interval allocation.

With 280 ps of interconnect skew and jitter on either side of the setup and hold window, the operating margin is exactly zero.

The write and read timing budgets shown previously in Tables 4-1 and 4-2 are templates to be filled in with actual interconnect timing, which will consume the remainder of the chapter.

4.5   GET TO KNOW YOUR IO

At the beginning of the design cycle, the temptation to stitch models together and start running simulations rears its ugly head. Management inevitably applies pressure to generate design rules using those simulations under a demanding schedule, but investing a few extra days in exploration of the design space always yields valuable knowledge on the current design and those to come. Many electrical parameters exert influence on a digital interface, but which ones are worth further exploration?

The parameters that affect the eye most strongly are likely candidates for optimization. These parameters, along with those that have a high degree of uncertainty, are also likely candidates for a failure mechanism during hardware bring-up. This knowledge comes with the application of physics to the problem at hand—and experience. (Some refer to the combination of these two factors as intuition, which is a bit of a misnomer.) After you have established the timing constraints, IO circuits are the next logical step in the process of exploring the design space.

Three important questions can help you gain confidence that the end hardware will function reliably as simulated. First, does the model match the IO circuit design? Given a transistor-level model extracted from the IO circuit physical design database, the answer to this question is likely to be yes. If the silicon vendor supplied an IBIS model, you can evaluate correlation between transistor-level simulations and behavioral simulations using the Golden Waveforms feature of IBIS.

Second, does the model match the hardware? In the past, lower performance interfaces had enough operating margin to absorb inaccurate models. That is no longer the case with interfaces running at 1 Gbps and above. It is in the best interest of the system or card designer to ask a component vendor to provide proof of model accuracy before issuing a purchase order.

Third, does the model match the interface specification? One would hope the answer is yes, but compliance with industry specifications can be elusive. Assuming accurate models, a few simple charts can shed light on the compliance of DDR2 IO circuits with JESD79-2, as follows:

1. Off-chip driver IV curves
2. On-die termination (ODT) IV curves
3. Rising and falling waveforms

4.6 OFF-CHIP DRIVER

DDR2 drivers operate in either full- or reduced-strength mode, corresponding roughly to 20 ohm and 40 ohm output impedances. Bit A1 of the DRAM extended mode register set (EMRS) determines the impedance state of the driver. The controller has its own register set. The DQ and DQS receivers are equipped with variable on-die termination to help manage reflections and optimize timing margins. The controller is responsible for sending a signal to the DRAM to enable and disable termination at the appropriate time during a read or write cycle. EMRS bits A2 and A6 control the value of the termination resistance. The four possible states of on-die termination are: 50ohm (parallel 100 ohm), 75 ohm (parallel 150 ohm), 150 ohm (parallel 300 ohm), and disabled. Therefore, there are eight (2×4) possible combinations of driver impedance and on-die termination.

We can check for compliance of the driver's output impedance by overlaying IV curves from the transistor-level model or IBIS model on top of the curves defined in Tables 30–33 of JESD79-2 (DDR2 SDRAM Specification). Figure 4-9 demonstrates the compliance of a reduced-strength driver model. The bold black lines delineate the boundaries of IV space within which a driver's curves must fit, and the light black lines are the minimum, typical, and maximum curves from the IBIS model.

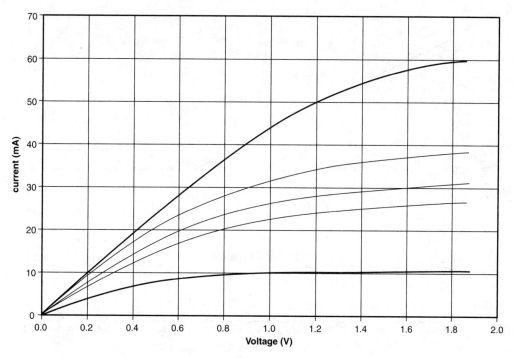

Figure 4-9   Reduced-strength driver pull-down curves.

4.7   ON-DIE TERMINATION

The characteristics of the on-die termination are just as important as those of the driver since the two work together to determine the dc levels and reflection behavior of the network. In the IBIS model, six curves represent the impedance of the termination circuit: power and ground clamps for minimum, typical, and maximum conditions. Figure 4-10 plots the power and ground clamp curves under typical conditions. If the termination network were two 150 ohm resistors, one to

VDDQ and one to VSSQ, we would expect to see two lines of slope 150 ohm crossing the x-axis VDDQ/2 or 0.9 V. The Thevenin equivalent circuit is represented by two 75 ohm lines, both of which go to zero current around 0.9 V. Both the parallel termination and the Thevenin equivalent circuit satisfy the conditions of JESD79-2, which defines on-die termination impedance in Table 18 by the current drawn at the high and low ac thresholds. If you are modeling switching noise, carefully consider the current path in the Thevenin network.

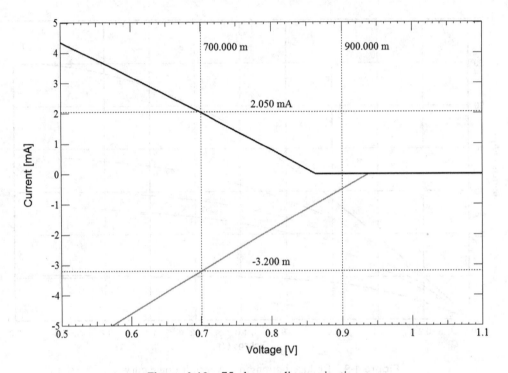

Figure 4-10   75 ohm on-die termination.

The ac thresholds sit at ±200 mV around VDDQ/2. Using the currents from the VIL(ac) and VIH(ac) crossing points, the termination impedance calculation gives 76.6 ohm under typical conditions. The impedance at minimum conditions (VDDQ = 1.7 V) is 80.5 ohm and it is 72.2 ohm under maximum conditions (VDDQ = 1.9 V). All three values fall within the JESD79-2 specification limits of 60 and 90 ohm.

$$Rtt(eff) = \frac{V_{IH}(ac) - V_{IL}(ac)}{I(V_{IH}(ac)) - I(V_{IL}(ac))} = \frac{1.100V - 0.700V}{2.025mA - (-3.199mA)} = \frac{0.400V}{5.224mA} = 76.6\Omega$$

Equation 4-2

The DDR2 case study in this chapter focuses on the half-strength driver and 75 ohm ODT combination, which works well for a point-to-point net running at 667 Mbps. The full-strength driver is too strong for a single load. Setting ODT to 50 ohm would improve the reflection coefficient, but it comes at the expense of higher power and lower dc swing.

4.8   RISING AND FALLING WAVEFORMS

The DDR2 specification also defines the allowable range for off-chip driver dV/dt measured into 25 ohm termination to VTT (VDDQ/2). A driver that is too slow consumes too much of the unit interval, and a driver that is unnecessarily fast generates excessive noise. According to the specification, dV/dt is a linear approximation defined where the waveform crosses VIL(ac) max and VIH(ac) min. In Figure 4-11, rising dV/dt = 2.3 V/ns under typical process, temperature, and voltage conditions, which is within the limits of 1.5 and 5.0 V/ns. Although the falling dV/dt is close to the same value, the falling edge takes significantly longer to round the corner. This asymmetry becomes noticeable in the eye diagram, which does not switch exactly at 0.9 V but slightly above. Pay close attention to the balance between high and low impedances and between rising and falling edges. A suboptimal driver design can degrade system performance.

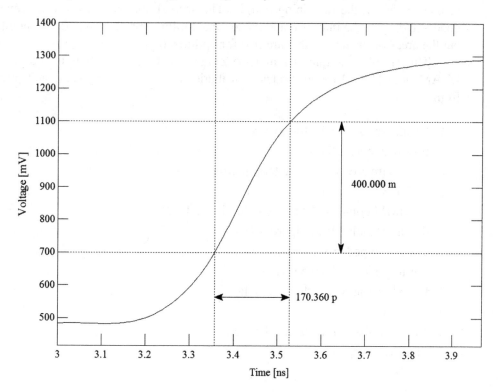

Figure 4-11   Reduced-strength driver typical dV/dt.

Instantaneous dV/dt is another important parameter, although no specification exists for it. Crosstalk is directly proportional to instantaneous dV/dt, and that makes it a valuable parameter to watch. The instantaneous value can vary from the linear approximation by as much as 20% to 30%. It also varies from location to location within a network. An unpackaged driver will likely give rise to the highest dV/dt, an interesting but unrealistic case. Attenuation degrades dV/dt as the wave propagates down a lossy transmission line, so pay attention to its value at the receiver and other important locations such as a DIMM connector.

4.9   INTERCONNECT SENSITIVITY ANALYSIS

After establishing the electrical characteristics of the IO circuits, you can begin the process of understanding how much of the unit interval the interconnect consumes. At least two approaches to this process are possible. You can throw all the models into one big pot, stir it, cook it, and find out how it tastes when it's done. This approach may or may not produce the desired results. If it does not, then it's back to the kitchen with another set of ingredients. An alternative approach involves starting with the simplest network imaginable—driver and termination—incrementally adding each interconnect model and stopping to assess the effects of that model on the remaining margin. The second approach leads to a deeper understanding of the interface and its dominant failure mechanisms. It also points out the areas of the design that are ripe for optimization.

This simple example of a memory controller driving one byte lane to a x8 DRAM on a DIMM demonstrates contributions to interconnect skew and jitter from

1. Conductor and dielectric losses
2. Impedance variations
   a. 75 ohm receiver on-die termination
   b. 2.5–3.5 pF receiver capacitance
   c. DIMM connector (labeled C in Figure 4-12)
   d. Vias (labeled V in Figure 4-12)
   e. Card impedance tolerance
3. Pin-to-pin capacitance variation
4. Length variations within a byte lane
5. Crosstalk
6. Vref ac noise and resistor tolerance
7. Slope derating factor

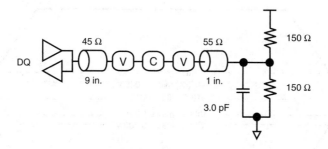

Figure 4-12   DQ net.

Before running any simulations, one question well worth asking is how much the driver's intrinsic rise and fall times degrade the eye. In the extreme case, if the rise and fall times were zero, they would not degrade the eye at all. Even if the rise and fall times were greater than zero but the DDR2 specification defined Vref as the timing threshold, the width of the eye would still stand at the unit interval, 1500 ps. However, the DDR2 specification defines the setup timing thresholds as VIL(ac) and VIH(ac) on the left side of the eye and the hold timing thresholds at VIL(dc) and VIH(dc) on the right side of the eye. This means that the eye is less open for slower rise and fall times.

If the DRAM driver dV/dt were centered exactly in the middle of the minimum and maximum values allowed by the specification, as shown in Figure 4-13, the eye would be 101 ps (62 + 39) smaller than the unit interval:

$$\Delta t_{SU} = \frac{\Delta V}{dV/dt} = \frac{0.200V}{3.25V/ns} = 62ps$$

Equation 4-3

$$\Delta t_{H} = \frac{\Delta V}{dV/dt} = \frac{0.125V}{3.25V/ns} = 39ps$$

Equation 4-4

If the DRAM driver dV/dt were at the low end of the specification, the eye would be 216 ps smaller than the unit interval, as shown in Figure 4-14—quite a difference.

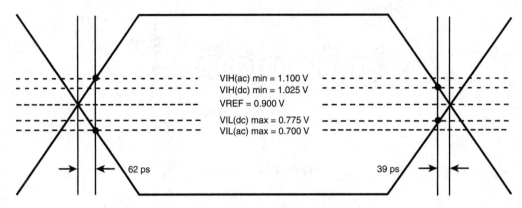

Figure 4-13   Eye opening at 3.25 V/ns.

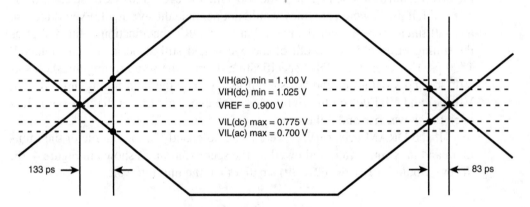

Figure 4-14   Eye opening at 1.5 V/ns.

The eye diagram in Figure 4-15 came from a nominal simulation of an IBIS driver model and an ideal 75 ohm termination with no interconnect. The eye opening is 1.380 ns, slightly less than the calculated 1.399 ps. Because the rise and fall times are not identical, the crossing points of the eye do not lie exactly at Vref.

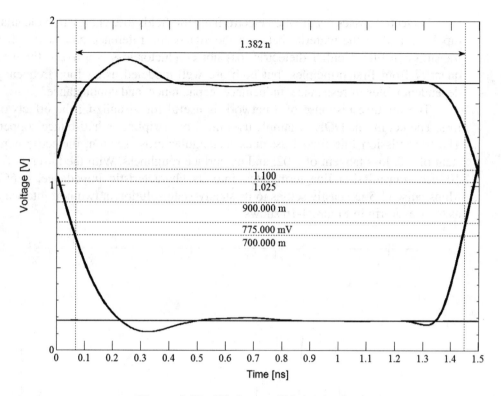

Figure 4-15 75 ohm parallel termination.

4.10 CONDUCTOR AND DIELECTRIC LOSSES

Energy loss, or attenuation, has at least one desirable effect: It rounds off the sharp and narrow waveform features introduced by reflections and noise. On the negative side, it can prevent a signal from reaching full swing when transmission lines are long and bit times are short. Therefore, it pays to understand its physical origins and independently quantify its effects on a particular net configuration.

Transmission line loss comes in two varieties: conductor loss and dielectric loss. When electromagnetic radiation encounters a metal surface, its amplitude dies out exponentially as it penetrates the surface of the metal. The amount of loss depends on the frequency content of the fields. At contemporary frequencies, the fields remain in the "skin" of the conductor, and so does the current. In this manner, skin effect or conductor loss drives the effective resistance of the wire much higher than it would be if the current were allowed to flow evenly throughout the cross-sectional area of the wire.

Dielectric losses occur when electromagnetic fields lose energy to molecular dipoles found in the material between the signal and reference conductors in a transmission line. Neither dielectric loss nor conductor loss is a trivial thing to quantify from first principles, but both are well-modeled using four frequency dependent tables of resistance, inductance, capacitance, and conductance.

The impulse response of a network is useful for visualizing the effects of loss. The net in this DDR2 example uses the 3 mil stripline in half-ounce copper. The transmission line model assumes rectangular cross section, dielectric constant of 4.0, loss tangent of 0.02, and no surface roughness. With the far-end of a 10-inch transmission line terminated into its characteristic impedance, a 667 Mbps pulse (1.5 ns) easily settles to its steady-state solution in two unit intervals, as the waveform in Figure 4-16 shows.

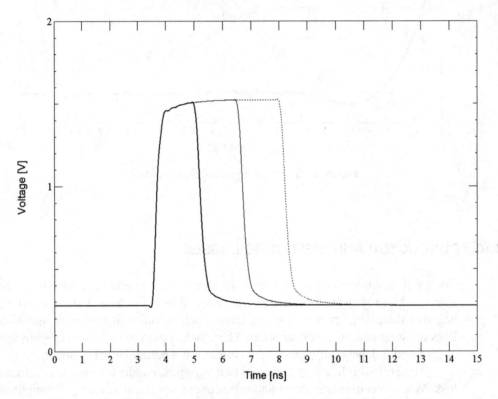

Figure 4-16   Pulse response of 10-inch lossy line with 50 ohm termination.

Given this impulse response, we would not expect to see much loss-induced deterministic jitter in a longer bit stream. In fact, jitter on a physical transmission line with conductor and dielectric losses is only 30 ps higher than jitter on an ideal lossless line (1.380 ns–1.350 ns).

0100 0110 0011 1000
1011 1001 1100 0111

This pattern corresponds to a simple DQ pattern with a maximum of three bits in the same state. The strobe would see a 0101 pattern. Although the DQS eye in Figure 4-18 is much cleaner than the DQ eye in Figure 4-17, the threshold crossings are nearly the same; DQ and DQS each contribute 30 ps of jitter in addition to the ideal transmission line. The sum of these contributions, 60 ps, appears in row 1 of Table 4-6, the final interconnect skew and jitter budget.

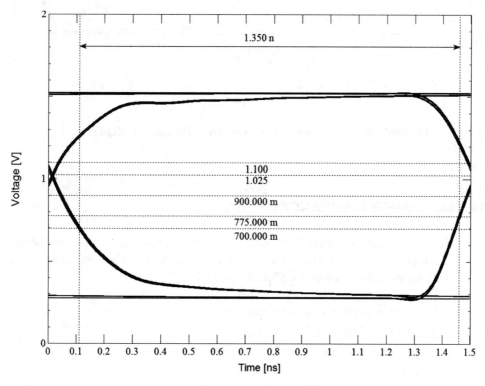

Figure 4-17   10-inch lossy line 50 ohm termination with DQ pattern.

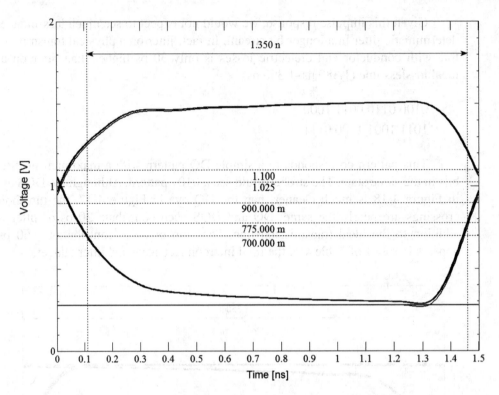

Figure 4-18   10-inch lossy line 50 ohm termination with DQS pattern.

4.11   IMPEDANCE TOLERANCE

This case study considers five sources of impedance discontinuity. For the sake of simplicity, it does not cover the effects of IC packaging, but you can extend the techniques in this chapter to IC packaging as well.

1. 75 ohm receiver on-die termination
2. 2.5–3.0 pF receiver capacitance
3. DIMM connector
4. Vias
5. Card impedance tolerance

Several factors play together to make this analysis more complicated than first meets the eye. The reflection coefficient determines the magnitude and direction of the reflection. The 75 ohm termination produces a positive reflection that is 20% of the original amplitude, while the receiver capacitance and vias produce a negative reflection whose size is a function of rise time. The dynamic impedance of the driver determines how much of the original reflection winds up back at the receiver. Because the final driver impedance (40 ohm) is less than that of the transmission line, the reflection becomes negative when it hits the driver and reverses direction.

The location of the discontinuity along the transmission line determines which part of the waveform the reflection of the reflection affects. If the reflection of the reflection superimposes on the waveform while it is switching, it closes the eye along the time axis. If it hits the waveform anywhere else, its effect is almost negligible on a signal whose amplitude at the receiver is 1.5 V.

Figure 4-19 demonstrates the effect of 75 ohm termination on waveforms that are one, two, and three bits long.

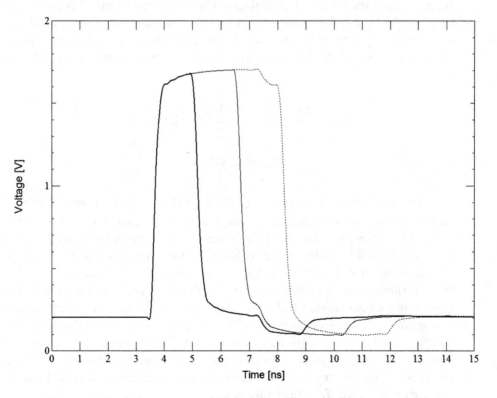

Figure 4-19   Pulse response of 10-inch lossy line with 75 ohm termination.

Compare Figures 4-16 and 4-19. When the far-end termination changes from 50 ohm to 75 ohm, the shape of the pulse does not change much, but the settling time increases to between three and four unit intervals.

The reflection from the rising edge of the two-bit pulse (light trace) bounces off the driver and returns to the receiver just as the waveform is finishing its transition from high to low. The reflection from the rising edge of the three-bit pulse (dashed trace) combines with the incident falling waveform just before it begins switching low. In simulations with a 10-inch line, the reflection causes very little timing degradation. However, lengthening the transmission line by a few hundred picoseconds places the reflection squarely in the edge of the pulse—where it affects eye width the most.

Two of the next three discontinuities are capacitive, and one is inductive. The receiver capacitance and the vias both generate negative primary reflections. The DIMM connector generates a positive primary reflection. The receiver capacitance and the vias tend to counteract the effect of the on-die termination, and if all three of these reflections reach the receiver at the same time, as they do in Figure 4-20, they reduce the overall jitter when compared to the effect of any one alone. By looking at the location in time of the secondary positive reflection from the receiver capacitance and the vias (around 7.8 ns), you can see how much to adjust the length of the first transmission line to create a worst-case jitter scenario.

$$\Delta L = \frac{\Delta t}{T_{pd}} = \frac{1}{2} \times \frac{920ps}{160ps/in.} = 2.9in.$$

Equation 4-5

Table 4-3 shows the cumulative effects on deterministic jitter (DJ) of adding each discontinuity to the simulation one at a time. To generate the DJ value in the last column, incrementally adjust the length of the first transmission line until the deviation from the nominal pulse width reaches its maximum value. At the point of maximum jitter, the reflection appears to be entirely absorbed into the edge of the waveform. It is important to note that this problem would have a different answer if each simulation had only one discontinuity and the effects were not cumulative. After understanding how the reflections from each individual discontinuity combine, you can then simulate a longer bit pattern and generate an eye diagram under silicon process, temperature, and voltage variations. It may be necessary to sweep the transmission line length around the nominal value to produce worst-case jitter at different silicon corners.

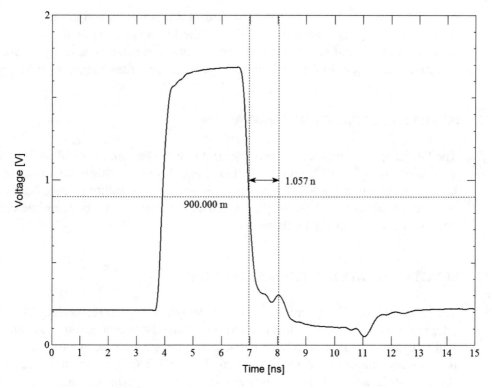

Figure 4-20   Pulse response with DIMM connector and vias.

Table 4-3   Pulse Width Variation Caused by Impedance Discontinuity

	Discontinuity	Length (in.)	Width (ns)	DJ (ps)
1	Baseline—50 ohm termination	10.000	3.140	0
2	75 ohm receiver on-die termination	9.100	3.110	−30
3	2.5 pF receiver capacitance	8.750	3.125	−15
4	3.0 pF receiver capacitance	8.750	3.130	−10
5	3.5 pF receiver capacitance	8.750	3.140	0
6	DIMM connector	6.700	3.155	+15
7	Vias	6.500	3.160	+20
8	Card impedance −10% / +10%	6.600	3.160	+20
9	Card impedance +10% / −10%	6.000	3.115	−25

The worst conditions on row 2 of Table 4-3 represent a nonphysical case: It is not possible to have an ideal 75 ohm on-die termination without also having some associated on-die capacitance. Row 9, then, represents a realistic worst case. This value goes in the final interconnect skew and jitter budget in Table 4-6.

4.12 PIN-TO-PIN CAPACITANCE VARIATION

The DDR2 specification allows any two DQ pins in the same DRAM package to vary as much as 0.5 pF from one another. Table 4-3 shows this to be a relatively minor effect—on the order of 5 ps. In the interest of not splitting hairs, both the setup and hold will get 5 ps for pin capacitance variation, and the final interconnect skew and jitter budget will get 10 ps.

4.13 LENGTH VARIATION WITHIN A BYTE LANE

The length of each DQS net must match the lengths of the corresponding DQ and DQM nets within some reasonable tolerance. Make the tolerance too small, and it will be extremely difficult to route the nets. Too large a tolerance will consume an unnecessary amount of the unit interval. For DDR2 667, a 100 mil tolerance seems to strike a good balance between routability and performance. For the final interconnect skew and jitter budget, we will round 16 ps to 15 ps for aesthetic value.

$$\Delta t = \Delta L \times T_{pd = 0.100in.} \times_{160ps/in.} = 16ps$$

Equation 4-6

When developing a new ASIC, keep package length tolerance in mind. Group the chip and package pins so that the DQ and DQM bits are near their DQS. Assign on-chip and package routing constraints that will allow the combination of chip and package to meet the timing specification targets at the package pins. Consider board routability when assigning pins.

These details are critical to the timing at the flip-flops inside the dashed box shown previously in Figure 4-4—that is, the primary inputs and outputs to the chip. When writing, the memory interface controller must phase align DQS with one DQ byte and DQM. When reading, it must shift DQS coming from the DRAM to the middle of the DQ/DQM eye. Routing skew between nets in this group on the chip, package, and board is a direct detractor from this fundamental timing relationship.

4.14   DIMM CONNECTOR CROSSTALK

There are several possible sources of crosstalk in a DDR2 interface: IC packages, resistor packages, printed circuit board wires, and DIMM connectors. Technically, each geometrically unique interconnect structure is a candidate for coupling, although some of them, such as ball grid arrays, do not contribute enough crosstalk to a typical DDR2 interface to bother tracking. IC vendors incorporate the effects of package crosstalk when writing timing specifications, but they have no way of knowing where crosstalk that reflects off their die will end up and when it will arrive there. Crosstalk analysis falls squarely in the domain of the signal integrity engineer. Determining the effects of crosstalk on timing is more complicated than it might appear, and this brief discussion focuses on the largest source of crosstalk in a DDR2 interface that uses striplines: the DIMM connector. Chapter 9, "PCI Express Case Study," covers crosstalk in more detail.

Analysis of DIMM connector crosstalk begins with a phone call to the customer engineer or manufacturer's representative. Ask for a multiline model and corresponding application note that documents modeling assumptions, how to use the model, and model-to-hardware correlation. The ideal application note will show a table of measured impedance, crosstalk, and propagation delay versus model predictions, as well as measured and simulated waveforms on the same set of axes. You will probably have to agree to release the manufacturer from any responsibility for the accuracy of the model or the success of your design. The second point is understandable, but the first has always seemed ironic since so much of a signal integrity engineer's ability to accurately predict system performance and reliability rests on accurate models.

The next step in assessing the impact of DIMM connector crosstalk on operating margin is to run a few simple simulations with only one pin switching and the quiet neighbors terminated into the printed circuit board characteristic impedance (see Figure 4-21). This exercise accomplishes several goals. It proves that the model runs and yields physical results consistent with the vendor's application note. A quick examination of quiet line crosstalk helps you to sort out any pin assignment problems; crosstalk should be strongest at the nearest neighbor. Finally, knowing the height and width of the crosstalk pulses allows you to make a first-order estimate of crosstalk-induced jitter. A more accurate simulation of crosstalk-induced jitter with both victim and aggressors switching requires an accurate estimate of the aggressor waveform time derivative at the point of coupling—the DIMM connector. This means using the correct DRAM and MIC driver models as well as realistic lossy transmission line models.

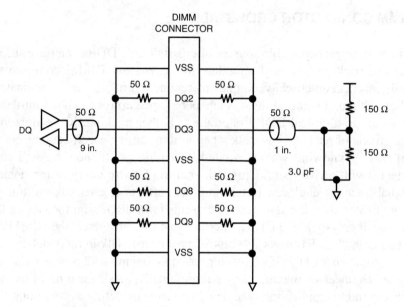

Figure 4-21   DIMM connector crosstalk simulation.

The general pin assignment pattern for the DDR2 DIMM connector is one reference followed by two signals: ref-sig-sig, ref-sig-sig, and so on. Two-to-one is a low signal-to-return ratio, and the coupling is manageable. The waveforms in Figure 4-22 show that the crosstalk signature from this particular DIMM connector follows a typical pattern for card-to-board connectors: Near-end crosstalk (NEXT, bold waveform) has the same sign as the aggressor waveform derivative, and far-end crosstalk (FEXT) has the opposite sign. Simulating with more than one unit interval between aggressor edges allows the crosstalk pulse to settle out before the aggressor switches again. This can be a useful technique for sorting out the effects of component coupling and reflected crosstalk from nonideal termination.

As expected, Table 4-4 shows that NEXT drops off strongly as the distance between victim and aggressor pins increases. Strangely, FEXT actually increases as distance between victim and aggressor increases. FEXT is a combination of electric and magnetic coupling; if they happen to be equal and opposite, FEXT will be nearly zero, even though NEXT may still be high. In this DDR2 DIMM connector, the electric and magnetic coupling between DQ2 and DQ3 are nearly balanced. Moving away from DQ3, magnetic coupling dominates and the waveform becomes more negative. Eventually it will begin dying off.

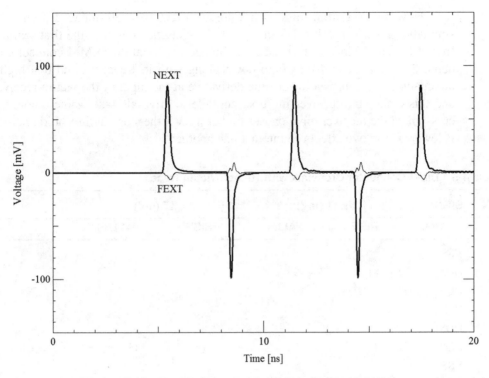

Figure 4-22   DIMM connector NEXT and FEXT waveforms.

Note that rising and falling derivatives are not necessarily identical for any given driver. The magnitude of the negative derivative on the falling edge exceeds that of the positive derivative on the rising edge by about 25%. This reinforces the previous discussions on the need to characterize the IO circuits before beginning timing and noise analysis.

Table 4-4   DIMM Connector Crosstalk with 1-Inch Victim Trace

PIN	SIGNAL	NEXT (mV)		FEXT (mV)	
		RISE	FALL	RISE	FALL
8	VSS				
9	DQ2	106	−133	−8	13
10	DQ3	VICTIM			
11	VSS				
12	DQ8	33	−41	−11	12
13	DQ9	19	−24	−15	18
14	VSS				

How much do transmission line losses affect crosstalk-induced jitter? The simulation results in Table 4-5 show that changing the length of the first segment from 1 inch to 10 inches decreases the time derivative at the DIMM connector by about 25%. NEXT and FEXT go down along with it. Keep in mind that higher attenuation also causes a lower time derivative at the input of the victim receiver, and this makes the receiver more susceptible to crosstalk—the same amount of crosstalk at the receiver input causes higher jitter if the slope at threshold crossing is lower. These two effects counteract one another.

Table 4-5   DIMM Connector Crosstalk with 10-Inch Victim Trace

PIN	SIGNAL	NEXT (mV)		FEXT (mV)	
		RISE	FALL	RISE	FALL
8	VSS				
9	DQ2	81	-99	-6	9
10	DQ3	VICTIM			
11	VSS				
12	DQ8	26	−31	−8	9
13	DQ9	15	−17	−12	13
14	VSS				

The DIMM connector crosstalk simulation shown previously in Figure 4-21, though useful in characterizing the connector, is still somewhat artificial. In the real net, the NEXT pulse sees 9 inches of lossy transmission line that attenuates it to 75% of the original value at the DIMM connector by the time it reaches the driver. Compare Tables 4.4 and 4.5 to see this effect. Assuming the worst-case NEXT to be the sum of the falling edge column in Table 4-5 (147 mV), the attenuated NEXT is down to 110 mV just before it reaches the driver.

If the victim driver happens to be active, the crosstalk pulse sees a 40 ohm output impedance rather than an ideal termination. The resistive (real) reflection coefficient at the driver is approximately 10%, and the capacitive (imaginary) component adds another 10% to the reflection coefficient. The pulse that travels back toward the receiver is only 22 mV high.

As the reflected NEXT pulse travels back toward the receiver, the transmission line robs another 20% of its amplitude. When it encounters the 75 ohm termination at the receiver, it gets a little boost from the positive reflection coefficient that is offset by the negative reflection coefficient of the 3 pF receiver capacitance. Just before it hits the receiver, the amplitude of the NEXT pulse is 18 mV.

Of course, the nets must be just the right length for the reflected NEXT to arrive at the receiver at the same time as a data edge and the FEXT. The sum of the worst-case FEXT and the reflected, attenuated NEXT is 49 mV. If each far-end crosstalk pulse and reflected near-end crosstalk pulse *does* arrive at the receiver while its input is switching, there is a well-defined relationship between crosstalk induced jitter (Δt), crosstalk magnitude (ΔV), and input edge rate (dV/dt). We will round this number down to 30 ps before entering it in the final interconnect skew and jitter budget.

$$\Delta t = \frac{\Delta V}{dV/dt} = \frac{0.049V}{1.5V/ns} = 33ps$$

Equation 4-7

Before proceeding to the next section, let's list the assumptions behind 33 ps of crosstalk-induced jitter:

1. DIMM connector is the only relevant source of crosstalk.
2. Only three DIMM connector pins contribute crosstalk.
3. Pins on opposite sides of DIMM do not contribute crosstalk.
4. 10 inches of 3 mil stripline.
5. 40 ohm DQ driver with nominal rise time.
6. Slowest rise time at receiver input (conflicts with previous assumption).
7. 75 ohm termination at receiver.
8. 3 pF driver and receiver capacitance.
9. FEXT and reflected NEXT arrive at receiver while input is switching.
10. All crosstalk peaks are concurrent.

Crosstalk is a multifaceted phenomenon. It takes a fair amount of thought to arrive at an estimate of crosstalk-induced jitter that is not fraught with excessive conservatism.

4.15 VREF AC NOISE AND RESISTOR TOLERANCE

The differential receiver in a DDR2 interface relies on a reference voltage, Vref, to discriminate between logic high and low levels at the receiver input. The nominal value for this reference is VDD/2 or 0.9 V, and the simplest implementation is a voltage divider with two low-tolerance resistors. This is a good place to spend money on more expensive components because it translates directly into improved operating margins. Just how much margin is required will become apparent when the final timing budget is complete. For now, assume a 1% tolerance on two 100 ohm resistors. Under worst-case resistor tolerance conditions, the reference voltage will drift 9 mV off its mark.

$$V_{ref} = \frac{R_1}{R_1 + R_2} \cdot VDD = \frac{99\Omega}{99\Omega + 101\Omega} \cdot 1.8V = 0.891V$$

Equation 4-8

Assuming a worst-case edge rate of 1.5 V/ns at the receiver input, the 9 mV shift in Vref effectively shortens or lengthens the bit by 6 ps.

$$\Delta t = \frac{\Delta V}{dV/dt} = \frac{0.009V}{1.5V/ns} = 6ps$$

Equation 4-9

What about ac noise on Vref? Conventional wisdom holds that the IC package will contain mid- and high-frequency noise. Therefore, if you want to design a DDR2 Vref solution, you need to consider noise in the neighborhood of 100 MHz and below. A good place to start would be holding Vref to 2.5% (45 mV) of its parent supply (1.8 V). This much ac noise would generate 30 ps of jitter. We will round the sum of these two numbers up to 40 ps.

$$\Delta t = \frac{\Delta V}{dV/dt} = \frac{0.045V}{1.5V/ns} = 30ps$$

Equation 4-10

How exactly do we keep Vref noise below 45 mV? Vref noise comes from several possible sources. The DRAM itself has four, eight, or sixteen differential amplifiers that are each trying to drag little spurts of current through a common Vref package pin. Also noise on the 1.8 V rail will inevitably find its way onto Vref by way of the voltage divider resistors. One theory says that it would be wise to make Vref and 1.8 V track each other if incoming DQ and DQM signals are referenced to 1.8V and ground. A supply collapse between 1.8 V and ground would affect the signal and its reference at the same time. However, a voltage divider made from two equal resistors tracks noise on 1.8 V at 50%, not 100%. This implies an ac noise budget of 5% on 1.8 V to achieve 2.5% on Vref, and the power distribution network must be optimized to meet this budget.

Pay close attention to the metal between the voltage divider and the DRAM. It should not be strongly coupled to any signal with high edge rates. Large shapes will couple noise from the planes on either side of them, which may be acceptable if the planes happen to be 1.8 V and ground. If you use transmission lines to bus Vref between the voltage divider and the DRAM, make sure both ends are terminated into their characteristic impedance or noise may build up on the line. Hanging a decoupling capacitor on the Vref node near the voltage divider destroys the ideal termination.

In the world of power distribution network analysis, rules of thumb are plentiful. Quantitative tools are scarce, complicated, and expensive. As always, dedication to controlled lab experiments is a sound investment.

4.16 SLOPE DERATING FACTOR

If there is a significant mismatch between the DQ and the DQS edges rate at the receiver, the DDR2 specification requires the user to apply a derating factor to the setup and hold times that ranges between 0 and 140 ps depending on the difference in edge rates. In a well-balanced design, this should not be an issue. Pay attention to DQ-DQS symmetry in the IO circuits, routing, and loading. This case study assumes a moderate imbalance: $\Delta tD = 25$ ps and $\Delta tH = 25$ ps. The sum of these two numbers goes in row 7 of Table 4-6.

4.17 FINAL READ AND WRITE TIMING BUDGETS

The end of the road is in sight. The DRAM and MIC timing parameters established the framework and defined how much of the unit interval remained for the interconnect. Table 4-6 and Figure 4-23 are the end result of chip-to-chip simulations and a compilation of items that fall under the category of interconnect delay.

Table 4-6   Final Interconnect Skew and Jitter Budget

	Parameter	ps
1	Rise and fall times	100
2	Conductor and dielectric losses	60
3	Impedance variation	25
4	Pin-to-pin capacitance variation	10
5	Length variation within a byte lane	15
6	DIMM connector crosstalk	30
7	Vref ac noise and resistor tolerance	40
8	Slope derating factor	50
9	DQS jitter	50
10	TOTAL	380

Experienced engineers will always argue with your numbers because their employers are paying them to think critically. That's what makes us successful engineers. A healthy argument can improve product quality levels. If you hadn't done the work to put these numbers on paper, the opportunity for improvement might not have presented itself until a field failure occurred. There is power in showing your work. When someone disagrees, you can point to the simulation results and calculations and have a rational, fact-based discussion. And if someone claims that your DDR2 interface design is broken because you did not include the effects of dielectric weave, you can respond with the obvious logical questions. What is the dominant failure mechanism associated with dielectric weave? Can we accurately model its behavior? Is there enough margin to cover it? How much money is the company willing to invest to quantify its effects on system performance and product quality? Is the investment worth the knowledge gained?

Rise and fall times, losses, and slope derating stand out as the top three contributors to 330 ps of total interconnect skew and jitter. This analysis presents a conundrum: Why do losses consume 60 ps of a 1500 ps DDR2 unit interval but less than 10 ps of a 400 ps PCI Express unit interval? (See Table 9.10.) The DDR2 case study assumed a 9-inch net, and the PCI Express example in Chapter 9 assumed a 16-inch net. Both interfaces use differential amplifiers as receivers, but DDR2 defines the input thresholds as Vref ±200 mV while PCI Express defines jitter at the zero Volt differential crossing. The 400 mV DDR2 window presents a big target for timing variations due to waveform slope.

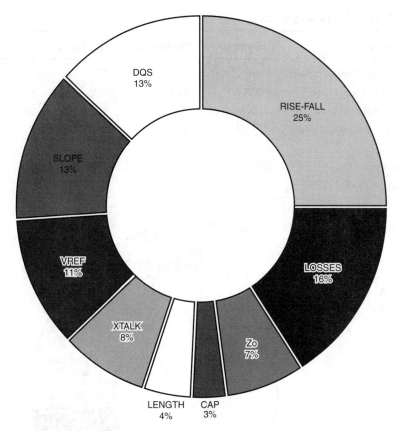

Figure 4-23   Interconnect skew and jitter for write transaction.

Table 4-7 and Figure 4-24 integrate the interconnect timing sum from Table 4-6 with the controller and DRAM timing from Table 4-1. We do not need to include Vref noise in the write timing budget because the DRAM input thresholds already account for Vref variations. The preliminary analysis assumed 825 ps for the interconnect and 0 ps of operating margin. The calculated interconnect skew and jitter sum is actually 230 ps, leaving 595 ps of margin. If the MIC actually met these specifications, DRAM write transactions would not be a performance limiting path.

Table 4-7   Final Write Timing Budget

	Parameter	Symbol	Setup (ps)	Hold (ps)
1	Memory interface controller	tMIC	200	200
2	DRAM input setup time	tDS	100	
3	DRAM input hold time	tDH		175
4	Interconnect skew and jitter		170	170
5	TOTAL		470	545
6	Unit interval	tCK/2	750	750
7	OPERATING MARGIN		280	205

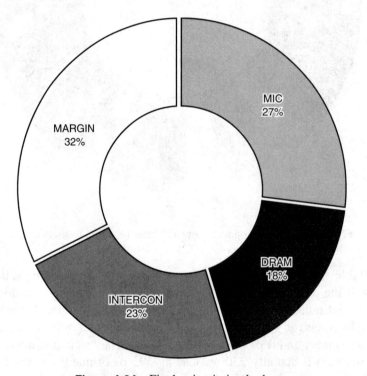

Figure 4-24   Final write timing budget.

Let's assume that the MIC input timing parameters are defined at the Vref crossing rather than more conservative DRAM thresholds. This implies that we need to account for Vref variations but not rise and fall times. We will also assume the MIC does not need a slope derating factor. (To prove this to yourself, find a SPICE model for a contemporary input buffer and investigate delay sensitivity to input slew rate.) Table 4-8 and Figure 4-25 show a 9% margin—a healthy amount. Given that engineers are prone to adding unspecified amounts of conservatism, any positive amount of calculated operating margin is healthy so long as the calculations were thorough.

Table 4-8 Final Read Timing Budget

	Parameter	Symbol	Setup (ps)	Hold (ps)
1	Controller input setup time	tDS	250	
2	Controller input hold time	tDH		250
3	DRAM		320	320
4	Interconnect skew and jitter		115	115
5	TOTAL		685	685
6	Unit interval	tCK/2	750	750
7	OPERATING MARGIN		65	65

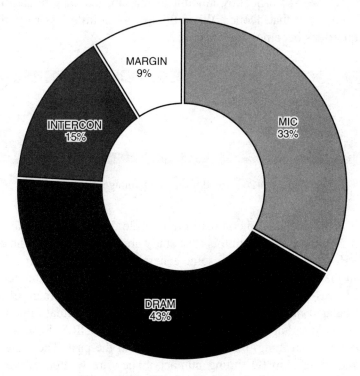

Figure 4-25 Final read timing budget.

4.18 SOURCES OF CONSERVATISM

During a recent project, our signal integrity team in Rochester, Minnesota, had the opportunity to compare margin predictions against actual hardware for a DDR2-667 interface involving a new ASIC and card. The ASIC DDR2 core had the capability to phase shift the read strobe relative to the incoming data at the primary flip-flops (those connected directly to the C4 pads) in 30 ps increments. We also shifted the reference voltage up and down in 100 mV increments. The white space in Figure 4-26 represents functional voltage-time points. Although the open eye certainly appears to be healthy, it begs the questions:

1. What did we expect to see?
2. Is this good enough?
3. Did we overachieve?

When we see a healthy amount of margin, it's easy to step back and say, "I'm happy with that. Done." When things look a little less rosy, the answers to these questions become more obscure.

VREF
1.30V
1.20V
1.10V
1.00V
0.90V
0.80V
0.70V
0.60V

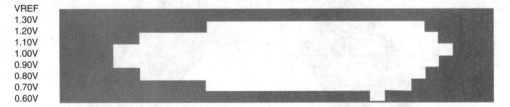

Figure 4-26   Read timing and voltage margins.

It is not possible to compare Figure 4-26 directly with Figure 4-25 because the MIC timing numbers in this case study are not the same as those in the actual ASIC. However, we can make some general statements.

In the case study, silicon accounts for the majority of the budget (76%) on paper. It was even more for our design. A thorough assessment of measured versus calculated margin would have to account for the actual effective gate length and threshold voltage conditions on the ASIC and DRAM chips that we tested. This not a trivial thing, but we have done it in the past. There are two common facts regarding chip IO timing numbers: They are written to maximize silicon yield, and they take a back seat to on-chip timing. This is strange because on-chip timing is for naught if you can't move data through the IOs. All these factors make it difficult to assess margins and weed out creeping conservatism.

Another source of conservatism lies in the potential for double-counting interconnect delays. Because DDR2 is a source-synchronous interface, skew between strobe and data aggravates timing margins. The people who designed the DRAM and the ASIC DDR2 circuits had to decide what process, voltage, and temperature conditions would result in worst-case skew between strobe and data. At IBM, we have more than three sets of corner models, but most chip vendors only distribute fast, nominal, and slow IO circuit models. This means that one of the "mixed corners" could be responsible for worst-case DQ-DQS skew while the interconnect skew and jitter at this corner are less than the fast or slow IO simulations predicted. This may or may not influence operating margins, but you must turn over the rocks and carefully inspect the underlying assumptions.

The numbers in Table 4-6 came from simulations under nominal process, voltage, and temperature conditions. This balances out some of the conservatism built into the DDR2 threshold voltages. Recall how narrow the threshold window was in Figure 2-18. Amplitudes have to shrink really low before the output of a receiver starts to become sensitive to input edge rate.

Is it really possible that all these things could happen at the same time—if the planets and stars were properly aligned? In theory, yes. Statisticians would probably give us a different answer if we could define the problem well enough. We definitely need to account for the fact that two layers of a printed circuit board may be at opposite impedance extremes. This happens in real life. Crosstalk assumptions have a strong potential for conservatism due to launch times, routing skew, and reflections. Chapter 9 covers this topic in more detail.

Based on our timing analysis and lab data, our team was confident that the DDR2-667 we designed would perform reliably over the lifetime of the product. It is possible that we overachieved, but we do not know this for certain because our program did not plan for the extra expense of tracing silicon process conditions. Time will tell.

CONCLUSION

Let's review some of the assumptions made in this analysis regarding physical implementation:

1. 6–9 inch stripline on memory carrier card
2. 1-inch stripline on DIMM
3. 3 mil line width
4. Package effects are negligible (simplifying assumption)
5. Receiver modeled as capacitor and 75 ohm Thevenin termination

Assumptions 1 through 3 form a rudimentary set of design rules. Although packaging effects are significant at 667 Mbps, this example does not address them in the interest of not bogging down the discussion. Procurement of high-quality package models for the MIC and the DRAM is well worth the effort.

If the analysis were to imply that operating margins were negative for this set of design rules, would the signal integrity engineer have any recourse? Let's examine the interconnect skew and jitter budget item by item:

1. Conductor and dielectric losses. Wider lines will improve jitter due to conductor losses at the cost of more difficult routing and a thicker printed circuit board. Advanced materials will decrease dielectric losses. Both options are costly. Figure 4-16 would lead you away from wider lines and advanced materials, but Figure 4-23 indicates loss is a top contributor to jitter. Having a complete timing budget makes it possible for you to make the cost-performance trade-offs for a specific application.

2. Impedance variations. Not much can be done about the impedance tolerance inherent to a printed circuit board manufacturing process. Optimizing antipad diameter has the potential to lower reflections from vias, and it is a low-cost option. It may be that the combination of net lengths in a given application does not produce a worst-case combination of secondary reflections. Study this carefully if margins are tight.

3. Pin-to-pin capacitance variation. This is a minor effect, which is fortunate since signal integrity engineers have no control over it.

4. Length variations within a byte lane. Cutting the routing tolerance in half could reduce this effect to less than 10 ps, which is probably not worth the effort. However, if there is significant routing skew in the package, there is an opportunity to make up the difference in the card wiring.

5. Crosstalk. The JEDEC DDR2 committee defined the mechanical features of the DIMM connector and its signal-to-return ratio, so the only opportunity for optimization lies in clean printed circuit board artwork on either side of the connector. Ask the connector vendor for model assumptions and lab correlation data.

6. Vref ac noise and resistor tolerance. Although Vref consumes only 13% of the total interconnect skew and jitter, it is ripe for optimization. Controlling Vref noise begins with a solid 1.8 V power distribution network (PDN) from dc up to 100 MHz or so. Well-decoupled printed circuit boards will not support higher frequencies than that.

7. Slope derating factor is an unavoidable artifact of JESD79-2 that should not present a major problem so long as there is good symmetry between DQ and DQS in the IO circuits and the interconnect between them, as there should be.

This analysis covers the read and write data paths, but four others need consideration: memory controller to address buffer, address buffer to DRAM, CK distribution, and DQS-DQS skew. The memory controller to address buffer net is lightly loaded and easy to time, but the address buffer to DRAM path is heavily loaded and strongly dependent on net topology. Designing the address buffer to DRAM nets is a traditional signal integrity problem involving optimization of net topology for minimum reflections. CK distribution usually involves a PLL buffer chip, which requires the usual attention to analog supply filtering, termination, and feedback loop length.

DQS-DQS skew can be a significant challenge depending on the internal architecture of the memory controller's DDR2 core. Because the DRAM launches DQ and DQS together, within the tolerance of tDQSQ and tQHS, the first job of the controller is to align DQS in the center of the data eye. The second and more difficult job is capturing eight or nine separate DQS streams from the DRAMs in the core clock domain. It is important to pay close attention to this part of the controller design and understand as much as possible about how it accomplishes this task. There are limits to how much skew between DQS signals the core logic can tolerate and still synchronize incoming data with the core clock. The controller datasheet should specify the maximum allowable DQS-DQS skew.

This chapter probably seems like a lot of work for someone doing it for the first time. It is a lot of work for someone who has done many similar exercises! Following the recipe from a design guide is always the preferred way to do signal integrity—if you can live with the conservatism inherent in every design guide. However, someone will inevitably ask whether you are willing to sign your name on a design that breaks one or more of the rules in the design guide. Then you will need a basis for making data-driven decisions. Doing your homework develops a high degree of trust between you and your colleagues. When the numbers are on the table, an engineering team can engage in a rational discussion about their relevance rather than a brainless argument in which strong opinions cover up lack of information. That is not to say this kind of analysis is completely objective and free from personal bias, but without a set of numbers, there is nowhere to begin the discussion.

5

Real-Time Measurements: Probing

Central to this book and the principal topic in this chapter are the challenge of data acquisition and the examination of the thorny issues of the ideal nonintrusive probe. Traditionally, probing has been the Cinderella of instrumentation. Probes are seen as an accessory, subservient to the enthralling world of state-of-the-art instrumentation. However, the opposite is true today. We are seeing a number of new and exciting advances in probing technology that are pushing the boundaries of approaches to the ideal probe and how we think about probing. As active probes evolve into advanced data acquisition systems, we have to rethink and question our concept of the ideal probe. For example, the current thought-provoking innovation of translating real probe acquisitions into virtual probe measurements emulating internal integrated circuit signals is changing our perception of data acquisition. Consequently, as embedded systems move toward higher levels of performance, instrument manufacturers have applied innovative technology to probing. On the other hand, probe manufacturers continue to hone conventional solutions to address the three classic probing issues:

- Connecting to high-density embedded systems acquiring analog signals and digital data and connecting to any one of a myriad of bus systems
- Minimizing probe effects on the operation of a circuit
- Maintaining signal fidelity in a measurement and limiting probe-induced artifacts, thereby measuring the circuit and not probe characteristics

At this stage, it is important to emphasize the difference between signal fidelity and signal integrity, since the terms are sometimes mistakenly interchanged. In fact, signal fidelity specifically refers to the ability of a measurement system (instrument plus probes) to reproduce a captured signal without producing distortion or aberrations. Signal integrity, on the other hand, refers to the effects that are the principal topic of this book, such as the unwanted signal anomalies and aberrations that occur within a high-performance circuit or system. Also, this chapter has to be carefully considered alongside the sections in the book that examine the relationship between fast-edge signal voltages and currents. Almost all high-performance probes are designed to acquire signal voltages, whereas current probes typically are designed to acquire power supply signals. A significant number of SI problems are related to signal current transients. Therefore, in many cases we capture and see voltages, but we need to think current.

If you have yet to discover probing, pay particular attention to sections 5.1 through 5.6. If you're an experienced SI engineer, start with section 5.7, which discusses advanced probing theory and applications.

5.1 THE ANATOMY OF A MODERN OSCILLOSCOPE PROBE

By way of an introduction, we can consider the anatomy of a modern high-performance oscilloscope probe, which is shown in Figure 5-1.

The following list describes each numbered component shown in Figure 5-1:

1. The interface box contains the circuitry required to control the probe and communicate with the oscilloscope. A probe interface has to allow for a quick and easy mechanical connection and robust electrical connections to an oscilloscope.
2. A common 50ohm coaxial cable carries signals from the probe amplifier to the oscilloscope. This cable includes both the signal path coax and power connection for the probe amplifier, so the coax cable must be carefully chosen for its electrical properties. As a cable bends, its electrical properties can shift slightly, leading to a trade-off between cable flexibility and the required electrical response. Faster probes often call for a stiffer cable to increase signal fidelity during high-performance measurement applications.

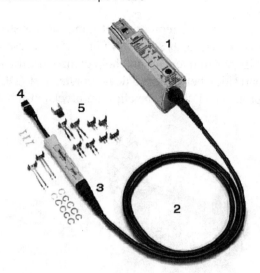

Figure 5-1   A modern oscilloscope probe.

3. This is the heart of the probe, where the probe body amplifier acquires a signal from the device under test and then buffers it to drive the 50ohm coax cable. The amplifier needs to have enough high-frequency gain to compensate for skin loss effects in the coax cable.

4. To allow a probe head to connect to differential signals, the probe head requires an attachment for two small coax cables. These small coax cables add flexibility and reach to a probe head so that it can be positioned in a variety of ways. The polarity of a differential probe connection is important, so the red ring on one of the cables is a visual indication of the positive probe input side. The unmarked cable is the negative probe input side. Plus and minus symbols are molded into the top and bottom of the probe head plastic to indicate polarity. The probe head also contains an attenuator network, allowing the probe amplifier to provide a usable input dynamic range for a variety of applications. Modern probe heads generally are made out of precision molded plastic to accommodate a variety of connection clip assemblies for connection to a range of devices.

5. The probe tip is the point of contact between the circuit under test and the probe. Because of this, the tip is the most likely probe component to be damaged. It can be expensive and time-consuming to repair or replace a broken probe tip. The answer to this issue is to provide inexpensive tips that can be easily interchanged. You also need a variety of circuit attachment styles to allow a number of different connection applications, such as solder-in probing, handheld probing, fixture probing, or probing on square pins. Also,

connection assemblies require input damping resistors that are integral to the probe architecture, as shown in Figure 5-2. These resistors reduce probe loading at higher frequencies and help damp parasitic resonances from interconnect parasitics between the device under test (DUT) and the probe tip. Like the probe head, the probe clips are precision-molded plastic parts.

Figure 5-2   Probe head resistors absorb some reflected signals caused by probing, which minimizes unwanted ringing in the measured signal. In this case, the longer flexible tip allows ease of use but limits the measurement to 4 GHz.

The advent of high-speed embedded systems buses, whether the traditional parallel type or the recently introduced ultra-high-speed serial buses with their associated signal-integrity issues, has led to a comprehensive rescheduling of the traditional design cycle. The digital design engineer now needs to think about testing and debugging much earlier in the design cycle, largely because of the new challenges involved in probing a modern embedded system. The designer now needs to consider where to probe and, just as importantly, what type of probe to use at the initial stages of the design cycle. Figure 5-2 is an example of the need to balance ease of connection with measurement fidelity.

A particular feature of established embedded system design is the customary use of parallel interconnection buses and standard peripheral buses. However, a traditional parallel copper bus has an upper data rate limit of about 1 gigabit per second (Gbps). As a result, most of today's high-performance embedded systems incorporate complex high-speed serial buses that present new test and debug challenges as data rates go beyond 10 Gbps and use low-voltage differential signals. A consequence of the migration to high-performance buses is the need to decide which signal to probe at the initial phase of the design cycle. We can no longer leave probing until the prototyping stage of a design. You can see in Figure 5-2 that connection pads have to be designed into a product in the same way as we think about printed circuit board interconnection sockets at the layout stage of a design.

5.2   A PROBING STRATEGY

Today the digital design engineer has to consider real-time test and debug connections at the initial phase of the design. This ensures real-time observation and, in some cases, the control of the final system under test. Of particular concern are the diverse bus configurations in a modern embedded system, where there is a trade-off between the relatively high cost of providing test connections and the necessities of test access. For the interconnection of an oscilloscope probe, we can consider the connection pads shown in Figure 5-2. However, a modern logic analyzer typically has 136 or more probe lines. The design engineer has to devise a cost-benefit connection strategy for testing and debugging a modern embedded system. As you will see in this chapter, a probe strategy needs to be developed at the onset of a design. In particular, the SI engineer needs to consider three initial questions:

- Do I care about the visibility of a particular microprocessor signal or bus, and what happens if particular signals cannot be seen?
- How do I minimize the test and debug risks, and will test visibility impact my ability to meet my schedule?
- What will be the impact of probing on the operation of my design?

Fortunately, test and measurement manufacturers have considered these questions and have provided a precise set of probing solutions, which we will examine in this chapter. Even so, the embedded system designer and SI engineer must consider a new test paradigm and critically decide what, where, and how they intend to probe their system.

5.3 MEASUREMENT QUALITY

Probes are the critical links to logic analyzer and oscilloscope measurements, but there is a subtle difference in the requirements for the two types of instruments. The purpose of the oscilloscope probe is to deliver the best representation of the analog signal. Logic analyzer probes, for which there could be hundreds of channels, have to deliver the best representation of a digital signal expressed as a discrete logic level. In addition to being vital to the process of carrying out measurements, probes are also critical to measurement quality for two important reasons:

- Connecting a probe to a circuit can affect the circuit's operation, either by causing the circuit to malfunction or by making it function when it is not supposed to.
- If the probe does not have sufficient fidelity, the best instrument in the world will not show the correct result.

It is therefore imperative that the probe have a minimal impact on the probed circuit and that it maintain adequate signal fidelity for the desired measurements. If the probe fails to maintain signal fidelity, if it changes the signal in any way, or if it changes how a circuit operates, the instrument that the probe is connected to sees a distorted version of the actual signal. As a result, the instrument can provide wrong or misleading measurements. In the real world, of course, all probes exert some load on the device under test, so some distortion of the waveform is inevitable. Ways in which this distortion can be kept to a minimum are discussed in later sections.

In essence, the probe is the first link in the measurement chain, and the strength of this measurement chain relies as much on the probe as on the test instrument. Weaken that first link with an inadequate probe or poor probing methods, and the entire chain is weakened.

5.4 DEFINING A PROBE

Basically, a probe makes a physical and electrical connection between a test point or signal source and a test instrument (see Figure 5-3). Depending on the user's measurement needs, this connection can be made with something as simple as a length of wire or with something as sophisticated as an active differential probe. At this point, it is sufficient to say that a probe is some sort of device or network that connects the signal source to the input of the test instrument. However, at the very start you must realize that even the most advanced instrument can be only as precise as the data that goes into it. A probe functions in conjunction with an instrument as part of the measurement system. Precision measurements start at the probe tip.

Figure 5-3   Probes make physical and electrical connections between test points and measuring instruments. Today a probe with a small form factor is essential for connecting to a high-density PCB.

Whatever the probe is in reality, it must provide a connection of adequate convenience and sufficient bandwidth between the signal source and the instrument's input.

The adequacy of connection has three key defining issues:

• The physical attachment whereby the instrument needs to make contact with the DUT, with implications for test point provision and signal access.
• The impact the probe has on the operation of the DUT. There is no such thing as a totally nonintrusive probe.
• The signal transmission path between the DUT and the test instrument, where the requirement is for a true representation of the observed signal.

At this stage, it is important to distinguish between the requirements for oscilloscope probes and logic analyzer probes. A logic analyzer captures a large number of digital signals at one time. The acquisition probe contains an internal comparator in which the input voltage is compared against a threshold voltage to determine the logic state of a digital signal—namely, a logic 1 or logic 0. The threshold value varies according to which type of circuitry is being used, such as TTL, CMOS, or ECL; it is set by the user.

5.5 OSCILLOSCOPE PROBES

To make an oscilloscope measurement, the user must first be able to physically get the probe to the test point. To make this possible, most probes have at least a meter or two of cable associated with them. Unless the signal being measured is very small, an X10 probe is used to minimize the unwanted loading effects of the lead, as shown in Figure 5-4. Normally an oscilloscope manufacturer provides a probe adjustment procedure in its manual for that particular probe and a test signal for the X10 probe lead adjustment.

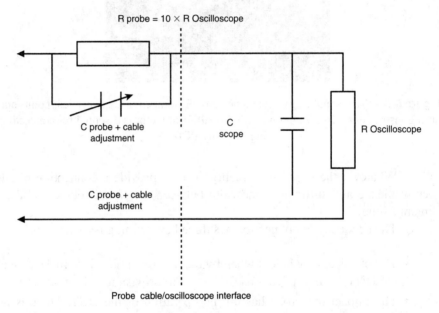

Figure 5-4   A simplified X10 probe circuit showing the capacitive probe and connection lead adjustment. Theoretically, when correctly adjusted, the X10 probe reduces the oscilloscope loading and unwanted cable-generated artifacts by a factor of 10.

A probe cable allows an oscilloscope to remain in a stationary position on a trolley or bench top while the probe is moved from test point to test point in the circuit being tested. However, this convenience has a trade-off, because the probe cable reduces the probe's bandwidth. The longer the cable, the greater the reduction in probe bandwidth. Also of significance are the length and position of the ground cable.

In addition to the length of cable, most probes also have a probe head, or handle, with a probe tip. The probe head allows the user to hold the probe while the tip is maneuvered to make contact with the test point. This probe tip is often in the form of a spring-loaded hook that allows the user to physically attach the probe to the test point. This physical attachment also establishes an electrical connection between the probe tip and the oscilloscope input. For useable measurement results, attaching the probe to a circuit must have a minimum effect on how the circuit operates. The signal at the probe tip must be transmitted with adequate fidelity through the probe head and cable to the oscilloscope's input.

These three issues—the physical attachment, the minimum impact on circuit operation, and adequate signal fidelity—encompass most of what goes into probe selection. Because probing effects and signal fidelity are the more complex topics, much of this chapter is devoted to those issues. However, the issue of physical connection should never be ignored, because difficulty in connecting a probe to a DUT or test point often leads to bad probing practices that reduce measurement fidelity. Fortunately, new developments in active probe technology, involving split attenuators and fine-pitch extension cables, are making these connection tasks much easier.

5.5.1 The Ideal Probe

In a perfect world, the ideal probe would have the following key attributes:

- Connection ease and convenience
- Absolute signal fidelity
- Zero signal source loading
- Complete noise immunity

5.5.1.1 Connection Ease and Convenience

Making a physical connection to the test point was mentioned as one of the key requirements of probing. With the ideal probe, a user should also be able to make the physical connection easily and conveniently. For miniaturized circuitry, such as high-density surface-mount devices, connection ease and convenience are promoted through subminiature probe heads and various probe-tip adaptors. These probes, however, are too small for practical use in applications such as industrial power circuitry, where higher voltages and larger gauge wires are common. Today's high-speed serial buses require high-performance differential probing systems; Figure 5-5 shows the type of probe that is used for high-bandwidth connections. In fact, no single ideal probe size or configuration is suitable for all applications. However, an advanced form of universal probe is discussed in section 5.7, where a single probe system is used to acquire both analog and digital signals for an interconnected logic analyzer and oscilloscope.

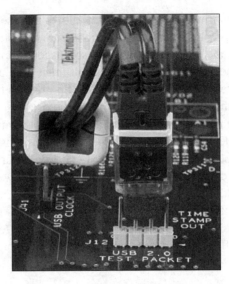

Figure 5-5   A modern high-performance differential signal probing system that can accurately acquire high-speed serial bus signals. For accurate measurement, test pin placement must be specified early in the design.

5.5.1.2   Absolute Signal Fidelity

The ideal probe should transmit any signal from probe tip to instrument input with absolute signal fidelity. In other words, the signal as it occurs at the probe tip should be faithfully duplicated at the oscilloscope input. For absolute fidelity, the probe circuitry from tip to oscilloscope input must have zero attenuation, infinite bandwidth, and linear phase across all frequencies. Not only are these ideal requirements impossible to achieve in reality, but they are impractical. For example, there is no need for an infinite bandwidth probe—or, for that matter, an infinite bandwidth oscilloscope—when dealing with, say, audio frequency signals. Nor is there a need for infinite bandwidth when 1 GHz will cover most high-speed digital, TV, and other commonly found oscilloscope applications. Nonetheless, within a given application, absolute signal fidelity is an ideal to be sought after. Measurement bandwidth is increasingly significant as embedded system buses migrate to ultra-high-speed serial data paths where the required measurement bandwidth is in the tens of gigahertz.

5.5.1.3   Zero Signal Source Loading

The circuitry behind a test point can be thought of, or modeled, as a signal source. Any external device, such as a probe, that is attached to the test point can appear as an additional load on the signal source behind the test point. The external device acts as a load when it draws signal current from the circuit, or the signal source. This loading, or signal current draw, changes the operation of the circuitry behind the test point and thus changes the signal seen at the test point. An ideal probe causes zero signal source loading. In other words, it does not draw any signal current from the signal source. This means that, for zero current draw, the probe must have infinite impedance, essentially presenting an open circuit to the test point. In practice, of course, a probe with zero signal source loading cannot be achieved, because it must draw some signal current, however small, to develop a signal voltage at the oscilloscope input. Therefore, you must always expect some signal source loading when using a probe, but your goal should be to minimize the amount of loading through appropriate probe selection.

5.5.1.4 Complete Noise Immunity

Electrical noise sources such as fluorescent lights and fan motors can pose a challenge to the correct functioning of probes. Induction effects can result in noise being induced onto the probes and probe cables, which add to the signals being measured. This susceptibility to induced noise means that a simple piece of wire is a less-than-ideal choice for an oscilloscope probe. The ideal probe should be immune to all noise sources so that the signal delivered to the oscilloscope has no more noise on it than what appeared at the test point.

In practice, the use of shielding allows probes to achieve a high level of noise immunity for most common signal levels. Nevertheless, noise can still be a problem for certain low-level signals. In particular, common-mode noise can present a problem for differential measurements, as discussed later.

5.5.2 The Realities of Probes

It is important to realize that any probe, even one consisting of just a simple piece of wire, is potentially a complex circuit. For DC signals (to be pedantic, a frequency of 0 Hz), a probe appears as a simple conductor pair with some series resistance and a terminating resistance, as shown in Figure 5-6.

However, for AC signals, the picture changes dramatically as signal frequencies increase, as shown in the bottom part of Figure 5-6. Any piece of wire has distributed inductance, and any pair of wires has distributed capacitance. The distributed inductance (L) reacts to AC signals by increasingly impeding AC current flow as the signal frequency increases. The distributed capacitance (C) reacts to AC signals with decreasing impedance to AC current flow as the signal frequency increases. The interaction of these reactive elements (L and C), along with the resistive elements (R), produces a total probe impedance that varies with signal frequency. Through good probe design, a probe's R, L, and C elements can be controlled to provide desired degrees of signal fidelity, attenuation, and source loading over specified frequency ranges. Even with good design, probes are limited by the nature of their circuitry. It is important to be aware of these limitations and their effects when selecting and using probes.

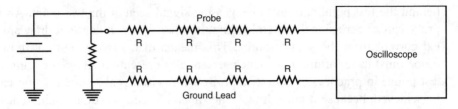

a. Distributed R for DC (0 Hz) signals.

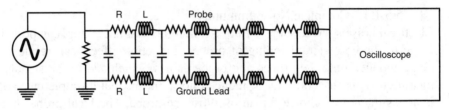

b. Distributed R, L, & C for AC signals.

Figure 5-6 Probes are circuits composed of distributed resistance (R), inductance (L), and capacitance (C).

5.5.3 Bandwidth and Rise Time Limitations

Bandwidth is the range of frequencies that an oscilloscope or probe is designed to convey. For example, a 100 MHz probe or oscilloscope is designed to make measurements within specification on all frequencies up to 100 MHz. Unwanted or unpredictable measurement results can occur at signal frequencies above the specified bandwidth, as shown in Figures 5-7 and 5-8.

In cases where the rise time (T_r) is not specified, it is possible to derive it from the bandwidth (BW) specification with this relationship:

$$T_r = 0.35/BW$$

Equation 5-1

where the bandwidth is that specified for the signal being measured.

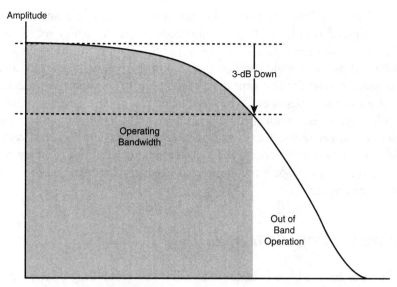

Figure 5-7   Probes and oscilloscopes are designed to make measurements to specification over an operating bandwidth. At frequencies beyond the 3 dB point, signal amplitudes become overly attenuated, and measurement results may be unpredictable.

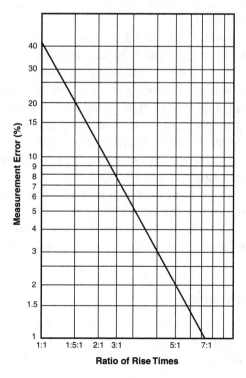

Figure 5-8   An oscilloscope and probe combination with a rise time three times faster than the pulse being measured (3:1 ratio) can be expected to measure the pulse rise time to within 5%. A 5:1 ratio would result in only 2% error.

Every oscilloscope and probe has defined bandwidth and rise time limits. When a probe is attached to an oscilloscope, it creates a new set of system bandwidth and rise time limits. Unfortunately, the relationship between system bandwidth and the individual oscilloscope and probe bandwidths is not a simple one. The same is true for rise times. To cope with this complexity, manufacturers of quality oscilloscopes specify bandwidth or rise time to the probe tip when the oscilloscope is used with specific probe types. This is important because the oscilloscope and probe together form a measurement system, and the system's bandwidth and rise time determine its measurement capabilities. Using a probe that is not on the oscilloscope's recommended list of probes may lead to unpredictable measurement results.

5.6 DYNAMIC RANGE LIMITATIONS

All probes have a high-voltage safety limit that should not be exceeded. For passive probes, this limit can range from hundreds to thousands of volts. However, for active probes, the maximum safe voltage limit is often in the range of tens of volts. To avoid personal safety hazards as well as potential damage to the probe, the user needs to be aware of the voltages being measured and the voltage limits of the probes being used.

In addition to safety considerations, there is also the practical consideration of measurement dynamic range. A typical oscilloscope may have an amplitude sensitivity range from 1 mV to 10 V per division. This allows the engineer to make reasonably accurate measurements on signals ranging from 4 mV peak-to-peak to 40 V peak-to-peak on a typical eight-division display. This assumes, at a minimum, a four-division amplitude display of the signal to obtain a reasonable measurement resolution. With a X1 probe, the dynamic measurement range is the same as that of the oscilloscope. For the example just mentioned, this would be a signal measurement range of 4 mV to 40 V.

Normally, however, an oscilloscope's dynamic range is shifted to higher voltages by using an attenuating probe. In the example, a X10 probe shifts the dynamic range to 40 mV to 400 V per division. It does this by attenuating the input signal by a factor of 10, which effectively multiplies the oscilloscope's scaling by 10.

5.6.1 Source Loading

As mentioned, a probe must draw some signal current to develop a signal voltage at the oscilloscope input. This places a load at the test point that can change the signal that the circuit or signal source delivers to the test point.

The simplest example of source loading effects is to consider measurement of a battery-driven resistive network. See Figure 5-9 and Equation 5-2.

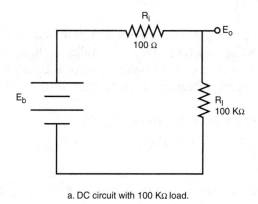

a. DC circuit with 100 KΩ load.

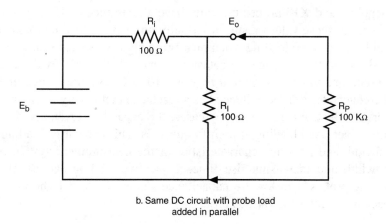

b. Same DC circuit with probe load
added in parallel

Figure 5-9   An example of resistive loading.

Before a probe is attached, the battery's DC voltage is divided across the battery's internal resistance (R_i) and the load resistance (R_l). For the values given in Figure 5-9, this results in the following output voltage:

$$E_o = \frac{E_b \times R_l}{R_i + R_l}$$

Equation 5-2

where E_o is the probe voltage and E_b is the battery voltage.

$E_o = 100 \text{ V} \times 100,000 / (100 + 100,000)$
$E_o = 10,000,000 \text{ V} / 100,100$
$E_o = 99.9 \text{ V}$

In the bottom part of Figure 5-9, a probe has been attached to the circuit, placing the probe resistance (R_p) in parallel with R_1. If R_p is 100 kilohm, the effective load resistance in the bottom part of Figure 5-9 is reduced by half to 50 kilohm. The loading effect of this on E_o is as follows:

$E_o = 100 \text{ V} \times 50,000 / (100 + 50,000)$
$E_o = 5,000,000 \text{ V} / 50,100$
$E_o = 99.8 \text{ V}$

This loading effect of 99.9 versus 99.8 is only 0.1% and is negligible for most purposes. However, if R_p were smaller—say, 10 kilohm—the effect would no longer be negligible.

To minimize such resistive loading, X1 probes typically have a resistance of 1 megohm, and X10 probes typically have a resistance of 10 megohm. For most applications these values result in virtually no resistive loading. Nonetheless, you should expect some loading when measuring high-resistance sources.

Usually, the loading of greatest concern is that caused by the capacitance at the probe tip, which is shown in Figure 5-10 as C_p at a high value of 100 pF. For low frequencies, this capacitance has a reactance that is very high, and there is little or no effect. As frequency increases, the capacitive reactance decreases. The result is increased loading at high frequencies. This capacitive loading affects the bandwidth and rise time characteristics of the measurement system by reducing bandwidth and increasing rise time. Capacitive loading can be minimized by selecting probes with low tip capacitance values. Table 5-1 shows some typical capacitance values for various probes.

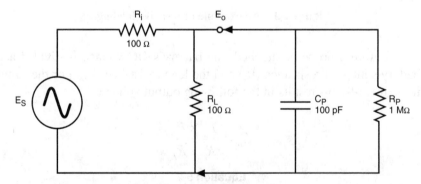

Figure 5-10   For AC signal sources, probe tip capacitance (C_p) is the greatest loading concern. As signal frequency increases, capacitive reactance (X_c) decreases, causing more signal flow through the capacitor and signal distortion.

Table 5-1   Resistive and Capacitive Probe Attenuation

Probe Type	Attenuation	Resistance	Capacitance
Low performance	X1	1 megohm	100 pF
Passive probes	X10	10 megohm	11 pF
High performance	X10	10 megohm	8 pF
Active probes	X10	1 megohm	" 1 pF

Because the ground lead is a wire, it has some amount of distributed inductance, as shown in Figure 5-11. This inductance interacts with the probe capacitance to cause ringing at a certain frequency that is determined by the L and C
values. This ringing is unavoidable, and it may be seen as a sinusoid of decaying
amplitude that is impressed on pulses. The effects of ringing can be reduced by
grounding the probe in such a way that the ringing frequency occurs beyond the
bandwidth limit of the probe and oscilloscope system.

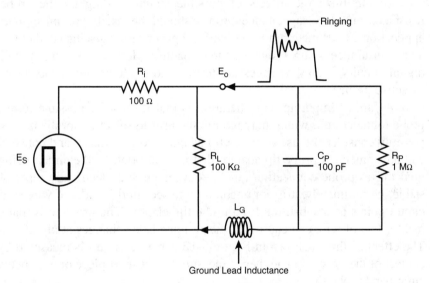

Figure 5-11   The probe ground lead adds inductance to the circuit. The longer the ground lead,
the greater the inductance and the greater the likelihood of generating unwanted ringing on the
edges of fast pulses.

5.6.2   Choosing the Right Probe

Because different instruments are designed for different bandwidth, rise time, sensitivity, and input impedance considerations, taking full advantage of the instrument's measurement capabilities requires a probe that matches the instrument's performance. The probe selection process should also include consideration of the user's measurement needs in terms of the parameters and amplitudes of the signals that are being measured.

For example, it is important to make sure that the bandwidth or rise time at the probe tip exceeds the signal frequencies or rise times that are being measured. Users should also remember that nonsinusoidal signals have important frequency components, or harmonics, that extend well above the signal's fundamental frequency. For example, to fully include the fifth harmonic of a 100 MHz square wave, the measurement system needs a bandwidth of 500 MHz at the probe tip. Similarly, the instrument system's rise time should be three to five times shorter than the signal rise times that are being measured.

To minimize the effects of possible signal loading by the probe, high-resistance and low-capacitance probes should be used. For most mainstream applications, a 10 megohm probe with 20 pF or less capacitance should provide ample insurance against signal source loading. However, for many high-speed digital circuits, it may be necessary to move to the lower tip capacitance offered by active probes.

A further important consideration is that it must be possible to attach the probe to the circuit, which may require the user to select the probe head size and possibly consider the use of probe tip adaptors to allow easy and convenient circuit attachment. A probe tip adaptor that is appropriate to the circuit being measured makes probe connection quick, convenient, and electrically repeatable and stable. Unfortunately, it is not uncommon to see short lengths of wire soldered to circuit points as a substitute for a probe tip adaptor. The problem is that even an inch or two of wire can cause significant impedance changes at high frequencies. The effect of this is shown in Figure 5-12, where a circuit is measured by direct contact of the probe tip and then is measured via a short piece of wire between the circuit and probe tip.

When performing performance checks or troubleshooting large boards or systems, it may be tempting to extend the probe's ground lead so that the probe can be freely moved around the system to look at various test points. However, the added inductance of an extended ground lead can cause ringing to appear on fast-transition waveforms, as shown in Figure 5-13. It shows waveform measurements made using both the standard probe ground lead and, in the second oscillogram, an extended ground lead.

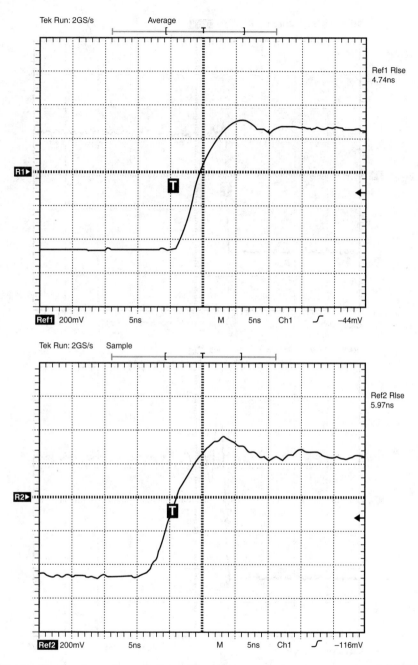

Figure 5-12   Even a short piece of wire soldered to a test point can cause signal fidelity problems. In this case, rise time has been changed from 4.74 ns (part a) to 5.97 ns (part b).

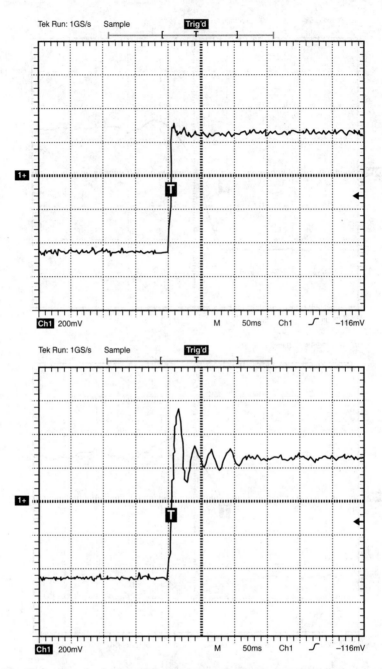

Figure 5-13 Extending the length of the probe ground lead can cause ringing to appear on pulses, which is clearly seen in the second oscillogram.

Most probes are designed to match the inputs of specific oscilloscope models. However, there are slight variations from oscilloscope to oscilloscope and even between different input channels in the same oscilloscope. To deal with this where necessary, many probes, especially attenuating X10 and X100 probes, have built-in compensation networks that need to be adjusted to compensate the probe for the oscilloscope channel being used.

5.6.3   Different Probes for Different Needs

The wide selection of oscilloscope models and capabilities is one of the fundamental reasons for the large number of available probe types. For example, a 500 MHz oscilloscope requires probes that will support the instrument's 500 MHz bandwidth, but those same probes would be uneconomical both in capability and cost for an oscilloscope that has a 100 MHz bandwidth. As a general rule, probes should be selected to match or exceed the oscilloscope's bandwidth whenever possible. However, bandwidth is only the beginning, because oscilloscopes can also have different input connector types and different input impedances. For example, most oscilloscopes use a simple BNC-type input connector, others may use an SMA connector, and still others may have specially designed integrated connectors to support features such as readout, trace ID, probe power, and control. Probe selection therefore must consider connector compatibility with the oscilloscope being used—whether for direct connection or via an appropriate adaptor.

Readout support is a particularly important aspect of probe and oscilloscope connector compatibility. When X1 and X10 probes are interchanged on an oscilloscope, the oscilloscope's vertical scale readout should reflect the X1 to X10 change to maintain the correct amplitude measurement values. Some generic or commodity probes may not support this readout capability for all oscilloscopes. As a result, extra caution is necessary when using generic probes in place of the probes specifically recommended by the oscilloscope manufacturer.

In addition to bandwidth and connector differences, various oscilloscopes have different input resistance and capacitance values. Typical oscilloscope input resistances are either 50 ohm or 1 megohm, but there can be great variations in input capacitance, depending on the oscilloscope's bandwidth specification and other design factors. For proper signal transfer and fidelity, it is important that the probe's resistance and capacitance match the resistance and capacitance of the oscilloscope the probe is to be used with. An exception to this one-to-one resistance matching occurs when attenuator probes, also known as divider or multiplier probes, are used. For example, a X10 probe for a 50 ohm environment has a 500 ohm input resistance, and a X10 probe for a 1 megohm environment has a 10 megohm input resistance.

Capacitance matching can be done by adjusting a probe's compensation network. However, it is only possible to capacitance-match an oscilloscope probe to an oscilloscope that has an input capacitance within the probe's compensation range. Therefore, it is not unusual to find probes with different compensation ranges to meet the requirements of different oscilloscope input capacitances.

5.6.4 Oscilloscope Probe Families

The most basic differentiator between probe types lies in the voltage ranges being measured. For example, millivolt, volt, and kilovolt measurements typically require probes with different attenuation factors, such as X1, X10, and X100. There are also many cases in which the signal voltages are differential, whereby the signal exists across two points or two wires, neither of which is at ground or common potential. See Figure 5-14.

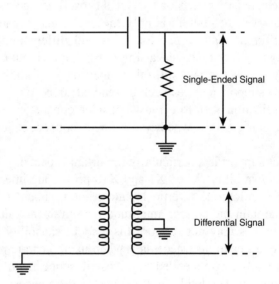

Figure 5-14   Single-ended signals are referenced to ground (part a), and differential signals are the difference between two signal lines or test points (part b).

Such differential signals are common in telephone voice circuits, computer disk read channels, multiphase power circuits, and the new high-speed serial buses mentioned at the beginning of this chapter. Measuring these signals requires yet another class of probes—differential probes. There are also many cases, particularly in power applications, in which current is as of much or more interest than voltage. Such applications, which fall outside the scope of this chapter, are best served with yet another class of probes that sense current rather than voltage.

As a preface to discussing various common probe types, it is important to realize that overlap occurs between different types. Certainly a voltage probe senses voltage exclusively, but a voltage probe can be passive or active. Similarly, differential probes are a special type of voltage probe, and differential probes can also be active or passive. Where appropriate, these overlapping relationships will be pointed out.

5.6.5   Passive Voltage Probes

Passive probes are constructed of wires and connectors. When needed for compensation or attenuation, they contain resistors and capacitors. The probe has no active components, so there is no need to supply power to the probe.

Because of their relative simplicity, passive probes tend to be the most rugged and economical kind. They are easy to use and also are the most widely used type of probe. However, simplicity of use or of construction belies the fact that high-quality passive probes are rarely simple to design. Passive voltage probes are available with various attenuation factors for different voltage ranges, of which X10 is the most commonly used probe. A X10 probe is the type typically supplied as a standard accessory with an oscilloscope.

For applications in which signal amplitudes are 1 V or less peak-to-peak, a X1 probe may be more appropriate or even necessary. Where there is a mix of low-amplitude and moderate-amplitude signals, within a range of say tens of millivolts to tens of volts, a switchable X1/X10 probe can be a great convenience. However, such a probe is essentially two different probes in one. Not only are their attenuation factors different, but their bandwidth, rise time, and impedance characteristics are also different. As a result, these probes do not exactly match the oscilloscope's input and do not provide the optimum performance achieved with a standard X10 probe.

Most passive probes are designed for use with general-purpose oscilloscopes. As such, their bandwidths typically range from less than 100 MHz to 500 MHz or more. However, a special category of passive probes provide much higher bandwidths. They are referred to variously as 50 ohm probes, Z_0 probes, or voltage divider probes. These probes are designed for use in 50 ohm environments in applications such as high-speed device characterization, microwave communication, and time-domain reflectometry (TDR). A typical 50 ohm probe for such applications has a bandwidth of several gigahertz and a rise time of 100 ps or less.

5.6.6 Active Voltage Probes

Active probes contain or rely on active components such as transistors for their operation. The active device in most oscilloscope probes is a field-effect transistor (FET), which offers the benefit of providing a very low capacitance input—typically a few picofarads (pF) down to less than 1 pF. Such ultra-low capacitance has several desirable effects. Because capacitive reactance is a probe's primary input impedance element, a low capacitance results in high input impedance over a broader band of frequencies. As a result, active FET probes typically have specified bandwidths ranging from 500 MHz to as high as 20 GHz. In addition to higher bandwidth, the high input impedance of active FET probes allows measurements at test points of unknown impedance with a much lower risk of loading effects. Moreover, longer ground leads can be used, because the low capacitance probe reduces ground lead effects.

The most important benefit, however, is that FET probes offer such low loading that they can be used on high-impedance circuits that would be seriously loaded by passive probes. With all these positive benefits, including bandwidths as wide as DC to 20 GHz, why should anyone bother with passive probes? The answer is that active FET probes do not have the voltage range of passive probes. The linear dynamic range of an active probe is generally anywhere from ±0.6 V to ±10 V, compared to a range from millivolts to tens of volts with a passive probe. In addition, the maximum voltage that an active probe can withstand can be as low as ±40 V (DC plus peak AC). Probes can even be catastrophically damaged by a static discharge. However, the high bandwidth of FET probes is a major benefit, and their linear voltage range covers many typical semiconductor voltages. Active FET probes therefore are often used for low signal level applications, including high-speed logic families.

5.6.7 Differential Probes

Differential signals, as shown in Figure 5-15, are signals that are referenced to each other instead of to an earth ground.

These include the signal developed across a collector load resistor, a disk-drive read channel signal, multiphase power systems, high-speed serial buses, and numerous other situations in which signals in essence "float" above the ground. Differential signals can be probed and measured in two basic ways: by using the invert and add feature of a dual-channel oscilloscope, or by using a dedicated differential probe, as shown in Figure 5-16.

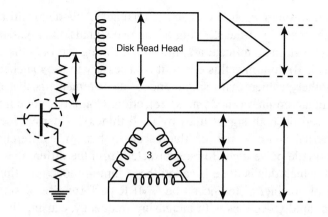

Figure 5-15   Examples of differential signal sources.

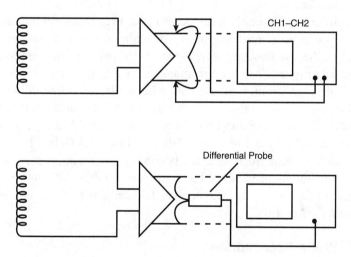

Figure 5-16   Differential signals can be measured using the invert and add feature of a dual-channel oscilloscope (part a), or preferably by using a differential probe (part b).

Using two probes to make two single-ended measurements is often done because a dual-channel oscilloscope with two probes is available. It is also usually the least desirable method of making differential measurements. Measuring both signals to ground, single-ended, and using the oscilloscope's mathematics functions to subtract one from the other seems like an elegant solution to obtaining the difference signal. But it is generally applicable only where the signals are of low frequency and have enough amplitude to be free from noise problems. Combining two single-ended measurements introduces several potential problems. One problem is that there are two long and separate signal paths down each probe and through each oscilloscope channel. Any delay differences between these paths results in time skewing of the two signals. On high-speed signals, this skew can result in significant amplitude and timing errors in the computed difference signal. To minimize this effect, it is necessary to use matched probes.

Another problem with single-ended measurements is that they do not provide adequate common-mode noise rejection. Common-mode noise is noise that is impressed on both signal lines by such things as nearby clock lines or noise from external sources such as fluorescent lights. In a differential system, this common-mode noise tends to be subtracted from the differential signal. The success with which this is done is called the common-mode rejection ratio (CMRR). Because of channel differences, the CMRR performance of single-ended measurements rapidly declines with increasing frequency, causing the signal to appear noisier than it actually would be if the common-mode rejection of the source had been maintained.

A differential probe, on the other hand, uses a differential amplifier to subtract the two signals, resulting in one differential signal for measurement by one channel of the oscilloscope (see Figure 5-16, part b). This approach provides a substantially higher CMRR performance over a broader frequency range. In addition, advances in circuit miniaturization have allowed differential amplifiers to be moved down into the actual probe head. In the latest differential probes, this has allowed a 12.5 GHz bandwidth to be achieved with CMRR performance ranging from 60 dB (1000:1) at 1 MHz to 30 dB (32:1) at 12.5 GHz. These levels of bandwidth and CMRR performance are becoming increasingly necessary as the data rates in ultra-high-speed serial buses surpass 5 Gbps, although it has to be said that differential probes are expensive. Differential probing is covered in more detail in a later section.

5.6.8 Probe Accessories

Most probes come with a package of standard accessories. These accessories often include a ground lead clip that attaches to the probe, a compensation adjustment tool, and one or more probe-tip accessories to aid in attaching the probe to various test points.

Probes that are designed for specific application areas, such as probing surface-mount devices, may include additional probe-tip adaptors in their standard accessories package. Various special-purpose accessories and application packages may be available as options for the probe.

5.6.9 Basic Probing Know-How

All signals, regardless of their type, have frequency components. DC signals have a frequency component of 0 Hz, and pure sinusoidal waveforms have a single frequency that is the reciprocal of the sinusoidal period. Fourier showed that almost all other signals contain multiple frequencies whose values depend on the signal's wave shape. For example, a symmetrical square wave has a fundamental frequency that is the reciprocal of the square wave's period and additional harmonic frequencies that are odd multiples of the fundamental. The fundamental is the foundation of the wave shape, and the harmonics combine with the fundamental to add structural detail such as the wave shape transitions and corners.

For a probe to convey a signal to an oscilloscope while maintaining adequate signal fidelity, the probe must have enough bandwidth to pass the signal's major frequency components with minimum disturbance. In the case of square waves and other periodic signals, this generally means that the probe bandwidth needs to be three to five times higher than the signal's fundamental frequency. This allows the fundamental and the first few harmonics to be passed without undue attenuation of their relative amplitudes. The higher harmonics also are passed, but with increasing amounts of attenuation, because these higher harmonics are beyond the probe's 3 dB bandwidth point. However, because the higher harmonics are still present to at least some degree, they still can contribute somewhat to the waveform's structure. The primary effect of bandwidth limiting is not only to reduce the displayed signal amplitude, but to distort the captured waveform.

The closer a signal's fundamental frequency is to the probe's 3 dB bandwidth, the lower the overall signal amplitude seen at the probe output. At the 3 dB point, amplitude is reduced by 30%. In addition, those harmonics or other frequency components of a signal that extend beyond the probe's bandwidth experience a higher degree of attenuation because of the bandwidth roll-off. The result of higher attenuation on higher-frequency components may be seen as a rounding of sharp corners and a slowing of fast waveform transitions, as shown in Figure 5-17.

Probe tip capacitance can also limit signal transition rise times, which in turn are related to signal-source impedance and signal-source loading.

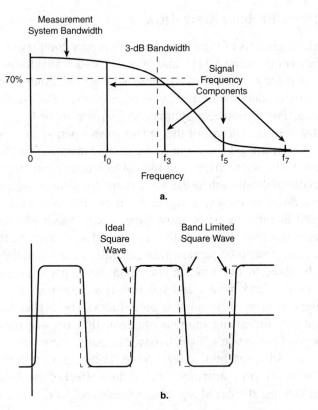

Figure 5-17 When major harmonic frequency components, such as f5 and f7, of a signal are beyond the measurement system bandwidth (part a), the result is loss of waveform detail through rounding of corners and lengthening of transitions (part b).

5.6.10 Signal-Source Impedance

Oscilloscope probe impedances combine with signal source impedances to create new signal load impedances that must have some effect on signal amplitude and signal rise times. When the probe impedance is substantially greater than the signal source impedance, the effect of the probe on signal amplitude is negligible. Probe tip capacitance, also called input capacitance, has the effect of stretching a signal's rise time. This is due to the time required to charge the probe's input capacitance from the 10% to the 90% level, which is 2.2 times the product of the source impedance and the probe capacitance.

5.6.11   Physically Connecting to the Signal

The location and geometry of signal test points can be a key consideration in probe selection. Is it enough to just touch the probe to the test point and observe the signal on the oscilloscope? Or will it be necessary to leave the probe attached to the test point for signal monitoring while making various circuit adjustments? For the former situation, a needle-style probe tip is appropriate. The latter situation requires some kind of retractable hook tip or fixed connection. Moreover, the SI engineer must consider such questions early in the design cycle. Hoping that a signal will be available at the test stage is often too late.

The size of the test point can also impact probe selection. Standard-size probes and accessories are fine for probing connector pins, resistor leads, and backplanes. However, for probing surface-mount circuitry and high-density printed circuit boards (PCBs), smaller probes with accessories designed for surface-mount applications and the latest connectorless probing systems are recommended. These probing systems have to be designed into the PCB early in the design cycle.

5.7   ADVANCED PROBING TECHNIQUES

Ground lead issues are a significant concern in oscilloscope measurements because of the difficulty in determining and establishing a true ground reference point for measurements. This difficulty arises from the fact that ground leads, whether on a probe or in a circuit, have inductance and become circuits of their own as signal frequency increases. In addition to being the source of ringing and other waveform aberrations, the ground lead can act as an aerial for receiving noise.

An interesting question is where in particular the engineer connects the probe ground lead when measuring ground signals or ground bounce. The answer for a digital circuit is often a solid logic zero, where the ground signal is referenced to logic zero.

Suspicion is the first defense against ground-lead problems. The SI engineer should always be suspicious of any noise or aberrations being observed on a signal's oscilloscope display. The noise or aberrations may be part of the signal, or they may be the result of the measurement process. Any probe ground lead has some inductance, and the longer the ground lead, the greater the inductance. When combined with probe tip capacitance and signal source capacitance, ground lead inductance forms a resonant circuit that causes ringing at certain frequencies. To see ringing or other aberrations caused by poor grounding, the following two conditions must exist:

- The oscilloscope system bandwidth must be high enough to handle the high-frequency content of the signal at the probe tip.
- The input signal at the probe tip must contain enough high-frequency information, such as a fast rise time, to cause the ringing or aberrations due to poor grounding.

Two main conclusions can be drawn from these examples. The first conclusion—stated in this chapter a number of times—is that ground leads should be kept as short as possible when probing fast signals. The second conclusion is that product designers can ensure higher effectiveness of product maintenance and troubleshooting by designing in product "testability." This includes using electronic circuit board (ECB)/probe-tip adaptors where necessary to better control the test environment and to avoid wrong adjustment of product circuitry during installation or maintenance, or—for ultra-high-bandwidth applications—the use of built-in surface-mount adaptors (SMAs).

For measurements on fast waveforms where an ECB/probe-tip adaptor is not installed, the probe ground lead needs to be kept as short as possible. In many cases, this can be done by using special probe-tip adaptors with integral grounding tips.

Noise is another type of signal distortion that can appear on oscilloscope waveform displays. As with ringing and aberrations, noise might actually be part of the signal at the probe tip, or it might appear on the signal as a result of improper grounding techniques. The difference is that the noise generally is from an external source, and its appearance is not a function of the speed of the signal being observed. In other words, poor grounding can result in noise appearing on any signal of any speed. Noise can be impressed on signals as a result of probing through two primary mechanisms. One is by ground-loop noise injection, and the other is by inductive pickup through the probe cable or probe ground lead.

Ground loop noise injection into the grounding system can be caused by unwanted current flow in the ground loop between the oscilloscope common, the test circuit power-line grounds, and the probe ground lead and cable. Normally, all these points are, or should be, at zero volts, and no ground current will flow. However, if the oscilloscope and test circuit are on different building system grounds, small voltage differences or noise could occur on one of the building ground systems. Figure 5-18 shows the complete ground circuit.

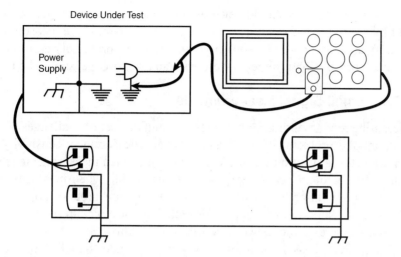

Figure 5-18   The complete ground circuit, or ground loop, for an oscilloscope, probe, and test circuit on two different power plugs.

The resulting current flow develops a voltage drop across the probe's outer cable shield. This noise voltage is injected into the oscilloscope in series with the signal from the probe tip. The result is that the observer sees noise riding on the signal of interest, or the signal of interest may be riding on noise. With ground loop noise injection, the noise is often mains frequency noise, 50 Hz or 60 Hz. It may also be in the form of spikes or bursts resulting from building equipment switching on and off.

Various steps can be taken to avoid or minimize ground-loop noise problems. The first approach is to minimize ground loops by using the same power circuits for the oscilloscope and circuit under test. The probes and their cables should also be kept away from sources of potential interference. In particular, probe cables should not lie alongside or across equipment power cables. If ground-loop noise problems persist, the ground loop may need to be opened by one of the following methods:

- Using a ground isolation monitor
- Using a power-line isolation transformer on either the test circuit or the oscilloscope
- Using an isolation amplifier to isolate the oscilloscope probes from the oscilloscope
- Using differential probes to make the measurement, hence rejecting common-mode noise

The SI engineer should never attempt to isolate the oscilloscope or test circuit by removing the safety three-wire ground system. If the measurements need to be floated, an approved isolation transformer, or preferably a ground isolation monitor specifically designed for use with an oscilloscope, should be used.

5.7.1 Small-Signal Measurements

Measuring low-amplitude signals presents a unique set of challenges. Foremost of these are the problems of noise and adequate measurement sensitivity. Ambient noise levels that would be considered negligible when measuring signals of a few hundred millivolts or more are no longer negligible when measuring signals of tens of millivolts or less. Consequently, minimizing ground loops and keeping ground leads short are imperative for reducing noise pickup by the measurement system. At the extreme, power-line filters and a shielded room may be necessary for noise-free measurement of very low-amplitude signals. However, before resorting to extremes, it is worth considering signal averaging as a simple and inexpensive solution to noise problems. If the signal being measured is repetitive and the noise is random, signal averaging can provide significant improvements in the signal-to-noise ratio (SNR) of the acquired signal (see Figure 5-19).

Signal averaging is a standard function of most digital storage oscilloscopes (DSOs). It operates by summing multiple acquisitions of the repetitive waveform and computing an average waveform from the multiple acquisitions. Since random noise has a long-term average value of 0, the process of signal averaging reduces random noise on the repetitive signal. The amount of improvement is expressed in terms of SNR. Ideally, signal averaging improves the SNR by 3 dB per power of two averages. Thus, averaging just two waveform acquisitions (2^1) provides up to 3 dB of SNR improvement, averaging four acquisitions (2^2) provides 6 dB of improvement, averaging eight acquisitions (2^3) provides 9 dB of improvement, and so on.

Secondly, the obvious solution to measuring a small signal is to increase the oscilloscope measurement sensitivity, which is a function of the oscilloscope input circuitry. The input circuitry either amplifies or attenuates the input signal for an amplitude-calibrated display of the signal on the oscilloscope screen. The amount of amplification or attenuation needed to display a signal is selected via the oscilloscope's vertical sensitivity setting, which is adjusted in terms of volts per display division (V/div). In addition to the requirement of adequate oscilloscope sensitivity for measuring small signals, it is also necessary to use an adequate probe. Typically, this is not the usual probe supplied as a standard accessory with most oscilloscopes. Standard accessory probes are usually X10 probes, which reduce oscilloscope sensitivity by a factor of 10.

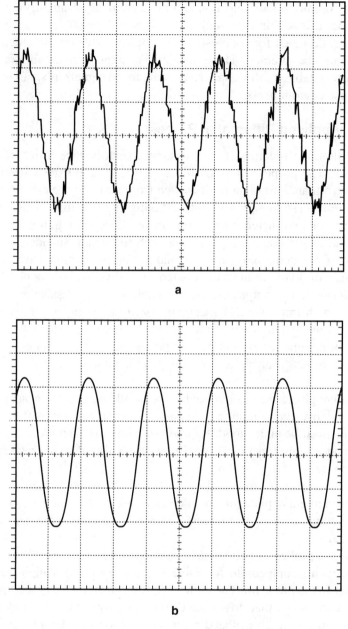

Figure 5-19   A noisy signal (part a) can be cleaned up by signal averaging (part b).

In cases where the small-signal amplitude is below the oscilloscope's sensitivity range, some form of pre-amplification is necessary. Because of the noise susceptibility of very small signals, differential pre-amplification is generally used. Differential pre-amplification offers the advantage of some noise immunity through common-mode rejection and the advantage of amplifying the small signal so that it is within the oscilloscope's sensitivity range. With differential pre-amplifiers designed for oscilloscope use, sensitivities on the order of 10 μV per division can be attained. These specially designed pre-amplifiers have features that allow useable oscilloscope measurements on signals as small as 5 μV, even in high-noise environments.

5.7.2 Applying Differential Probes

An oscilloscope specifies its measurement performance at its front-panel input connector. The purpose of an oscilloscope probe is to extend specified measurement performance from an oscilloscope front panel to the circuit under test with the highest possible signal fidelity. Making high-fidelity measurements with a probe one or two meters away from an oscilloscope requires that the probe be connected to the oscilloscope with a high-frequency shielded cable. Since the coaxial cables commonly used to make this connection have a characteristic impedance of 50 ohms and a distributed capacitive load of about 30 pF per foot, a probe design must compensate for this cable loading. Passive probes, which connect to a high-impedance (1 megohm) oscilloscope input, compensate for cable loading by using a high-impedance probe tip attenuator. The high-impedance X10 attenuator commonly used in passive probes generally limits probe bandwidth to 500 MHz or less. For higher-frequency applications, active probes generally are used.

Although an ideal differential probe would be able to measure a signal with perfect fidelity and without any loading of the signal, any real probe has limitations that must be understood. By understanding the limitations of practical probe performance, a user can more readily relate measured signal response to real signal response.

By choosing a probe and oscilloscope whose performance is adequate for the measurement application and then using the probe in a manner that minimizes its limitations, a user gets the best match between measured and real signal performance.

Active probes, which are designed to connect to a broadband 50-ohm oscilloscope input, compensate for cable loading by using a high-frequency buffer amplifier in the probe tip. The buffer amplifier in an active probe tip has a 50-ohm output driver stage that drives the distributed probe cable loading as a transmission line to provide broadband frequency response. Like all transmission lines, however, the probe cable is susceptible to electrical cable loss. To guarantee signal fidelity, the probe buffer amplifier must compensate for probe cable frequency-dependent loss effects, up to the probe's frequency limit. The use of a high-frequency buffer amplifier in the probe head also reduces probe loading effects at the probe tip. The addition of a high-impedance attenuator in front of the buffer amplifier reduces the effect of the buffer amplifier input capacitance and increases input dynamic range. These input attenuator advantages, however, are traded off against reduced SNR.

A differential probe provides differential measurement capability at the probe tip and connects to a single oscilloscope channel input. Typically, the probe has an active hybrid circuit in the probe head containing a high-impedance laser-trimmed attenuator and differential buffer amplifier. By placing the buffer amplifier in the probe head, very close to the probe input pins, probe loading effects are minimized and CMRR can be maximized. The probe head is miniaturized to allow its use in physically restrictive environments and to minimize parasitics in the probe attachment path for best signal fidelity.

An ideal differential probe could be attached to a differential circuit and make measurements without disturbing the circuit under test. Any real probe, however, has a finite load impedance that may need to be accounted for in measurements, particularly at high frequencies. Figure 5-20 shows a first-order load model for an active differential probe.

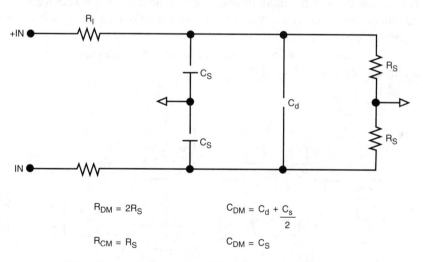

$$R_{DM} = 2R_S \qquad\qquad C_{DM} = C_d + \frac{C_s}{2}$$

$$R_{CM} = R_S \qquad\qquad C_{DM} = C_S$$

Figure 5-20   A first-order differential probe load model.

As expected, the probe load model shows a balanced structure, which to the first order is identical for each of the two inputs. The input resistor is a damping resistor with a typical value of about 130 ohms. The common-mode input capacitance is half the capacitance that would be measured with the inputs shorted together. For the probe illustrated it has a typical value of about 0.4 pF. The differential-mode input capacitance is the capacitance measured differentially between the two inputs. It is the sum of CD and the series combination of the two common-mode input capacitors. For the probe under consideration, which has a typical value for CD of about 0.1 pF, this results in a differential-mode input capacitance of about 0.3 pF. The common-mode input resistance for the probe is 50 kilohms, which results in a differential-mode input resistance of 100 kilohms.

This first-order probe load model provides useful information on probe loading at lower frequencies. From DC up to about 1 MHz, the probe differential-mode input impedance is primarily resistive, and for the probe shown it's constant at 100 kilohms. Above about 1 MHz, the probe differential-mode input capacitance begins to reduce the probe input impedance until it reaches the probe's damping resistance of 130 ohms for frequencies above 1 GHz. Although this discrete component model does provide some useful information at lower frequencies, it does not include any probe input inductance. Neither intrinsic inductance in the probe hybrid nor attachment inductance from the probe pins or connection adaptors is included in the first-order model. Probe input inductance adds resonant effects to this first-order probe load model. The addition of probe interconnect adaptors, which generally add more inductance to the relatively small probe pin inductance, increases probe resonant overshoot and may require additional damping resistance at the measurement node.

Figure 5-21 shows a distributed component model for a probe with a bandwidth specification of 3.5 GHz.

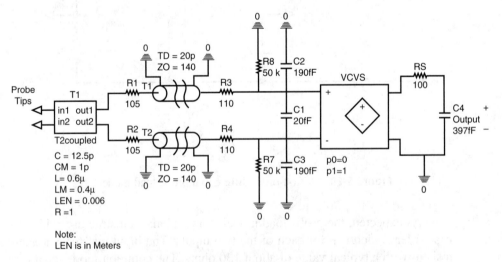

Figure 5-21   A distributed model for a modern differential probe.

The distributed model contains a coupled transmission line input and a simplified amplifier output model with a single-pole roll-off response. This distributed component model provides accurate modeling of probe loading effects up to the limit of the probe frequency response. Although it's optimized for probe loading effects, the distributed component model also provides first-order transient and frequency response performance for the probe.

To measure an electrical signal with high fidelity, the gain of a probe amplifier must be carefully controlled over a wide frequency range. The gain of an ideal probe amplifier would be constant from DC out to frequencies near its specified bandwidth limit and then would drop off fairly rapidly. The gain roll-off characteristic of this ideal probe must also be controlled to minimize pulse response aberrations. A real probe amplitude response, although designed to be flat over a broad frequency range, will exhibit some gain variation, particularly at frequencies above 1 GHz.

Although the gain of a typical probe is relatively flat from DC to about 1 GHz (see Figure 5-22), the probe gain accuracy deviates noticeably from the DC specification at frequencies above 1 GHz. For a probe with a specified bandwidth limit of 5 GHz, for example, the gain may be "rolled off" by as much as 3 dB, which is a reduction in gain of about 30% from the DC voltage gain. The DC voltage gain of this probe is specified as 0.16, which represents a X6.25 attenuation of the input signal. This X6.25 probe attenuation results from the combination of the probe input attenuator and the fixed probe amplifier gain. Figure 5-23 shows a simplified diagram of the probe architecture, which illustrates the probe input attenuator and buffer amplifier.

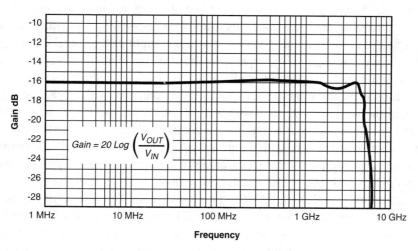

Figure 5-22   Probe gain characteristics.

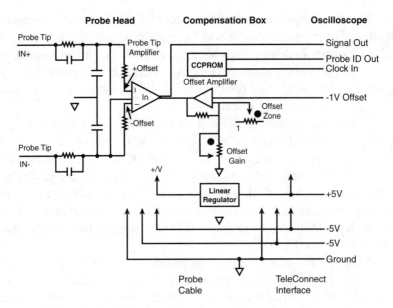

Figure 5-23   A block diagram illustrating differential probe architecture.

With modern probes and oscilloscopes, the probe's DC gain usually is transparent to the user. The oscilloscope automatically scales its internal gain control to match the oscilloscope vertical scale factor with the specified probe gain. Information about the probe DC gain is stored in nonvolatile memory in the probe and is read by the oscilloscope when the probe is attached. It is also possible to optimize the accuracy of the system gain and offset of the probe and oscilloscope.

Probably the most important amplitude response specifications for an active probe are its differential and common-mode input range specifications. The differential mode input range represents the effective dynamic range of an input differential signal. To prevent inadvertent overdriving of the probe amplifier, some oscilloscopes show temporary annunciation markers at the dynamic range limit when the vertical scale or position is adjusted near the range limit. The probe's differential-mode input range is then communicated to the oscilloscope to allow the instrument to set its dynamic range markers correctly. The common-mode input range represents the DC voltage range with respect to ground that can be applied to both the probe input pins without limiting the probe response. In a modern probe with high CMRR, a common-mode DC signal within the +6.25 V to –5.0 V common-mode input range will be reduced to a millivolt level DC offset voltage. While the bandwidth and rise time specifications for many probes are measured with small input signals, they generally can give usable speed for large signals as well.

It is possible to extend a probe's differential mode input range by applying a differential signal offset to the probe. In this case, the DC offset signal from the oscilloscope is buffered by a single-ended amplifier and passed to the offset input of the probe head amplifier. The amplifier then converts the single-ended offset signal to a complementary differential offset signal that drives the ends of the input attenuator. The differential offset signal effectively cancels out differential DC voltages applied to the input pins. This technique is particularly useful for single-ended measurements, which are made by grounding the probe's negative input pin. If a single-ended DC voltage is present at the probe's positive input pin, it is effectively converted to a DC differential mode voltage:

$$V_{DIFF} = V_+ - V_- = V_{DC} - 0 = V_{DC}$$

This DC differential mode voltage can be nulled using the differential offset control, if it is within the ±1.25 V differential offset range. By nulling this DC differential mode voltage, the probe's dynamic range window is effectively expanded, although the ±2.5 V differential signal range limit still applies within the expanded dynamic range window.

The CMRR specification for a high-performance probe normally is specified at several frequencies from DC to 1 GHz. CMRR is defined as the ratio of differential-mode gain to common-mode gain. It's effectively a measure of how well the two differential signal paths of the probe input have been designed and manufactured. Path mismatch, particularly at higher frequencies, can cause some of the differential-mode signal to be converted into a common-mode signal, which reduces CMRR. Because of increasing parasitic effects with increasing frequency, CMRR drops off with frequency, as shown in Figure 5-24.

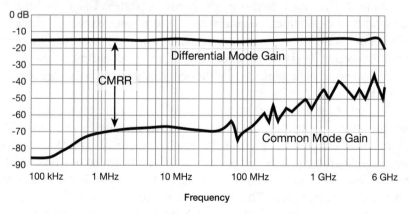

Figure 5-24   CMRR response/frequency graph.

On this logarithmic scale display of differential-mode and common-mode gain, CMRR can be calculated by simply subtracting differential-mode gain from common-mode gain at any frequency:

$$CMRR = A_{DM}(dB) - A_{CM}(dB)$$

A probe's CMRR performance typically is measured with a network analyzer, which has a very controlled impedance environment. The CMRR that is observed in an actual differential measurement application is affected by signal source and load impedance mismatch as well as differential signal skew and other routing variations.

5.7.3   Differential Probe Attachment Issues

One of the critical probing issues in making accurate measurements of high-speed signals is the probe attachment to the circuit under test. Any high-bandwidth active probe, whether differential or single-ended, provides full bandwidth performance only when the measurement nodes are connected to the probe input pins with short leads. The use of long connection leads on a probe input adds parasitic inductance and capacitance that either reduces the probe bandwidth or produces resonant effects. Attaching a probe to a circuit using long connection leads thus causes the signal being measured to appear distorted. Part of the distortion seen may actually be real, caused by disturbance of the measured circuit by the loading of probe and connection parasitics. Even if the probed circuit is not affected noticeably by the parasitic measurement load, the interconnect parasitics may still distort the measurement of the probed signal.

These parasitic problems from long connection leads become worse as the rise time of the measured signals becomes faster. Even the use of higher-bandwidth probes, which generally have lower input parasitics at the probe tips, will not help much if significant parasitics are added to the measurement by using long connection leads. The attachment of probe connection adaptors may also increase measurement parasitics, although it may be the only way that contact can be made to some circuits.

Another important probe attachment issue is the measurement's reliability and repeatability. In the case of single-ended measurements, the quality of the ground lead connection is often the primary factor in connection repeatability. Since single-ended probes generally are designed with an asymmetrical structure, the connection focus is often on making a solid contact between the probe input pin and the circuit signal. The probe ground requirement for low-speed signals is to simply locate a convenient ground point somewhere on the circuit board. For high-speed signals, however, the ground connection can be as critical as the signal connection for good measurement fidelity. A high-speed ground connection should be as short as possible and should be attached to a low inductance ground reference close to the signal connection. A short ground connection is important not only to minimize interconnect parasitics, but also to reduce inductive noise pickup by the probe grounding loop. The sensitivity of single-ended measurements to ground noise can sometimes result in measured signal variation when different ground nodes are contacted near the measured signal node. Single-ended measurements are also sometimes affected by the physical location of the hand-hold on the probe, since body capacitance contributes to the overall probe interconnect parasitics. The issue of probe lead mechanical attachment compliance is usually handled by spring action in the ground lead. This can be done using a "pogo pin" ground lead when the correct probe spacing between signal and ground points on the probe has been designed into test points on the circuit board. For the more general case, a flexible z-ground lead can be used, or a customizable ground lead can be soldered to a ground reference near the signal point and then plugged into the probe ground socket.

Differential probe attachment has two signal inputs as well as a ground-lead connection point on the side of the probe head. Because of the high CMRR of a differential probe, the requirements for a ground connection are very different than for a single-ended probe. In general, it is not necessary to make a ground-lead connection to a differential probe when making measurements.

In fact, the addition of a ground lead connection from a differential probe to the circuit ground may actually inject noise into the measurement. When a differential probe is used to make a single-ended measurement, the circuit ground connection should be made with the differential probe negative input pin rather than the probe head ground-lead connection. The only application where a differential probe head ground connection is necessary is with earth-ground isolated circuits such as battery-powered devices.

One of the more challenging attachment issues in the use of differential probes is probe-lead connection compliance. Because differential probes have two identical input pins, making reliable connections to circuit contacts with both pins at the same time generally is more challenging than with single-ended probes. As noted earlier, a single-ended probe structure usually is asymmetrical, making the addition of connection compliance to the ground pin an easier task than with a differential probe, where symmetry must be preserved for good CMRR. Although some lower-performance differential probes have sockets on their inputs, the present generation of high-performance differential probes use solid male pins as input leads. Input pin sockets generally provide more flexibility in measurement applications by allowing the easy attachment of different adaptors. Male input pins are used on higher-performance probes, however, because of their reduced parasitics compared to pin sockets. The problem of attaching adaptors to male input pins has been solved by the use of elastomeric contacts in the probe tip adaptors.

Because of the connection compliance problem with differential probes, subtle measurement errors can be introduced. Figure 5-25 shows the kind of error that can occur because of poor differential probe pin contact.

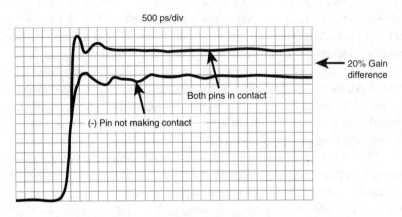

Figure 5-25   Differential probe contact error.

The two waveforms show that it is possible to make a measurement that appears to have the correct waveform shape when only one of the two differential input pins makes solid contact with the circuit. The single pin contact measurement, however, shows a large gain error, a slower rise time, and long-term signal distortion compared to a correct two-pin contact measurement. One method that can help with differential probe attachment compliance is the use of probe positioners incorporating coarse and fine position adjustment controls and flexible probe retention rings. These rings, which attach the probe to the probe holder arm, also provide some spring action in holding the probe in solid contact with the circuit.

The use of solid male pins on differential probe inputs also encourages planning for measurement test points with the correct probe pin spacing on circuit boards. Ideally, critical measurement nodes should be identified and pads added for probing to give the best measurement performance. Realistically, however, not all measurement nodes can be identified in advance. It becomes increasingly difficult to add probing pads to high-speed circuits without introducing significant parasitic loading to the circuit, even without the probe attached. Trying to probe a pair of differential traces having less than 10 mil width and spacing with differential probe pins having a fixed 100 mil spacing is, of course, impossible.

A variety of probe tip adaptors have been developed to address this probe attachment challenge, including variable spacing, square pin, and solder-in types. Electrical connection between the probe pins and the adaptor pins is made with elastomeric contacts inside the adaptor housing. These adaptors are also polarized, requiring alignment between the positive and negative input pins on the differential probe tip housing and the matching markings on the adaptor housing.

5.7.4   Alternative Differential Measurement Methods

Although it might seem obvious that single-ended measurements should be made with single-ended probes and differential measurements should be made with differential probes, alternative measurement methods can be used. It is certainly possible, for example, to make single-ended measurements with a differential probe. It is similarly possible to make differential measurements using a pair of single-ended probes and oscilloscope waveform mathematics. It is also possible, in a 50-ohm signaling environment, to make single-ended or differential measurements without probes. The signals in a 50-ohm signaling environment are terminated by the 50-ohm inputs on the measurement equipment. They can be directly cabled to the 50-ohm inputs available on all high-performance oscilloscopes, both real-time and sampling. The differential signal response is then computed by subtracting the two channel inputs using the oscilloscope's waveform processing feature. Probes have also been developed that contain a differential input termination network as well as a high-performance differential buffer amplifier. These devices can be used to make a differential measurement in a 50-ohm signaling environment on a single oscilloscope channel.

5.7.5   Single-Ended Measurements with a Differential Probe

One of the common misconceptions about the use of differential probes is that they can only be used to measure differential signals. In fact, the high CMRR of a differential probe means that a differential probe generally can be used to make single-ended measurements with high signal fidelity. The low-input capacitance of high-performance differential probes results in reduced circuit loading and may actually provide superior signal fidelity compared to single-ended probes of similar bandwidth. In addition, because differential probes make their single-ended measurement ground connection with one of their high-impedance input pins rather than a low-impedance ground lead, they seem to be less susceptible to noise injected into the measurement ground loop. This common-mode noise rejection characteristic of a differential probe also results in less sensitivity to the probe's physical placement and handling.

Single-ended measurements of differential signals are very important in differential signal testing. They can identify possible asymmetry between the two differential signal pairs that cannot be separated in a differential measurement alone. When a differential probe is used to measure an ideal complementary differential signal, the signal's common-mode component is a DC voltage, which is largely rejected by the probe's high DC CMRR. When the differential probe is then used to make a single-ended measurement of one of the complementary signals, the signal's common-mode component has both an AC and DC response. When measured with a differential probe, the common-mode component of a single-ended signal is equal to one-half the signal amplitude:

$$V_{CM} = (V_+ + V_-)/2$$

where V_+ is the single-ended signal and $V_- = 0$ V (ground).

If the CMRR of the differential probe is not large enough, some of the common-mode signal "bleeds through and distorts the output signal. This common-mode bleed-through problem is worse at high frequencies, where parasitic mismatches tend to reduce the CMRR. The differential-mode component of the single-ended measurement is also only half that of a differential measurement on a complementary signal. As a result of this smaller signal, the error from common-mode bleed-through appears relatively larger as well.

Even though a common-mode error term may be present in a single-ended measurement made with a differential probe, it will be small for a probe with good CMRR. In addition, the error term may be smaller than the ground noise error signal present in a single-ended measurement using a single-ended probe.

5.7.6 Pseudo-Differential Measurements

Differential measurements made using two single-ended probes and oscilloscope waveform mathematics are commonly called "pseudo-differential" measurements. They can be made with either real-time or sampling oscilloscopes. Depending on the measurement signaling environment, pseudo-differential measurements can be made with either cables or probes. Cables can be used for serial data compliance testing, where the transmitted signal is in a 50-ohm signaling environment and the signal path can be broken and terminated at the ground-referenced 50-ohm oscilloscope input. High-impedance probes must be used where the signal path cannot be broken or where the signal must be debugged at any of a number of places along the signal path. Because of the difficulty of matching the signal paths in two separate oscilloscope channels, the CMRR for pseudo-differential measurements tends to be somewhat poorer than that of a good differential probe. Differential probes are carefully designed for differential signal path matching to optimize CMRR. As a result, pseudo-differential measurements may exhibit less signal fidelity than measurements made with a good differential probe, although other factors such as amplitude response and circuit loading can also affect the measurement's fidelity. Pseudo-differential probing, of course, requires more oscilloscope channels per differential signal than needed by a differential probe.

Sampling oscilloscopes are often chosen for pseudo-differential measurements made in a 50-ohm signaling environment because of the high bandwidth of the sampling system. The measurement fidelity of a sampling system also tends to be superior, with fewer signal reflection problems and excellent noise performance. A sampling oscilloscope, however, requires a periodic signal, which makes it more difficult to use for debugging transient problems. It is also impossible to use a sampling oscilloscope to capture and analyze a real-time data stream—something that is required for the jitter testing of data signals in some of the serial data standards such as PCI Express. A sampling oscilloscope also generally requires an external trigger signal, such as a data rate clock, which is not always readily available. High-performance real-time oscilloscopes are now available with high-enough bandwidth and sampling rate to be effectively used in high-speed serial data measurement applications. Real-time oscilloscopes also provide triggering on the input data signal, including the capability to trigger on a specific serial data pattern, including internal clock-recovery triggering.

Because pseudo-differential measurements are made using two different measurement signal paths, the measurement path amplitude response and timing delays must be closely matched for high measurement fidelity. Amplitude response mismatch and signal delay variation between the two oscilloscope channels usually set the fundamental CMRR limit for a pseudo-differential measurement. It is important to ensure that interconnect mismatch between the signals to be measured and the oscilloscope channels is minimized. If the two signal input paths are not carefully matched, interconnect path differences distort the response. Identical cables or probes, with matched response and time delays, should be used for pseudo-differential measurement interconnections. When cables are used for signal interconnect, delay variations are primarily determined by variations in cable length, with a typical signal delay of about 150 ps per inch for commonly used 50-ohm coaxial cables.

Probe delay variations are affected by differences in both probe cable length and probe head amplifier delay. Real-time oscilloscope channels also exhibit signal-path delay variations from the input to the acquisition sampler. There can be a timing variation of several hundred picoseconds between channels of an oscilloscope, as well as a similar delay variation between attenuator paths of a single channel. Sampling oscilloscopes have delay variation between channels, particularly channels on different sampling modules, but they have no attenuator path delay variations because they have no input attenuators.

5.7.7 Measurement Channel Deskew

Both sampling and real-time oscilloscopes require a channel deskew procedure to compensate for delay differences between the two measurement channels and the attached cables in a pseudo-differential measurement. Deskewing two sampling oscilloscope channels requires two sampling modules. Even where the electrical sampling modules are dual-channel, the sampling architecture provides only one trigger strobe per module. Although this architecture provides excellent timing alignment between the intramodule channels, only intermodule channels can be deskewed with time-aligned channels.

Channels on a high-performance real-time oscilloscope can be deskewed on any set of channels without concern for time-aligned waveform problems. Many high-performance real-time oscilloscopes, however, have maximum real-time sampling rate limitations that may affect the choice of channels used to make pseudo-differential measurements. Since the greatest signal fidelity is obtained with the maximum sampling rate, pseudo-differential measurements should use channel combinations that sample at the highest rate.

The deskew procedure requires a high-speed signal edge that is connected in common to both differential signal interconnects. Where two probes are used, it is possible to add a probe deskew fixture that generates a fast edge signal and convenient probing points. With cables, obtaining a pair of tightly matched delay signals generally is more difficult, since a fast edge signal in a 50-ohm environment is susceptible to serious distortion problems unless care is taken in splitting the signal between the two paths. Although a 50-ohm power splitter is often used in this application, the use of a dual TDR sampling module provides a more controlled set of matched-delay fast edge signals. No matter what signal source is used for the deskew signals, the oscilloscope should be configured to the vertical and horizontal scale settings that will be used in the measurement application. Changing the vertical scale factor or horizontal scale factor after deskewing the channels can result in a significant variation in timing accuracy between the channels, which will almost certainly increase the channel skew.

5.7.8 Differential, Sub-Miniature Version A (SMA) Input Probes

Probes are available with architecture that is optimized for serial data compliance testing and other high-speed differential measurement applications using differential to single-ended signal conversion in a 50-ohm signaling environment with a single oscilloscope channel input. The use of SMA connectors provides a reliable, repeatable signal attachment method for compliance testing, where the signal path can be interrupted and terminated at the measurement input. In addition, through the use of an embedded differential amplifier, the SMA probe can have its differential measurement interface at the end of a 1.2 m cable rather than at the oscilloscope front panel. Bringing the differential measurement interface closer to the circuit under test in this fashion minimizes frequency-dependent cable inter-connect losses.

The probe architecture shown in Figure 5-26 includes dual SMA connector inputs, a dual 50-ohm resistor termination network, a common-mode DC bias connection to the termination network, and an embedded differential amplifier. The SMA inputs provide a reliable connection interface for high-frequency 50-ohm signal paths to the embedded differential amplifier and input termination net-work. The probe input termination network is implemented in a shielded module that uses laser-trimmed hybrid circuit technology to provide high bandwidth, good power dissipation, and excellent CMRR performance.

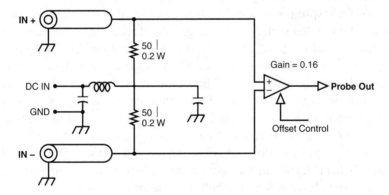

Figure 5-26   SMA probe architecture.

The input termination resistors are coupled with a common-mode termina-tion network that is designed to provide a low-impedance path to ground. The DC bias connection to the common-mode termination network is made from a user-supplied external DC power supply through a dual "banana plug" connector on the probe. The input termination network is designed to provide broadband 50-ohm terminations for both common-mode signal pairs of a differential signal. For an ideal complementary differential signal, there is only a DC common-mode component, and the termination resistors alone should terminate the signal with minimal reflections. A practical differential signal, however, has both amplitude and timing mismatch, which results in an AC common-mode component. The input termination network includes a common-mode capacitance at the node between the termination resistors, which provides a common-mode termination for high-frequency signals. The common-mode node impedance breakpoint frequency can also be shifted all the way down to DC by driving the DC bias input from a low-resistance DC source.

When used to measure a high-speed complementary differential signal that has no need for a common-mode DC termination voltage, the SMA probe can be used with the DC bias port open. This is generally the safest configuration, because there is less risk of exceeding the 500 mW power limit of the 50-ohm termination resistors. Some common high-speed logic circuits, however, are designed to operate properly only with their termination resistors connected to a specified DC termination voltage. The LVPECL logic family, for example, when operated from a 3.3 V power supply is designed to operate with its output termination resistor connected to a 1.3 V pull-down voltage. The conventional method for characterization testing of an LVPECL device requires the test board to be powered from split power supplies (such as +2.0 V and –1.3 V) so that the outputs could be terminated with 50 ohms to ground at an oscilloscope channel input. The DC bias input of the SMA probe should allow testing of LVPECL circuits without the use of split power supplies. An SMA probe can similarly be used to test a CML logic output, where the DC bias input provides the CML termination resistor pull-up voltage. Care should be taken when using the DC bias input of the SMA probe so as not to exceed the 500 mW power rating of the probe termination resistors. Since the termination power is a function of both the probe input signal voltage and the probe DC bias voltage, a maximum DC bias voltage specification is insufficient. In the special case where the probe input signal is a DC voltage, the voltage difference between the probe SMA input and the DC bias input should be kept to less than 5.0 V so as not to exceed the 500 mW power rating of the termination resistors.

Although designed and specified primarily for use in making differential signal measurements, these SMA probes can be configured to make single-ended measurements. Single-ended measurements can be made using the SMA probe by connecting the single-ended signal to the probe's positive input and terminating the negative input with a 50-ohm SMA termination resistor. When used to measure a high-speed serial data signal with limited low-frequency power below 10 MHz, the common-mode capacitance effectively terminates the single-ended signal input. As a result, the DC bias port can be left open. Most gigabit serial data signals with 8B/10B encoding meet this low-frequency spectral requirement and thus can be used without a DC bias connection. The exception to this general rule is where the signal driver requires a common-mode DC termination voltage.

Although the timing skew between the inputs of the probe at the SMA connectors typically is less than 1 ps, additional skew in the differential signal path due to interconnect cables must be carefully controlled. Low-loss, matched-delay SMA cables are recommended for general-purpose interconnect use.

It should be expected that a skew of ±10 ps might have a more noticeable effect on other differential signal measurements, such as the crossover point of a high data rate eye pattern. If the skew of the matched delay cable set is not small enough to meet an application's requirements, it is possible to manually deskew the cables using a pair of mechanical SMA phase adjusters. Before doing this, however, it is prudent to measure the actual skew of the cables, because it may be smaller than the guaranteed specification of less than 10 ps. Measured results showing a low-enough cable skew may remove the need for manual deskew of the cables, except perhaps when very low skew characterization must be done.

Each of the alternative measurement methods described for making differential measurements has advantages and disadvantages:

- Pseudo-differential measurements with a sampling oscilloscope might be the method of choice because of the need for the highest bandwidth available.
- Pseudo-differential measurements with a high-performance real-time oscilloscope might be the method of choice because of the need for a long record length real-time acquisition and the availability of a pair of high-bandwidth single-ended probes.
- The use of a pair of SMA probes might be the method of choice because of the need to make lane-to-lane skew measurements between several differential serial data lanes.

One factor that should be considered in evaluating the use of these alternative measurement methods is their CMRR performance. As described earlier, CMRR response, not just at DC, but over the full frequency range of operation, is one of the critical parameters for making differential measurements with high signal fidelity.

The large variety of possible channel combinations and the need for deskew over a wide range of vertical and horizontal settings makes specifying CMRR for oscilloscopes very difficult. The CMRR performance for a specific pseudo-differential measurement setup, however, certainly can be measured. It is relatively easy to make DC CMRR measurements of a pseudo-differential setup, because a DC signal can be split without reflection problems. AC CMRR measurements are more difficult to make, however, particularly for bandwidths greater than 1 GHz. One possible method for making AC CMRR measurements of a pseudo-differential measurement setup uses a dual TDR pulse source that can generate both differential-mode and common-mode pulse signals. This can be used to produce both differential and common-mode responses in the pseudo-differential measurement setup. By differentiating the differential-mode and common-mode pulse response waveforms and performing an FFT on the resulting impulse response, it is possible to produce frequency-domain conversions of the response. Taking the ratio of the differential-mode gain and the common-mode gain over frequency then provides the AC CMRR response, which can be evaluated for acceptability.

5.8 LOGIC ANALYZER PROBING

As outlined at the beginning of this chapter, a logic analyzer connects to, acquires, and analyzes digital signals. The large number of signals that the analyzer can capture at one time is what sets it apart from the oscilloscope. This means that the physical nature of the logic-analyzer probe is very different from that of the oscilloscope probe. In particular, the probe has an internal comparator that compares the input voltage to a predefined threshold voltage and makes a decision about the signal's logic state. The user can set the threshold voltage. This depends on which type of logic family is being tested, such as TTL, CMOS, or ECL.

Logic-analyzer probes come in many physical forms:

- "Clip-on" probes intended for point-by-point troubleshooting.
- High-density, multichannel probes requiring dedicated connectors on the circuit board are shown in Figure 5-27. These probes can acquire high-quality signals, and they have a minimal impact on the system under test.
- High-density compression probes that use a connectorless probe attachment are shown in Figure 5-28. This type of probe is recommended for applications involving higher signal density or for which the connectorless probe attachment mechanism provides quick and reliable connections to the system under test.

Figure 5-27 High-density logic analyzer probes.

Figure 5-28 A connectorless probe, showing pad layout and a socketless connector.

The impedance of the logic analyzer's probes, capacitance, resistance, and inductance becomes part of the overall load on the circuit being tested. All probes exhibit loading characteristics, but, just as with oscilloscope probes, the logic-analyzer probe should introduce minimal loading on the system under test and provide an accurate signal to the logic analyzer.

Probe capacitance tends to "roll off" the edges of signal transitions, as shown in Figure 5-29, and this roll-off slows down the edge transition.

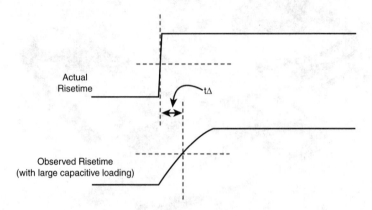

Figure 5-29   The impedance of the logic analyzer's probe can affect signal rise times and measured timing relationships.

Because this slower edge crosses the logic threshold of the circuit later, it introduces timing errors in the system under test. This is a problem that becomes more severe as clock rates increase. Indeed, in high-speed systems, excessive probe capacitance can potentially prevent the system under test from working. It is therefore essential to choose a probe with the lowest possible total capacitance. It is also important to note that probe clips and lead sets increase the capacitive loading on the circuits they are connected to. This effect can be minimized by using a properly compensated adaptor whenever possible.

During hardware and software debugging and system integration, the logic analyzer is used to obtain both state and timing information, which ideally should be captured simultaneously and correlated. For example, a problem such as a setup and hold timing violation may initially be detected as an invalid state on the bus. If the logic analyzer cannot capture both timing and state data simultaneously, isolating the problem becomes difficult and time-consuming. Some logic analyzers require separate timing and state probes and separate acquisition hardware to be connected to acquire both sets of information. This arrangement is known as "double probing," and it can compromise the impedance environment of the acquired signals. Using two probes at once "loads down" the signal, with resultant degradation of rise and fall times, amplitude, and noise performance of the system under test.

It is therefore best to acquire timing and state data simultaneously through the same probe so that one connection, one setup, and one acquisition provide both timing and state data. This arrangement simplifies the probes' mechanical connection, reduces problems, and allows the logic analyzer to capture all the information needed to support both timing and state analysis. There is no second step, so therefore there's less chance of the errors and mechanical damage that can occur with double probing. The single probe's effect on the circuit is lower, ensuring more accurate measurements and less impact on the circuit's operation. This results in higher timing resolution, enabling more details to be seen and triggered on, thereby increasing the chance of finding any problems.

5.8.1   The Socketless Probe: A Modern Probing Application

For the latest digital designs, test and debug access is based on a PCB direct probing system, where the logic analyzer or oscilloscope is connected to the PCB via a standardized socketless probing technique. The SI engineer needs to adopt new design paradigms and determine where to probe during the initial stages of the design process. A modern example is where to site modern socketless PCB access points, which have to be placed at an early stage in the design cycle—ideally, at the initial architectural design stage. Moreover, with modern high-performance systems the effects of probing on the performance of the system under test have to be considered at the simulation phase of the development cycle. Probing a fast edge is a significant issue in terms of maintaining the signal integrity of a modern design. Irrespective of a signal's frequency, the overriding issue today is the edge rate, the rise time, and what effect the test probe has on the edges. In particular, the following questions need to be asked:

- Will the probe impedance generate a logic error in the system?
- Will the probe characteristics affect the fidelity of a measurement?

Both these issues can be addressed in the early stages of the design by including the probe models within the schematic capture and logic simulation phases of the design cycle. Equally, any signal-integrity uncertainties caused by the PCB probe pads can be addressed at the PCB simulation phase of the design cycle by entering the PCB probe pad models into the layout simulator. Logic analyzer manufacturers supply both the probe simulation models and CAD models for PCB socketless pad layouts.

Fast signal edges demand high-impedance probing, and manufacturers have responded by providing active probes in modern socketless packages using silicon-germanium technology to achieve a capacitance of less than 0.5 pF. An important point to note is the high density of the PCB pads required for a socketless probe, making it even more important to consider probing at the early stages of the design. A significant factor in the high density of the probes is the small overhead in PCB area and the alleviation of a number of signal integrity issues with the application of small connection stubs. Also, the benefit of a standard logic analyzer connection system is important to systems designers for a number of reasons. It is fortunate that major instrument manufacturers have now established such a standardized design for socketless probes. The investments during the initial phases of a design are critical in the long term, and the decision to incorporate a socketless PCB into a design is now independent of the analyzer manufacturer, and any dependence on a single source of instrument manufacturer is alleviated.

Incorporating socketless connectors into a design offers a number of additional benefits. The low overhead in the new compact PCB socketless footprints allows the developer to provide a low-cost connection for both development and production test. Moreover, with the latest logic analyzers and oscilloscopes, any 4 logic analyzer probes, typically out of 136, can be multiplexed to an oscilloscope to allow the developer to view both digital and corresponding analog signals on a logic analyzer display. This is an important facility for detecting obscure signal integrity faults and debugging fast-edge state-of-the-art embedded systems. Further benefits are achieved by incorporating a differential clock probe within the socketless probe to avoid any imbalance in the differential clock signals. These probe developments give the systems designer a significant number of probing and signal measurement benefits. The fact that it is possible to carry out single-point digital and analog probing, with the analog probe outputs always "live," is an important element in detecting modern signal integrity issues such as ground bounce, crosstalk, and impedance mismatches.

CONCLUSION

This chapter has demonstrated that probing is a key issue in ensuring signal fidelity of measurements. It also is a major factor in ensuring that the tests needed to ensure signal integrity in a design are carried out accurately and reliably. In addition, the complexity of the issues surrounding probing technology has re-emphasized the need for designers to think about probing early in their designs.

Testing and Debugging: Oscilloscopes and Logic Analyzers

This chapter reviews signal integrity testing from both an analog and digital viewpoint, since at the high frequencies encountered in today's designs the two are inextricably linked. The emphasis is on frequencies over 100 MHz, where the measurement tools of choice are the digital oscilloscope and the logic analyzer. Practical examples included in the chapter show how detailed observation of analog and digital waveforms, using integration tools that demonstrate their interaction side by side, can provide a close understanding of signal integrity challenges—such as setup and hold violations, data skew, and metastable states—and how they can be tackled.

As earlier chapters have shown, signal integrity is a key issue to anyone concerned with high-performance digital system design under real-world conditions, where time constraints, cost limitations, quality requirements, and manufacturability concerns can all be impacted.

As we have seen, the term *signal integrity* encompasses the analog factors that affect both the performance and the reliability of digital designs. As system speeds increase, signal integrity becomes a greater challenge as it is ever more difficult to maintain clean pulse edges, low noise, and nominal amplitude and timing characteristics. When these parameters are flawed, a rigorous regime of signal integrity measurements can trace these problems to their root causes.

6.1   FUNDAMENTALS OF SIGNAL INTEGRITY

The digital bandwidth race requires innovative thinking. Increasing a system's operating rate is not a matter of simply designing a faster clock. As frequency increases, the traces on a circuit board become more than simple conductors. At lower frequencies (such as the clock rate of an older digital system) the trace exhibits mostly resistive characteristics. As frequencies increase, the trace begins to act like a capacitor. At the highest frequencies, the trace's inductance plays a larger role. All these characteristics can adversely affect signal integrity.

At clock frequencies in the hundreds of megahertz and above, every design detail is important—in particular the following:

- Clock distribution
- Signal path design
- Stubs
- Noise margins
- Impedances and loading
- Transmission line effects
- Signal path return currents
- Termination
- Decoupling
- Power distribution

All these considerations impact the integrity of the digital signals that must carry clocks and data throughout a system. An ideal digital pulse is cohesive in time and amplitude, free from aberrations and jitter, and has fast, clean transitions. As system speeds increase, it becomes ever more difficult to maintain ideal signal characteristics.

That is why signal integrity is a critical issue. A pulse rise time may be adequate in a system clocked at 10 MHz or 20 MHz, but will not suffice at clock rates of 100 MHz and above.

6.2   SIGNAL INTEGRITY CONCEPTS

The notion of signal integrity pertains to noise, distortion, and anomalies that can impair a signal in the analog domain. At frequencies in the hundreds of megahertz range, a host of variables can affect signal integrity: signal path design, impedances and loading, transmission line effects, and even power distribution on the circuit board. It is the designer's responsibility to minimize such problems during the design phase and correct them when they appear.

When considering testing and measurement related issues, it is important to distinguish between the terms *signal integrity*, which applies to the device under test, and *signal fidelity*, which applies to the measurement equipment. Factors such as probe loading and measurement system bandwidth determine signal fidelity. Meaningful signal integrity and compliance measurements require tools with good signal fidelity.

Two fundamental sources of signal degradation are as follows:

- **Digital issues:** Typically timing-related. Bus contentions, setup and hold violations, metastability, and race conditions can cause erratic signal behavior on a bus or device output.
- **Analog issues:** Low-amplitude signals, slow transition times, glitches, overshoot, crosstalk, and noise. These phenomena may have their origins in circuit board design or signal termination, but other causes may also need to be investigated.

Not surprisingly, there is a high degree of interaction and interdependence between digital and analog signal integrity issues. For example, a slow rise time on a gate input can cause the output pulse to be delayed, in turn causing a bus contention in the digital environment further downstream. A thorough solution for signal integrity measurement and troubleshooting involves both digital and analog tools.

6.2.1   Impedance Measurements: The Foundation of Signal Integrity

Seasoned engineers know that good signal integrity is the result of constant vigilance during the design process. It is all too easy for signal integrity problems to be compounded as a design evolves, and to become more difficult to track down.

A tiny aberration that goes unnoticed in the first prototype board can bring the whole system to a crashing halt when the board is merged with others to form a complete system.

Given these realities, where does signal integrity work begin? Designers working on the most critical high-speed technologies often start their signal integrity work at the very beginning—with the raw, unpopulated circuit board. Most high-speed protocols require a 50-ohm impedance; for example, PCI Express specifications call for a 50-ohm transmission line with 10-ohm tolerance. Analyzing transmission line principles is covered in more detail in Chapter 3, "Signal Path Analysis As an Aid to Signal Integrity"; suffice it to say, however, that the tolerances are critical to high-fidelity signal transmission. And less deviation is always better.

Modern layout tools implement the applicable impedance rules for high-speed protocols, but physics, manufacturing variation, circuit board materials, and human error can introduce unforeseen departures. As a result, many developers have learned that a rigorous process of verifying impedance characteristics can help them detect and correct problems early. Design choices can be reconsidered, if need be, before quantity orders are placed with a vendor.

6.2.1.1   Impedance Measurement Techniques

The tool of choice for measuring impedances is a sampling oscilloscope equipped with a time domain reflectometer (TDR) module, which permits the signal transmission environment to be analyzed in the time domain, just as the signal integrity of live signals will be analyzed in the time domain.

Time domain reflectometry measures the reflections that result from a signal traveling through transmission environments such as circuit board traces, cables, or connectors. The TDR instrument sends a fast step pulse through the medium and displays the reflections from the observed transmission environment. Figure 6-1 is a simplified block diagram of this scheme.

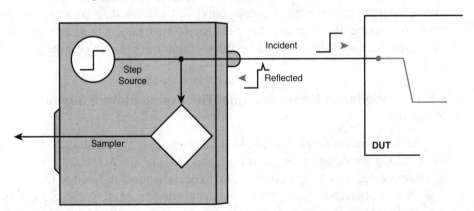

Figure 6-1   Block diagram of a TDR acquisition setup using a TDR module.

The TDR display is a voltage waveform that includes the incident step and the reflections from the transmission medium. The reflections increase or decrease the step amplitude depending on whether the nature of the discontinuity is more inductive or capacitive, respectively.

A reflection from an impedance discontinuity has a rise time equal to or (more likely) slower than that of the incident step. The physical spacing of any two discontinuities in the circuit determines how closely their reflections will be positioned relative to one another on the TDR waveform.

Two neighboring discontinuities may be indistinguishable to the measurement instrument if the distance between them amounts to less than half the system rise time. The quality of the incident step pulse is critical, especially when measuring short traces. In addition to its fast rise time, the step must be accurate in terms of amplitude and free from aberrations.

6.2.2 True Differential TDR Measurements

Many of today's high-speed serial standards rely on differential transmission techniques using complementary signals. Two "wires" or printed-circuit board (PCB) traces carry simultaneous mirror images of the signal. Though more complex than the single-ended approach, differential transmission is less vulnerable to external influences such as crosstalk and induced noise, and generates less of the same.

Differential paths require TDR measurements just as other transmission environments do. The incident pulse must be sent down both sides of the differential pair and the reflections measured. This can be done in two ways:

- The "virtual" or "calculated" differential TDR method, in which the TDR sends out two positive-going incident steps alternating in time, and corrects both the polarity and the alignment before the measurement waveform is displayed on the oscilloscope screen.
- The true differential TDR method, in which the TDR sends out complementary signals that are accurately and correctly aligned in time. The device under test (DUT) receives a differential stimulus signal more like those it will encounter in its end-user application, potentially producing better insights into the device's real-world response. The TDR system does not need to manipulate the displayed step placements.

Even though the true differential TDR delivers an inverted (negative-going or complementary) pulse simultaneously with its positive counterpart, the resulting TDR impedance display shows a seemingly uninverted trace, with all values on the trace positive. Why is this?

Consider what the display is showing: impedance in ohms. It is comparable to measuring the resistance of a potentiometer with a digital multimeter. You expect a range of resistance values beginning at zero and increasing positively toward some maximum. The same is true when using a TDR to observe a passive element such as a circuit board trace: The impedance is always a positive construct with the lowest possible value being near zero ohms. There are no negative ohms or Rho values.

6.2.2.1 ASIC Verification: Ensuring Signal Integrity at the Source

TDR measurements on the unpopulated board are the first critical step in the hardware verification process.

The active devices that make the raw PCB into a server finished system are the next candidates for evaluation as the design moves ahead.

"Off the shelf" semiconductor devices are usually tested by their manufacturers, but many designers prefer to confirm performance claims rather than risking an entire design on unproven components. More important, early samples of custom devices must be verified. Short-run ASIC prototypes (used to fabricate prototypes of the end-user product itself) will be tested, and ultimately the device tests might be released to production.

The verification process may include up to four types of tests:

- **DC parametric:** The most basic measurements to ensure that the device provides proper logic levels under loading, adequate current at each binary state, and very low leakage in the high-impedance "compliant" state.
- **Analog functional:** In mixed-signal devices, this test verifies the analog features. Again, simple digital commands set up the device and execute the gain and range changes unique to analog components.
- **Digital functional:** A simple command set initializes the device in its various modes, and test vectors (binary data) are applied at a low data rate. The resulting outputs are compared with expected data.
- **At-speed functional:** In effect, a digital functional test at the device's maximum clock and data rates.

The DC parametric and analog functional tests are beyond the scope of this chapter, but the digital functional tests examine timing details that can impact signal integrity.

The issue of signal access is a challenge when testing complex ASIC devices. Particularly in the case of the at-speed tests, fixturing must be designed to connect—with the least possible signal degradation—oscilloscopes and high-speed signal sources to a sequence of selected device pins. Most designers also characterize the fixture itself (using TDR as explained in the previous section) to prevent impedance-related signal distortions.

It is common practice to design fixtures with relay matrixes to distribute oscilloscope inputs and signal source outputs to respective pins on the DUT. Test programming environments such as National Instruments LabVIEW® provide an expedient means to automate measurements and control signal routing on the fixture. If properly designed, test programs developed for validation and evaluation can be ported to production test applications with minor modifications.

6.2.2.2 Digital Verification

Setup and hold timing values are essential specifications for any clocked digital device, and the heart of the ASIC digital functional verification measurements. *Setup time* is defined as the length of time for which the data must be present (and in a stable, valid state) before the clock edge occurs. *Hold time* defines the amount of time that the data state must be maintained after the clock edge. In high-speed digital devices used for computing and communications, both values may be as low as a few hundred picoseconds.

Setup and hold violations can cause signal integrity problems in the form of transients, edge aberrations, glitches, and other intermittencies. Figure 6-2 depicts a typical setup and hold timing diagram.

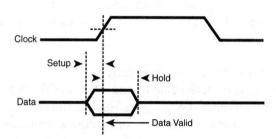

Figure 6-2   Setup and hold timing diagram.

In this example, the data envelope is narrower than the clock. This emphasizes the fact that, when using today's high-speed logic, transition times and setup and hold values can be very brief even when the cyclical rate is relatively slow.

There are two common approaches to evaluating the setup and hold time performance (as well as other timing parameters) of a device:

- **Low-speed tests:** Depending on the requirements foreseen in the end-user application, a fairly coarse functional verification procedure may be adequate. In some cases, it is not necessary to take quantitative measurements of the actual setup and hold values. If the device will tolerate a broad clock placement range, the timing "test" may be as simple as running a low-speed functional data pattern, adjusting the clock edge's position relative to the data, and observing the results on an oscilloscope. The oscilloscope's trigger is set to a 50% level. The device tends to become metastable as it exceeds its setup and hold timing limitations. *Metastability* is an unpredictable state in which the device output may switch to either "1" or "0" without regard to the logical input conditions. Similarly, excessive jitter may appear on the output when setup and hold tolerances are violated.

- **At-speed tests with "burst" data:** The burst functional test exercises the device at rates approximating its intended operational frequency. Modern signal sources can deliver a block of data to the device under test at rates much higher than the basic functional test just explained. The process is still one of empirically finding a range of setup and hold values and specifying the system's clock placement accordingly. Using the data generator to drive all the device's inputs, repetitive data patterns help to isolate the recurring skew problems that cause repeated setup and hold violations.

The ASIC evaluation step enables custom and programmable IC devices to be isolated from the system environment and qualified for installation. Moreover, it helps to detect and avoid circumstances that cause metastability, jitter, and noise, which can cause signal integrity problems that impact other system elements.

6.2.2.3   Basic Functional Verification

With PCB impedance measurements and ASIC verification completed, it is time to assemble a prototype and begin "live" measurements. The next few steps in the design process are where the majority of functional problems emerge and must be solved. Many of these problems stem from signal integrity issues that are impossible to detect until all the components are in place and working together.

As explained earlier, escalating data rates and edge rates are at the heart of many signal integrity problems in today's designs. In most modern bus protocols, the data rate is the basic currency of a device's throughput and efficiency. Raw bit rates extend from about 1 Gbit/s to more than 8 Gbit/s depending on the protocol, and multilane topology is often used to increase net throughput. Some standards use a single lane; for FibreChannel 4X and Serial ATA, the data throughput is equal to the signaling rate. Multilane configurations scale their throughput in direct proportion to the number of lanes used. InfiniBand, PCI Express, and XAUI all have 4X variants that provide good optimization for 10 Gbit/s, but InfiniBand and PCI Express carry the concept even further.

PCI Express offers up to 32 lanes (80 Gbit/s) on the circuit board and 16 lanes (40 Gbit/s) at the connector. Note that the raw data rate is not the same as the data transfer rate, which is expressed in gigabytes per second. Since the data transfer rate depends so much on processes occurring in higher layers of the architecture, a physical-layer comparison is not useful.

6.2.2.4   Operational Validation, Fault Detection, and Debugging

A key element in evaluating any new design is operational validation: the first look at how the overall system will accept, transfer, process, and store information as intended.

Any complex design is almost certain to encounter some errors during development. These may include incorrectly placed components, logic design problems, improper terminations, and more. A bus channel or memory signal whose analog characteristics such as rise time and amplitude are flawed can cause an error in a high-level system instruction. Conversely, fast-changing digital switching on a bus can impact the analog behavior of signals on nearby traces. The engineer is left with the challenge of detecting and correcting issues like these quickly to stay on schedule.

Quickly identifying the real cause of the problem involves tools and troubleshooting methods that address both analog and digital domains. The favored solution for such challenges is a pairing of familiar instruments: a high-bandwidth digital storage oscilloscope (DSO) or digital phosphor oscilloscope (DPO) and a logic analyzer. The DSO/DPO is the best tool for observing individual voltage/time events such as glitches, as well as distortions, transition times, and critical setup and hold timing values. The logic analyzer, on the other hand, captures logic signals in their elemental form—binary 1s and 0s with associated timing information—as they move through the system. Capturing the interaction between the two domains is key to efficient troubleshooting.

6.3   VERIFICATION TOOLS: OSCILLOSCOPES

For the verification and troubleshooting steps in this part of the design process, the real-time DSO and DPO are the tools of choice. Both types of instruments offer not only the necessary performance (bandwidth, sample rate, and so on), but also a wealth of triggering choices, probing options, application-specific software packages, and more. Most important, these real-time platforms make it easy to probe a series of test points and reliably acquire waveforms ranging from power supply noise to multigigahertz data streams. Automated setup routines can be used to find, scale, and display the signal, while cursors and built-in automated measurements simplify analysis.

The following sections describe these classes of instruments in more detail.

6.3.1   Digital Oscilloscopes

A digital oscilloscope uses an analog-to-digital converter (ADC) to convert the measured voltage into digital information. It acquires the waveform as a series of samples and stores these samples until it accumulates enough samples to describe a waveform. The digital oscilloscope then reassembles the waveform for display on the screen. The digital approach means that the oscilloscope can display any frequency within its range with stability, brightness, and clarity. For repetitive signals, the bandwidth of the digital oscilloscope is a function of the analog bandwidth of the front-end components of the oscilloscope, commonly referred to as the -3 dB point. For single-shot and transient events, such as pulses and steps, the bandwidth can be limited by the oscilloscope's sample rate.

Digital oscilloscopes can be classified into digital storage oscilloscopes (DSOs), digital phosphor oscilloscopes (DPOs), and sampling oscilloscopes.

6.3.1.1   Digital Storage Oscilloscopes

A conventional digital oscilloscope is known as a digital storage oscilloscope (DSO). Its display typically relies on a raster-type screen rather than luminous phosphor.

DSOs allow users to capture and view events that may happen only once—known as *transients*. Because the waveform information exists in digital form as a series of stored binary values, it can be analyzed, archived, printed, and otherwise processed, within the oscilloscope itself or by an external computer. The waveform need not be continuous; it can be displayed even when the signal disappears. Unlike analog oscilloscopes, DSOs provide permanent signal storage and extensive waveform processing.

However, DSOs typically have no real-time intensity grading; therefore, they cannot express varying levels of intensity in the live signal.

Some of the subsystems that comprise DSOs are similar to those in analog oscilloscopes. However, DSOs contain additional data-processing subsystems that are used to collect and display data for the entire waveform. A DSO employs a serial-processing architecture to capture and display a signal on its screen, as shown in Figure 6-3.

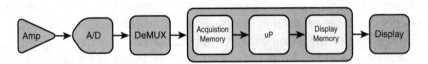

Figure 6-3   The serial-processing architecture of a DSO.

Like an analog oscilloscope, a DSO's first (input) stage is a vertical ampli-fier. Vertical controls allow the user to adjust the amplitude and position range at this stage.

Next, the ADC in the horizontal system samples the signal at discrete points in time and converts the signal's voltage at these points into digital values called *sample points*. This process is referred to as digitizing a signal. The horizontal system's sample clock determines how often the ADC takes a sample. This rate is referred to as the *sample rate* and is expressed in samples per second (S/s).

The sample points from the ADC are stored in acquisition memory as wave-form points. Several sample points may comprise one waveform point. Together, the waveform points comprise one waveform record. The number of waveform points used to create a waveform record is called the *record length*. The trigger system determines the start and stop points of the record. The DSO's signal path includes a microprocessor through which the measured signal passes on its way to the display. This microprocessor processes the signal, coordinates display activities, manages the front-panel controls, and more. The signal then passes through the display memory and is displayed on the oscilloscope screen.

Depending on the capabilities of the oscilloscope, additional processing of the sample points may take place to enhance the display. Pretrigger may also be available, enabling the user to see events before the trigger point. Most of today's digital oscilloscopes also provide a selection of automatic parametric measure-ments to simplify the measurement process.

A DSO provides high performance in a single-shot, multichannel instru-ment. DSOs are ideal for low-repetition-rate or single-shot, high-speed, multi-channel design applications.

6.3.1.2   Digital Phosphor Oscilloscopes

The digital phosphor oscilloscope (DPO) offers an alternative approach to oscillo-scope architecture. While a DSO uses a serial-processing architecture to capture, display, and analyze signals, a DPO employs a parallel-processing architecture to perform these functions, as shown in Figure 6-4.

The DPO architecture dedicates unique ASIC hardware to the acquisition of waveform images, delivering high waveform capture rates that result in a higher level of signal visualization. This performance increases the probability of wit-nessing transient events that occur in digital systems, such as runt pulses, glitches, and transition errors. A description of this parallel-processing architecture fol-lows.

A DPO's input stage is similar to that of an analog oscilloscope—a vertical amplifier—and its second stage is similar to that of a DSO—an ADC. However, the DPO differs significantly from its predecessors following the analog-to-digital conversion process.

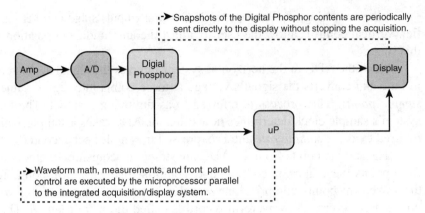

Figure 6-4   The parallel-processing architecture of a DPO.

For any oscilloscope—analog, DSO, or DPO—there is always a hold-off time during which the instrument processes the most recently acquired data, resets the system, and waits for the next trigger event. During this time, the oscilloscope is blind to all signal activity. The probability of seeing an infrequent or low-repetition event decreases as the hold-off time increases.

It should be noted that it is impossible to determine the probability of capture by simply looking at the display update rate. If a user relies solely on the update rate, it is easy to make the mistake of believing that the oscilloscope is capturing all pertinent information about the waveform when, in fact, it is not.

Because the traditional DSO processes captured waveforms serially, the speed of its microprocessor imposes a bottleneck because it limits the waveform capture rate. The DPO overcomes this limitation by rasterizing the digitized waveform data into a digital phosphor database. Every 1/30th of a second—about as fast as the human eye can perceive it—the content of this database is "pipelined" directly to the display system. This direct rasterization of waveform data, and the direct copying to the display memory from the database, removes the data-processing bottleneck inherent in other architectures. The result is an enhanced "lively" display. Signal details, intermittent events, and dynamic characteristics of the signal are captured in real time. The DPO's microprocessor works in parallel with an integrated acquisition system for display management, measurement automation, and instrument control, so that it does not affect the oscilloscope's acquisition speed.

A DPO faithfully emulates the best display attributes of an analog oscilloscope, displaying the signal in three dimensions: time, amplitude, and the distribution of amplitude over time, all in real time. The continuously updated digital phosphor database has a separate "cell" of information for every pixel on the oscilloscope's display. Each time a waveform is captured—in other words, every time the oscilloscope triggers—it is mapped into the digital phosphor database's cells. Each cell that represents a screen location and is touched by the waveform is reinforced with intensity information, while other cells are not. As a result, intensity information builds up in cells where the waveform passes most often.

When the digital phosphor database is fed to the oscilloscope's display, the display reveals intensified waveform areas, in proportion to the signal's frequency of occurrence at each point. The DPO also allows the display of the varying frequency-of-occurrence information on the display as contrasting colors. Hence it is easy to see the difference between a waveform that occurs on almost every trigger and one that occurs, say, every 100th trigger.

DPOs are equally suitable for viewing high and low frequencies, repetitive waveforms, transients, and signal variations in real time. Their capability to provide the additional intensity axis in real time also makes them ideally suited to communication mask testing, digital debug of intermittent signals, repetitive digital design, and timing applications.

6.3.1.3   Digital Sampling Oscilloscopes

When measuring high-frequency signals, a standard oscilloscope may not be capable of collecting enough samples in one sweep to reproduce the signal adequately. A digital sampling oscilloscope is an ideal tool for accurately capturing signals whose frequency components are much higher than the oscilloscope's sample rate. This class of oscilloscope is capable of measuring signals up to an order of magnitude faster than any other oscilloscope. It can achieve bandwidth and high-speed timing ten times higher than other oscilloscopes for repetitive signals. For example, sequential equivalent-time sampling oscilloscopes are available with bandwidths of up to 100 GHz.

In contrast to the digital storage and digital phosphor oscilloscope architectures, the architecture of the digital sampling oscilloscope reverses the position of the attenuator/amplifier and the sampling bridge, as shown in Figure 6-5.

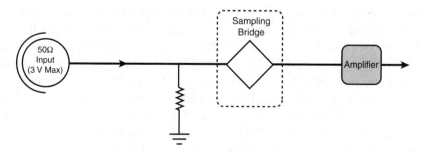

Figure 6-5   The architecture of a digital sampling oscilloscope.

The input signal is sampled before any attenuation or amplification is performed. A low bandwidth amplifier can then be utilized after the sampling bridge because the signal has already been converted to a lower frequency by the sampling gate, resulting in a much higher bandwidth instrument.

The trade-off for this high bandwidth, however, is that the sampling oscilloscope's dynamic range is limited. Since no attenuator/amplifier is in front of the sampling gate, there is no facility to scale the input. The sampling bridge must be capable of handling the full dynamic range of the input at all times. Therefore, the dynamic range of most sampling oscilloscopes is limited to about 1 V peak-to-peak. Digital storage and digital phosphor oscilloscopes, on the other hand, can handle hundreds of volts when used with the appropriate probe. In addition, protection diodes cannot be placed in front of the sampling bridge because this would limit the bandwidth. This reduces the safe input voltage for a sampling oscilloscope to about 3 V, as compared to 500 V available on other oscilloscopes.

6.3.2   Which Oscilloscope Should You Use?

In the context of signal integrity testing, the DSO is ideal for low- or high-repetition rate signals with fast edges or narrow pulse widths. The DSO also excels at capturing single-shot events and transients, and it is the best solution for multichannel acquisition at high bandwidths and sample rates.

The DPO, on the other hand, is the right tool for digital troubleshooting, for finding intermittent signals, and for many types of eye diagram and mask testing. The DPO's extraordinary waveform capture rate overlays sweep after sweep of information more quickly than other types of oscilloscope, presenting frequency-of-occurrence details—generally differentiated on the display in color—with unmatched clarity.

One important point to consider in any discussion of signal integrity testing, particularly where high-speed serial data is concerned, is that of bandwidth. Many of the key measurement processes—in particular the use of eye diagrams, discussed in detail in Chapter 8, "Signal Analysis and Compliance"—are critically dependent on bandwidth. In fact, some standards groups such as the Peripheral Component Interface, Special Interest Group (PCI SIG) now specify the bandwidth that must be used in characterization tests. As a result, instrument manufacturers are constantly striving for ever-higher bandwidths using techniques such as digital signal processing (DSP) to enhance performance. At the time of writing, a 20 GHz bandwidth oscilloscope for DPOs and 50 GS/s sampling for sampling oscilloscopes is the state of the art. The application of these high-performance serial data analyzers is discussed in more detail in the consideration of compliance testing in Chapter 8.

6.4   VERIFICATION TOOLS: LOGIC ANALYZERS

The logic analyzer is an indispensable tool for digital signal validation and troubleshooting. Equipped with accessories ranging from probes and fixturing to bus support packages and software disassemblers, the logic analyzer captures information on a bus and displays both timing diagrams and state information. It can trigger on a particular data word or an error such as a setup and hold violation. Alternatively, it can receive an external trigger and record synchronized cycle-by-cycle state data.

6.4.1   Data Capture

A logic analyzer captures data across many channels, detecting logic threshold levels and displaying them as "1" or "0" values (see Figure 6-6).

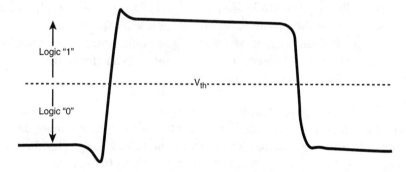

Figure 6-6   A logic analyzer determines logic values relative to a threshold voltage level.

A logic analyzer's waveform timing display is similar to that of a timing diagram found in a datasheet or produced by a simulator. All the signals are time-correlated so that setup and hold times, pulse width, and extraneous or missing data can be viewed. In addition to their high channel count, logic analyzers offer important features that support digital design verification and debugging. Among these are

- Sophisticated triggering that allows the user to specify the conditions under which the logic analyzer acquires data
- High-density probes and adapters that allow connection to many channels within a relatively small area
- Analysis capabilities that translate captured data into processor instructions and correlate it to source code

Logic analyzers are designed to capture data from multipin devices and buses. There are two types of data acquisition, or clock modes:

- *Timing acquisition* captures signal timing information. In this mode, a clock internal to the logic analyzer is used to sample data. The faster that data is sampled, the higher the resolution of the measurement will be. No fixed timing relationship exists between the target device and the data acquired by the logic analyzer. This acquisition mode is primarily used when the timing relationship between signals is of primary importance.
- *State acquisition* is used to acquire the "state" of the system under test. A signal (or signals) from the design is used to define the sample point (when and how often data will be acquired). The signal used to clock the acquisition may be the system clock, a control signal on the bus, or a signal that causes the system to change states. A complex clocking scheme can be used, typically using multiple signals to generate a master sample point. Data is sampled on the active edge, and it represents the condition of the system when the logic signals are stable. The logic analyzer samples when, and only when, the chosen signals are valid. What transpires between clock events is not of interest here.

Which type of acquisition is used is determined by how the user wants to look at the data. For capturing a long, contiguous record of timing details, timing acquisition using the internal (or asynchronous) clock is right for the job. Alternatively, it may be necessary to acquire data exactly as the system under test sees it, in which case state (synchronous) acquisition is the appropriate choice.

With state acquisition, each successive state of the system is displayed sequentially in a "listing" window. The external clock signal used for state acquisition may be any relevant signal.

6.4.2 Triggering

Unlike oscilloscopes, whose triggering capabilities respond to binary conditions, logic analyzers have powerful triggering that allows a variety of logical (Boolean) conditions to be evaluated to determine when the instrument triggers. The purpose of the trigger is to select which data is captured by the logic analyzer by tracking logic states in the system under test and triggering when a user-defined event occurs.

The event may be a simple transition, intentional or otherwise, on a single signal line. It may be a glitch, the moment when a particular signal such as "increment" or "enable" becomes valid, or it may be the defined logical condition that results from a combination of signal transitions across an entire bus. In all cases, however, the event is something that appears when signals change from one cycle to the next.

Many conditions can be used to trigger a logic analyzer. For example, the analyzer can recognize a specific binary value on a bus or counter output. Other triggering choices include

- **Words:** Specific logic patterns defined in binary, hexadecimal, and so on
- **Ranges:** Events that occur between a low and high value
- **Counter:** The user-programmed number of events tracked by a counter
- **Signal:** An external signal such as a system reset
- **Glitches:** Pulses that occur between acquisitions
- **Timer:** The elapsed time between two events or the duration of a single event, tracked by a timer
- **Analog:** Using an oscilloscope to trigger on an analog characteristic and to cross-trigger the logic analyzer

With all these trigger conditions available, it is possible to track down system errors using a broad search for state failures and then refine the search with increasingly explicit triggering conditions.

During hardware and software debug (system integration), it is helpful to have correlated state and timing information. A problem may initially be detected as an invalid state on the bus. This may be caused by a problem such as a setup and hold timing violation. If the logic analyzer cannot capture both timing and state data simultaneously, isolating the problem becomes difficult and time-consuming.

Some logic analyzers require a separate timing probe and separate acquisition hardware to acquire the timing information. One probe connects the system under test to a timing module, while a second probe connects the same test points to a state module: a technique known as *double probing*. However, this arrangement can compromise the impedance environment of the signals. Using two probes at once will "load down" the signal, degrading the system's rise and fall times, amplitude, and noise performance.

It is best to acquire timing and state data simultaneously, through the same probe at the same time. One connection, one setup, and one acquisition provide both timing and state data. This simplifies the mechanical connection of the probes and reduces the risk of the problems mentioned previously. With simultaneous timing and state acquisition, the logic analyzer captures all the information needed to support both timing and state analysis. There is no second step, and therefore less chance of errors and mechanical damage that can occur with double probing. The single probe's effect on the circuit is lower, ensuring more accurate

measurements and less impact on the circuit's operation. The higher timing resolution with this approach means that the user can see and trigger on more details,
increasing the chances of finding any problems.

6.4.3   Real-Time Acquisition Memory

The logic analyzer's probing, triggering, and clocking systems exist to deliver
data to the real-time acquisition memory. This memory is the heart of the instrument—the destination for all the sampled data and the source for all the instrument's analysis and display functions.

Logic analyzers have memory capable of storing data at the instrument's
sample rate. The memory can be envisioned as a matrix having channel width and
record length, as shown in Figure 6-7.

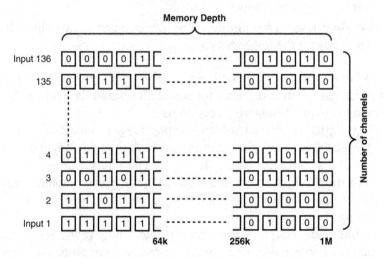

Figure 6-7   The logic analyzer stores acquisition data in deep memory with one full-depth
channel supporting each digital input.

The instrument accumulates a record of all signal activity until a trigger
event or until the user tells it to stop. The result is an acquisition—essentially a
multichannel waveform display that lets the user view the interaction of all the
acquired signals with a high degree of timing precision.

Logic analyzers continuously sample data, filling up the real-time acquisition memory, and discarding the overflow on a first-in/first-out (FIFO) basis.
There is thus a constant flow of real-time data through the memory. When the
trigger event occurs, the "halt" process begins, preserving the data in the memory.
The placement of the trigger in the memory is flexible, allowing the user to capture and examine events that occurred before, after, and around the trigger event.
This is a valuable troubleshooting feature since the logic analyzer can be set up to

store data preceding the trigger and capture the fault that caused the symptom. The analyzer can also be set up to store a certain amount of data after the trigger to see what subsequent effects the error might have caused.

6.4.4   Analysis and Display

The data stored in the real-time acquisition memory can be used in a variety of display and analysis modes. Once the information is stored within the system, it can be viewed in formats ranging from timing waveforms to instruction mnemonics correlated to source code.

6.4.4.1   Waveform Display

The waveform display is a multichannel detailed view that shows the time relationship of all the captured signals, much like the display of an oscilloscope (see Figure 6-8), which shows a D-type flip-flop (FF) with clock (CLK) and data (D) inputs, and an output (Q) that are all asynchronously sampled by the internal logic analyzer clock.

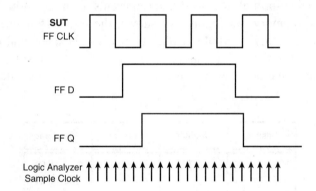

Figure 6-8   Logic analyzer waveform display (simplified).

The waveform display is commonly used in timing analysis, and it is ideal for diagnosing timing problems in hardware or verifying correct hardware operation by comparing the recorded results with simulator output or datasheet timing diagrams. It is also invaluable for measuring hardware timing-related characteristics, such as race conditions, propagation delays, and the absence or presence of pulses, and for analyzing glitches. At the time of writing, timing sample rates are up to 50 GHz (20 ps resolution) on each channel.

6.4.4.2   Listing Display

The listing display provides state information in user-selectable alphanumeric form. The data values in the listing are developed from samples captured from an entire bus and can be represented in hexadecimal or other formats.

Imagine taking a vertical "slice" through all the waveforms on a data bus, as shown in Figure 6-9, where a state acquition captures data bus lines D0-D3 when the external (device under test) clock signal enables an acquisition.

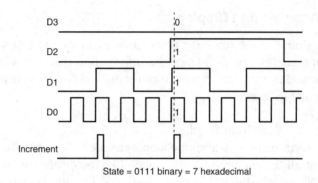

State = 0111 binary = 7 hexadecimal

Figure 6-9   State acquisition captures a "slice" of data across a bus.

The slice through the 4-bit bus represents a sample stored in the real-time acquisition memory. As Figure 6-9 shows, the numbers in the dotted slice are what the logic analyzer would display, typically in hexadecimal form.

The listing display in Figure 6-10 shows the state of the system under test with the information flow shown exactly as the system sees it: a stream of data words.

Sample	Counter	Counter	Timestamp
0	0111	7	0 ps
1	1111	F	114.000 ns
2	0000	0	228.000 ns
3	1000	8	342.000 ns
4	0100	4	457.000 ns
5	1100	C	570.500 ns
6	0010	2	685.000 ns
7	1010	A	799.000 ns

Figure 6-10   Listing display.

State data is displayed in several formats. The real-time instruction trace disassembles every bus transaction and determines exactly which instructions were read across the bus. It places the appropriate instruction mnemonic, along with its associated address, on the logic analyzer display (see Figure 6-11).

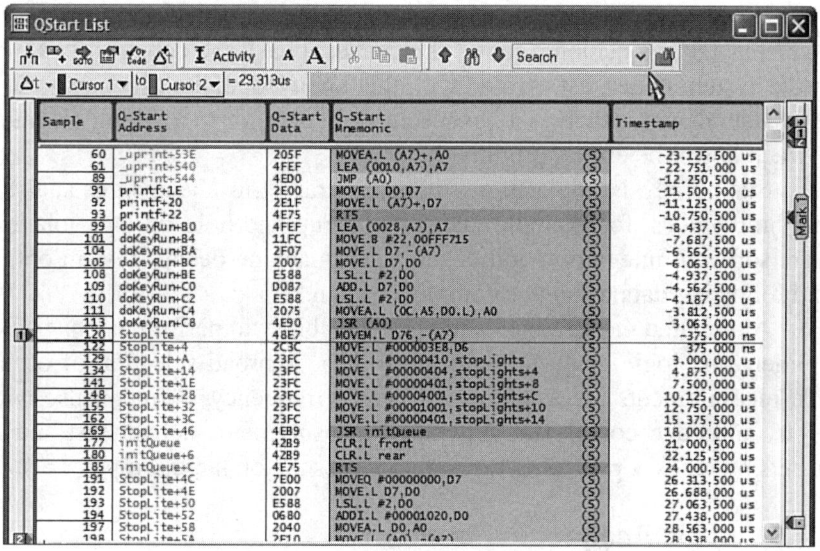

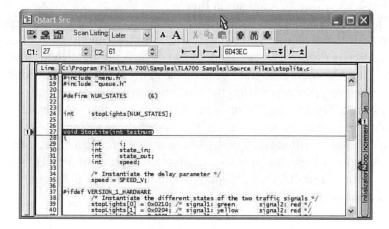

Figure 6-11   Real-time instruction trace display.

An additional display, the source code debug display, makes debug work more productive by correlating the source code to the instruction trace history. It provides instant visibility of what is actually going on when an instruction executes. Figure 6-12 shows a source code display correlated to the real-time instruction trace shown in Figure 6-11 where line 27 in this display is correlated with sample 120 in the instruction trace display of Figure 6-11.

Figure 6-12   Source code display.

With the aid of processor-specific support packages, state analysis data can be displayed in mnemonic form. This makes it easier to debug software problems in the system under test. Armed with this knowledge, the user can go to a lower-level state display (such as a hexadecimal display) or to a timing diagram display to track down the error's origin.

State analysis applications include parametric and margin analysis (setup and hold values, for example), detecting setup and hold timing violations, hardware/software integration and debug, state machine debug, system optimization, and following data through a complete design.

Automated measurements provide the ability to perform sophisticated measurements on logic analyzer acquisition data. A broad selection of oscilloscope-like measurements are available, including frequency, period, pulse width, duty cycle, and edge count. The automated measurements deliver fast and thorough results by quickly providing measurement results on large sample sizes.

6.4.5   Logic Analyzer Performance

The logic analyzer has a number of quantitative indicators of its performance and effectiveness, with several of these related to its sample rate. This is the measurement frequency axis that is analogous to the bandwidth of a digital storage oscilloscope. Certain probing and triggering terms will also be familiar to the DSO user, but many attributes are unique to the logic analyzer's digital domain. Because the logic analyzer is not attempting to capture and reconstruct an analog signal, issues such as channel count and synchronization (clock) modes are critical while analog factors such as vertical accuracy are secondary.

6.4.5.1   Timing Acquisition Rate

The logic analyzer's most basic mission is to produce a timing diagram based on the data it has acquired. If the device under test is functioning correctly and the acquisition is properly set up, the logic analyzer's timing display will be virtually identical to the timing diagram from the design simulator or data book.

However, this depends on the resolution of the logic analyzer—in effect, its sample rate. Timing acquisition is asynchronous, in that the sample clock is free-running relative to the input signal. The higher the sample rate, the more likely it is that a sample will accurately detect the timing of an event such as a transition. For example, a logic analyzer with a sample frequency of 50 GHz would have 20 ps resolution, and the timing display would reflect edge placements within 20 ps of the actual edge in the worst case.

6.4.5.2   State Acquisition Rate

State acquisition is synchronous. It depends on an external trigger from the device under test to clock the acquisitions. State acquisition is designed to assist engineers in tracing the data flow and program execution of processors and buses.

A typical logic analyzer may offer a state acquisition frequency of 450 MHz, with a setup/hold window of 625 ps across all channels to ensure accurate data capture.

Note that this frequency is relevant to the bus and I/O transactions that the logic analyzer will monitor, not the internal clock rate of the device under test. Though the device's internal rate may be in the multigigahertz range, its communication with buses and other devices is of the same order as the logic analyzer's state acquisition frequency.

6.4.5.3   Record Length

Record length is another key logic analyzer specification. A logic analyzer capable of storing more "time" in the form of sampled data is useful because the symptom that triggers an acquisition may occur well after its cause. With a longer record length, it is often possible to capture and view both, greatly simplifying the troubleshooting process.

Some logic analyzers can be configured with various record lengths. It is also possible to concatenate the memory from up to four channels to quadruple the available depth. This provides a means to build massive record lengths (up to 256 Mbit) when needed, or to get the performance of a long record length from a smaller, lower-cost configuration.

6.4.5.4   Channel Count and Modularity

The logic analyzer's channel count is the basis of its support for wide buses and/or multiple test points throughout a system. Channel count also is important when reconfiguring the instrument's record length: Two or four channels are required to double or quadruple the record length, respectively.

With today's trend toward high-speed serial buses, the channel count issue is as critical as ever. A 32-bit serial data packet, for example, must be distributed to not one, but 32 logic analyzer channels. In other words, the transition from parallel to serial architectures has not affected the need for channel count. Today's modular logic analyzers can accommodate a variety of acquisition modules and can be connected together for very high channel counts: ultimately up to thousands of acquisition channels.

6.4.5.5   Triggering

Triggering flexibility is the key to the fast and efficient detection of unseen problems. In a logic analyzer, triggering is concerned with setting conditions that, when met, will capture the acquisition and display the result. The fact that the acquisition has stopped is proof that the condition occurred (unless a timeout exception is specified).

Today, triggering setup is simplified by drag-and-drop operation for easier setup of common trigger types. These triggers spare the user from the need to devise elaborate trigger configurations for everyday timing problems. As the application examples later in this chapter demonstrate, logic analyzers also allow powerful specialization of these triggers to address more complex problems.

Logic analyzers also provide multiple trigger states, word recognizers, edge/transition recognizers, range recognizers, timer/counters, and a snapshot recognizer in addition to the glitch and setup/hold triggers.

6.4.6 Logic Analyzer Measurement Examples

The following series of examples illustrate several common measurement problems and their solutions. The explanations are simplified to focus on some basic logic analyzer acquisition techniques and the display of the resulting data.

6.4.6.1 General-Purpose Timing Measurements

Ensuring the proper timing relationships between critical signals in a digital system is an essential step in the validation process. A wide range of timing parameters and signals must be evaluated: propagation delay, pulse width, setup and hold characteristics, signal skew, and more.

Efficient timing measurements can be carried out with a logic analyzer having high-resolution acquisition across numerous channels with minimal loading on the circuit being measured. The analyzer must have flexible triggering capabilities that help the designer to quickly locate problems by defining explicit trigger conditions. In addition, it must provide display and analysis capabilities that simplify the interpretation of long records.

Timing measurements are commonly required when validating a new digital design. The example shown in Figure 6-13 illustrates a timing measurement on a D-type flip. In the real world, a timing measurement might simultaneously acquire hundreds or even thousands of signals. But the principle is the same in either case, and as the example proves, timing measurements are fast, easy, and accurate.

To initiate a timing measurement, it is necessary to first set up the triggering and clocking. This example uses the "if anything, then trigger" setup and internal (asynchronous) clocking. After executing a run operation to acquire the signal data, the horizontal position control or the memory scroll bar is used to position the on-screen data such that the trigger indicator is in view.

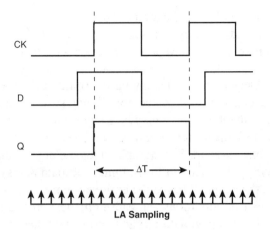

Figure 6-13   D-type flip-flop example of sample rate in relation to resolution.

Measurement cursors are then used to define the beginning and edge of the signal of interest, and, since the Y axis of the display denotes time, the subtractive difference between Cursor 2 and Cursor 1 is the time measurement. The result of 52 ns appears in the "delta time" readout on the display shown in Figure 6-14.

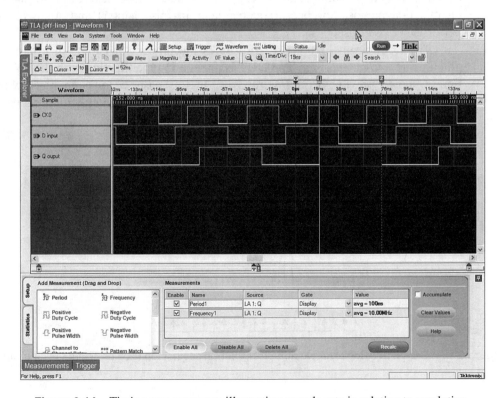

Figure 6-14   Timing measurement, illustrating sample rate in relation to resolution.

The resolution of the measurement depends on the sample rate; in Figure 6-14 it is 2 ns, as indicated by the ticks on the sample track. Note that the "delta time" measurement cannot have a resolution greater than the sample rate.

6.4.6.2   Detecting and Displaying Intermittent Glitches

Glitches are a constant annoyance for digital system designers. These erratic pulses are intermittent, and they may be irregular in amplitude and duration. They are inevitably difficult to detect and capture, yet the effects of an unpredictable glitch can disable a system. For example, a logic element can easily misinterpret a glitch as a clock pulse. This in turn might send data across the bus prematurely, creating errors that ripple through the entire system.

Any number of conditions can cause glitches: crosstalk, inductive coupling, race conditions, timing violations, and more. Glitches can elude conventional logic analyzer timing measurements simply because they are so brief in duration. A glitch can easily appear and then vanish in the time between two logic analyzer acquisitions.

A logic analyzer with very high timing resolution (that is, a high clock frequency when running in its asynchronous mode) is best suited to capture these brief events. Ideally, the logic analyzer automatically highlights the glitch and the channel. Recently, similar glitch capture facilities have also been available on mixed-signal oscilloscopes (covered later in this chapter) and some high-performance oscilloscopes.

The example in Figure 6-15 illustrates the process of capturing a narrow glitch using a logic analyzer. The DUT is again a D-type flip-flop with the signal timing shown previously in Figure 6-14. In this case, a logic analyzer with enhanced timing resolution is used to detect and display the glitch with great precision.

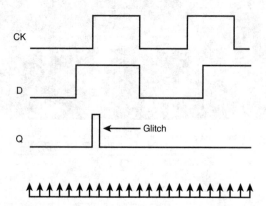

Figure 6-15   D-type flip-flop example of high-resolution acquisition (125 ps Timing resolution).

This operation relies on the use of a "glitch trigger" option, which captures glitches and displays them in the waveform window.

The acquisition is shown in Figure 6-16. This screen includes several channels that have been added (by means of a separate setup step that does not require a second acquisition) to display the contents of the high-resolution acquisition.

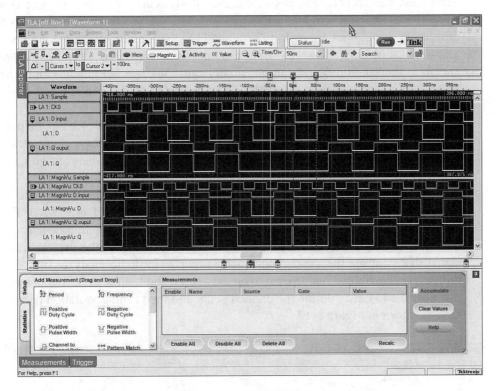

Figure 6-16   Glitch trigger with high-resolution acquisition.

On the Q output waveform trace, the flag to the left of (earlier than) the trigger indicator announces that a glitch has been detected somewhere in the area between the trigger sample point and the immediately previous data sample point (normally highlighted in a different color). The Q output's high-resolution channel (bottom trace) reveals exactly where the glitch occurred. At this point, the timing of the glitch is known, and the instrument's zoom and cursor features can be used to measure the pulse width.

6.4.6.3   Capturing Setup or Hold Violations

Setup time is defined as the minimum time that input data must be valid and stable prior to the clock edge. *Hold time* is the minimum time that the data must be valid and stable after the clock edge occurs.

Digital device manufacturers specify setup and hold parameters, and engineers must take great care to ensure that their designs do not violate the specifications. However, today's tighter tolerances and the widespread use of faster parts to drive more throughput are making setup and hold violations ever more common. These violations can cause the device output to become unstable (a condition known as metastability) and can potentially cause unexpected glitches and other errors. Designers need to examine their circuits closely to determine whether violations of the design rules are causing setup and hold problems.

In recent years, both setup and hold requirements have narrowed to the point where it is difficult for most conventional general-purpose logic analyzers to detect and capture the events. The only real answer is a logic analyzer with subnanosecond sampling resolution.

In this context, synchronous acquisition mode, which relies on an external clock signal to drive the sampling, is used. Once again, the DUT is a D-type flip-flop with a single output, but the example is equally applicable to a device with hundreds of outputs. In this example, the DUT itself provides the external clock signal that controls the synchronous acquisitions. The logic analyzer drag-and-drop trigger capability can be used to create a setup and hold trigger. In this mode, it is possible to easily define the explicit setup and hold timing violation parameters, as shown in Figure 6-17.

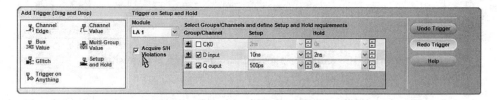

Figure 6-17   Setup and Hold event display.

Additional submenus in the setup window are available to refine other aspects of the signal definition, including logic conditions and positive- or negative-going terms.

When the test runs, the logic analyzer actually evaluates every rising edge of the clock for a setup or hold violation. It monitors millions of events and captures only those that fail the setup or hold requirements. Figure 6-18 shows the resulting display.

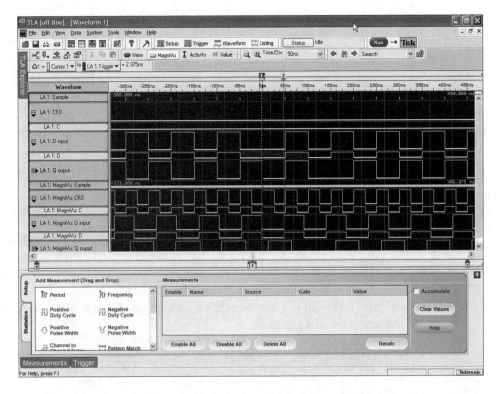

Figure 6-18   Resulting display shows setup and hold timing.

Here the setup time is 2.375 ns, far less than the defined limit of 10 ns.

6.4.6.4   Maximizing Usable Record Length

Sometimes the device under test puts out a signal that consists of occasional clusters of events separated by long intervals of inactivity. For example, certain types of radar systems drive their internal D/A converters with widely separated bursts of data.

This is a problem when using conventional logic analyzer acquisition and storage techniques. The instrument uses one memory location for every sample interval, a method appropriately named "store all." This can quickly fill the acquisition memory with unchanging data, consuming valuable capacity needed to capture the actual data of interest—the bursts of the active signal.

An approach known as *transitional storage* solves the problem by storing data only when transitions occur.

Figure 6-19 depicts the concept. The logic analyzer samples when, and only when, the data changes. Bursts that are seconds, minutes, hours, or even days apart can be captured with the full resolution of the logic analyzer's main sample memory. In between, the instrument "waits out" the long dormant periods. Note that these long spans of inactivity are not "ignored." On the contrary, they are constantly monitored. But they are not recorded.

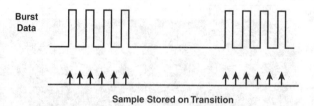

Figure 6-19   Transitional storage technique stores data only when transitions occur.

In this instance, the versatile "if/then" triggering algorithm is the best tool for distinguishing the unique circumstances that prompt the transitional storage.

The logic analyzer is set to select "transitional" rather than "all" events, and the "if channel burst = high, then trigger" mode is selected.

Running the test with these conditions specified produces a screen display similar to the one shown in Figure 6-20.

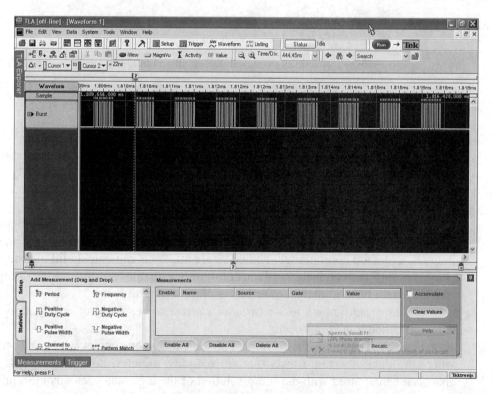

Figure 6-20   Display demonstrating transitional storage technique.

Here, the burst display contains nine groups of eight pulses, 22 ns in width, with the groups separated by quiescent intervals of 428 ns. Transitional storage has allowed the instrument to capture all 16 of these burst groups, including the seven remaining offscreen, only consuming a record length of 256 memory words. The time window represents almost 3.8 ms of acquisition time, where the groups repeat every 2 ms. In contrast, the "store all" acquisition mode would capture only one of the burst groups using 2,000 times the 256 memory space requires 512K memory words. The allocated memory would fill up in about 1 μs, with much of the space being occupied by "blank" inactive cycles. Transitional storage makes it possible to gather much more usable information every time an acquisition is run.

6.4.6.5   Field Programmable Gate Arrays (FPGAs)

The phenomenal growth in design size and complexity makes the process of design verification a critical bottleneck for today's FPGA systems. Limited access to internal signals, advanced FPGA packages, and printed circuit board (PCB) electrical noise all contribute to making FPGA debugging and verification the most difficult process of the design cycle. A designer can easily spend more than 50% of the design cycle time on debugging and verifying the design.

One of the key choices that needs to be made in the design phase is deciding which FPGA debug methodology to use. Ideally, the methodology should be portable to all the users' FPGA designs, provide insight into both FPGA operation and system operation, and provide the ability to pinpoint and analyze difficult problems.

There are really two basic in-circuit FPGA debug methodologies: The first is the use of an embedded logic analyzer, and the second is the use of an external logic analyzer. The choice of which methodology to use depends on the debug needs of the project.

Each of the FPGA vendors offers an embedded logic analyzer core. These intellectual property blocks are inserted into the FPGA design and provide both triggering capability and storage capability. It is important to note that FPGA logic resources are used to implement the trigger circuit and FPGA memory blocks are used to implement the storage capability. Joint Test Action Group (JTAG) is the name used for the standard boundary scan architecture for test access parts that is typically used to configure the operation of the core and is then used to pass the captured data to a personal computer (PC) for viewing. Because the embedded logic analyzer uses internal FPGA resources, this approach is most often used with larger FPGAs that can better absorb the overhead of the core. As with any debug methodology, the embedded logic analyzer has a number of trade-offs, as listed in Table 6-1.

Table 6-1   Trade-offs with Embedded Logic Analyzers

Advantages	Disadvantages
Fewer pins required	Size of core limits use to large FPGAs.
Simple probing	Internal memory must be given up.
Relatively inexpensive	State mode analysis only.
	Limited speed.
	No correlation between FPGA trace data and other system traces.

Because of the limitations of the embedded logic analyzer methodology, many FPGA designers have adopted a methodology that uses the flexibility of the FPGA and the power of an external logic analyzer. In this methodology, internal signals of interest are routed to pins of the FPGA, which are then connected to an external logic analyzer. This approach offers deep memory, which is useful when debugging problems where the symptom and the actual cause are separated by a large amount of time. It also offers the ability to correlate the internal FPGA signals with other activity in the system. As with the embedded logic analyzer methodology, there are trade-offs to consider (see Table 6-2).

Table 6-2   Trade-offs with External Logic Analyzers

Advantages	Disadvantages
Uses few, if any, FPGA logic resources	Requires more pins on FPGA.
Uses no FPGA memory	Moving probe points can require a recompile of the design.
Operates in both State and Timing modes	Requires manual update of signal names on LA.
Correlation between FPGA signals and other system signals	

Both methodologies can be useful, depending on the situation. The challenge is to determine which approach is appropriate for a particular design. For example, if the anticipated problems can be isolated to functional problems within the FPGA, then the use of an embedded logic analyzer may be all the debug capability that is required. If, however, the designer anticipates larger debug problems that may require timing margins to be verified, internal FPGA activity to be correlated with other activity on your board, or more powerful triggering capability to isolate the problem, the use of an external logic analyzer is more appropriate.

In essence, the external logic analyzer approach makes use of the "P" in FPGA to reprogram the device as needed to route the internal signals of interest to what is typically a small number of pins. This is a useful approach, but it does have limitations. Every time you need to look at a different set of internal signals, it may be necessary to change the design (either at the RTL-level or by using an FPGA editor tool) to route the desired set of signals to the debug pins. This in itself is time-consuming, but if it requires a recompile of the design, it will take even more time and can potentially hide the problem by changing the timing of the design.

There are typically a small number of debug pins, and the 1:1 relationship between internal signals and debug pins limits visibility and insight into the design. To overcome these limitations, FPGA debug software tools have been developed that deliver all the advantages of the external logic analyzer approach while removing its primary limitations. This software, when combined with a high-performance logic analyzer, allows the designer to see inside the FPGA design and correlate internal signals with external signals. Productivity is increased because the time-consuming process of recompiling the design is eliminated, and there is access to multiple internal signals per debug pin. In addition, this approach can handle multiple test cores in a single device: something that is useful when there is a need to monitor different clock domains inside the FPGA. It can also handle multiple FPGAs on a JTAG chain.

6.4.6.6   Memory

Dynamic random access memory (DRAM) has evolved over time, driven by faster, larger, and lower-powered memory requirements and smaller physical sizes. The first step was to synchronous dynamic RAM (SDRAM), which provided a clock edge to synchronize its operation with the memory controller. The data rate was then increased by using double data rate (DDR) RAM, and then—to overcome signal integrity issues—DDR2 SDRAM and DDR3 SDRAM evolved to go even faster.

To keep pace with the more complex and shorter design cycles, memory designers need a variety of test equipment to check out their designs. For looking at impedance and trace length, sampling oscilloscopes are used. To look at the electrical signals—from power to signal integrity to clocks, jitter, and so forth—digital phosphor oscilloscopes are the appropriate tool. For looking at commands, protocols, and sequencing, logic analyzers are necessary to verify the operation of the memory system.

6.5   COMBINING ANALOG AND DIGITAL MEASUREMENTS

So far, this chapter has looked separately at analog testing using oscilloscopes and digital debugging using logic analyzers. However, in some situations the interaction of analog and digital effects needs to be examined.

6.5.1   Mixed-Signal Oscilloscopes

One recent approach to this challenge is provided by the mixed-signal oscillo-scope, which, as its name implies, has the capability to capture and measure both analog and digital signals in a single instrument.

Although a modern digital oscilloscope can use up to four channels for probing serial data, several of the common serial protocols require three wires or more. Moreover, engineers often need the ability to decode and display multiple serial buses at the same time, and observe their timing correlation. The mixed-signal oscilloscope has been developed to address these challenges by combining the serial trigger and decode capabilities of the digital oscilloscope with additional digital channels (typically 16). As a result, engineers can probe and decode multiple serial buses along with custom parallel buses, all at the same time.

Figure 6-21 shows a screenshot from a mixed-signal oscilloscope carrying out simultaneous probing on an SPI bus (1), an I^2C bus (2), a 3-bit parallel bus (3), and an analog input. In this instance, the instrument is configured to take a single acquisition that triggers on particular I^2C activity. Setting the record length to 1M points guarantees that all the useful information around an event on the I^2C bus is accurately captured.

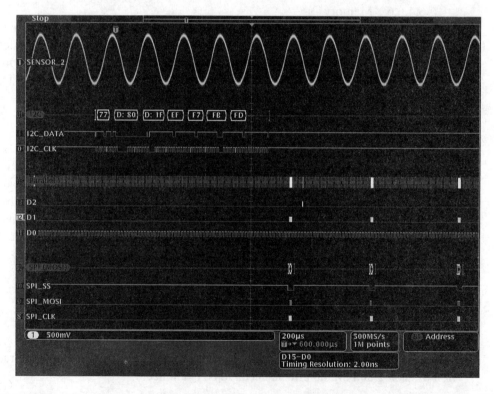

Figure 6-21   A mixed-signal oscilloscope display showing simultaneous acquisition of I^2C, SPI, and parallel buses with an analog signal on channel 1.

Figure 6-22 shows the same acquisition where a zoom facility is used to highlight the details of the SPI and parallel buses. In this particular instance, the ability to view and decode all the signals of interest aids the location of a software bug causing a signal malfunction.

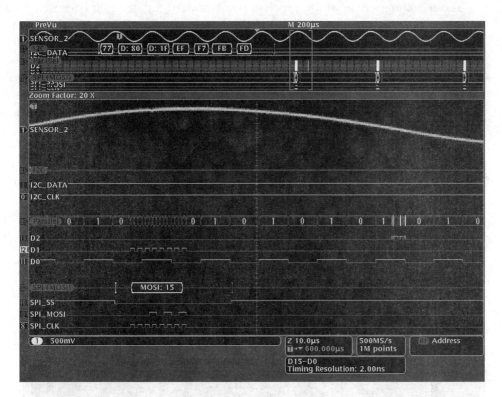

Figure 6-22   Zooming in on the SPI and parallel buses shows packet detail and highlights potential problems.

While the mixed-signal oscilloscope is a powerful tool for engineers developing and debugging embedded designs, it is important to understand that it is a hardware tool. In particular, the fact that it lacks external clocking means that, unlike a fully featured logic analyzer, it cannot be used for software analysis. The logic analyzer, of course, is totally dedicated to the digital environment, but, as this section shows, the oscilloscope still has a role to play in tracking down some types of problems.

Consider, for example, an application in which two separate buses—A3 (a 4-bit bus) and A2 (an 8-bit bus)—have to be verified. Initially the two buses seem to be transferring data just as the design models predicted, but after a few cycles of operation, errors begin to appear on the A3 bus. A value that should be "8" shows up as a "0"—not just once, but repeatedly. It is not a "stuck bit" or mis-routed signal, either of which would cause the same error to occur continuously. The erratic nature of the problem implies that some intermittent event is being mistaken for a legitimate data bit, altering the value of the hexadecimal results. The repeating nature of the problem points toward a glitch that is caused by an error in the layout or assembly of the prototype board. And the hexadecimal value, 8 instead of 0, implies a problem in the most significant bit of the A3 bus.

What are the circumstances under which the A3 bus error appears?

The true nature of the error begins to reveal itself after an acquisition with the logic analyzer's glitch capture trigger and display, which activates when it detects glitches and then flags their location in the timing display. The logic analyzer defines a glitch as an occurrence of more than one signal transition between sample points.

Figure 6-23 depicts the resulting display, which shows two types of waveforms. Each of the two buses, A2 and A3, is summarized with a bus waveform that reflects the word value on the respective bus. Bus waveforms provide an "at a glance" indication of the state of many individual signals, saving time when troubleshooting. In addition, the display can be configured to break out each individual signal line and again flag the glitch locations.

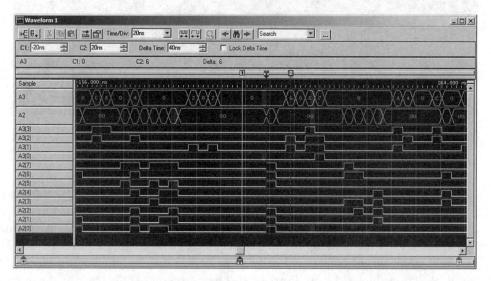

Figure 6-23   The logic analyzer has triggered on the glitch and flagged the individual glitch locations. (color version of figure available at www.informit.com/title/0131860062.)

In Figure 6-23, the period between clocks is 4.00 ns. The logic analyzer captures the signal values at 125 ps intervals, and this information can be displayed as a separate, high-resolution view. The high-resolution data is captured at the same time and through the same probe as the main timing data.

Figure 6-24 shows the display with the high-resolution acquisition traces added. Here, both the 125 ps clock ticks and the more detailed view of signal A3-3 are shown along with the bus waveform views. The signal shows a brief transition in the latter half of the cycle. Since it is already known that the cycle is producing an incorrect bus value, this transition is the likely cause of the error. An explanation of the cause of the invalid transition requires an additional examination of the analog characteristics of the signal, as described in the following section.

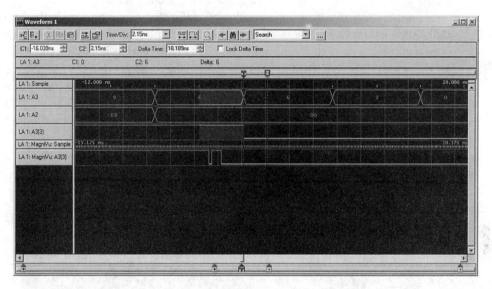

Figure 6-24   The logic analyzer's high-resolution acquisition display reveals a glitch in the A3(3) signal.

6.5.2   The Analog Perspective

Although logic analyzers and oscilloscopes have long been the tools of choice for digital troubleshooting, not every designer has seen the dramatic benefits that come with integrating these two key instruments. As we have seen, logic analyzers speed up debugging and verification by wading through the digital information stream to trigger on circuit faults and capture related events. Oscilloscopes, on the other hand, peer behind digital timing diagrams and show the raw analog waveforms, quickly revealing signal integrity problems.

To combine the virtues of both classes of instrument to create an integrated analog/digital debugging environment, new measurement tools have been created that incorporate analog multiplexing hardware with display and analysis software to allow logic analyzers and oscilloscopes to be used side by side. With this approach, the analog and digital signals are acquired together via a single logic analyzer probe, and a single time-correlated display shows the analog waveforms alongside digital logic signals. While the logic analyzer acquires and displays a signal in digital form, the attached oscilloscope captures the same signal in its analog form and displays it on the logic analyzer screen. Seeing these two views simultaneously makes it easy to see, for example, how a timing problem in the digital domain can be the result of a glitch in the analog realm.

Figure 6-25 illustrates the results of using this technique with the earlier application example and shows the display that results when analog signal A3(3) is aligned onscreen with its digital equivalent. This picture tells the whole story: At the exact moment of the digital glitch, the analog signal's amplitude is degraded in the area of the logic threshold. It apparently dips below the threshold voltage for an instant, creating a momentary "low," or logic "0" level. It then increases just enough to cross above the threshold and return to the "high" or logic "1" level before switching to logic "0" again at the cycle boundary.

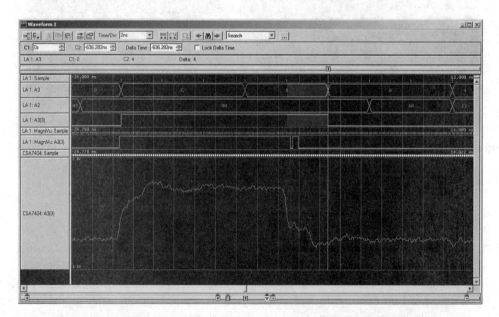

Figure 6-25   Integrated logic analyzer/oscilloscope display shows the analog behavior underlying the digital glitch on A3(3).

This analog behavior is the origin of the glitch and causes the error in the hexadecimal output on the bus. The instability is such that it does not affect every falling edge in the same way; many pulses pass without errors.

The result is that the designer has to review the design models to determine when the valid edge should occur: before or after the unstable portion of the waveform in this bus cycle.

The experienced engineer will recognize clues in this distorted waveform. A degraded logic level such as this is usually the result of a reflection coming back from an improperly terminated transmission line. In this particular example, the signal's fast edges have encountered a missing termination resistor at the signal's destination. The result is an erratic but damaging erosion of the falling edge.

To summarize, troubleshooting with the logic analyzer/oscilloscope combination is a matter of proceeding from a high-level, global view to a zoomed-in close-up of individual signals using the various signal formats available:

- The bus waveform gives an at-a-glance indication of problems occurring somewhere on the bus.
- The deep timing waveform reveals exactly which signal line is involved.
- The high-resolution timing waveform pinpoints the time placement of the error.
- The analog waveform, provided by the linked oscilloscope, captures the specific analog characteristics of the signal, revealing potential causes of the error.

6.6   EYE DIAGRAM ANALYSIS

Although the subject of eye diagrams is dealt with in more depth in the discussion of serial bus compliance testing in Chapter 8, it is worth pointing out that analog eye diagram analysis is also a valuable aid to troubleshooting using the integrated logic analyzer/oscilloscope combination.

The eye diagram is a visual tool for observing the "data valid" window and general signal integrity on clocked buses. It is a required compliance testing tool for many of today's buses, particularly serial types, but any signal line can be viewed as an eye diagram.

Using appropriate software tools, it is possible to speed troubleshooting by incorporating up to hundreds of eye diagrams into a single view that encompasses the leading and trailing edges of both positive-going and negative-going pulses.

Figure 6-26 depicts the result, showing how 12 signals (the entire signal content of the A3 and A2 buses) can be superimposed. The benefit of seeing all 12 at the same time is clear. Moreover, this approach is easily extended to 32-bit and 64-bit buses.

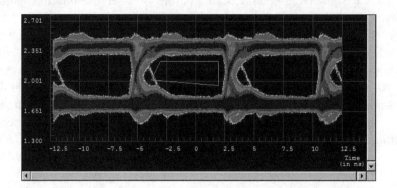

Figure 6-26   Using the eye diagram format to display multiple signals simultaneously.

Because an eye diagram presents all possible logic transitions in a single view, it can also provide a fast assessment of a signal's health. It reveals analog problems such as slow rise times, transients, attenuated levels, and more. Some engineers start their evaluation by looking first at the eye diagrams and then track down any aberrations.

The eye in Figure 6-26 reveals an anomaly in the signal: a thin line whose characteristics (normally represented by different colors on today's instruments) indicate a relatively infrequent transition. It proves that at least one of the 12 signals has an edge outside the normal range.

A mask feature helps to locate the specific signal that is causing the problem. By drawing the mask in such a way that the offending edge penetrates the mask area, the relevant signal can be isolated, highlighted, and brought to the front layer of the image. The result is shown in Figure 6-27, in which the flawed signal has been brought to the front and is highlighted in white.

In this example, the aberrant edge indicates a problem on the A3 (0) signal. The origin of the problem is crosstalk, with the edge change being induced by signals on an adjacent trace on the motherboard.

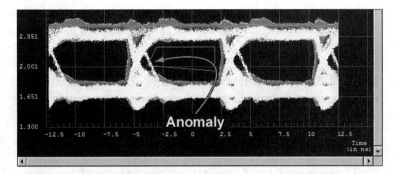

Figure 6-27   Using software analysis to bring an errant waveform to the front and highlight it for easy evaluation.

CONCLUSION

This chapter highlighted the need for expert, efficient solutions that can speed the process of validation and debugging.

The tool of choice for measuring the electrical portion of the physical layer is a high-performance real-time oscilloscope. State-of-the-art digital storage oscilloscopes and digital phosphor oscilloscopes can capture continuous, contiguous data in the multigigabit range with very good signal integrity. The real-time oscilloscope makes an ideal platform for physical layer electrical data validation, debugging, and compliance testing. These instruments are complemented by high-speed sampling oscilloscopes for looking at very high-speed analog signal integrity problems and for TDR signal path analysis.

The preferred tool for measuring the digital representation of the signal (as opposed to individual pulse characteristics) is the logic analyzer. Unlike both real-time and sampling oscilloscopes, logic analyzers capture binary data and express it in terms of clocks, cycles, and edge transitions. The purpose of the logic analyzer is to simplify acquisition and analysis of the purely digital aspects of the data. To carry out a serial bus debug mission, it must deliver features consistent with the needs of high-speed buses: high sample rate, deep memory, flexible triggering and synchronization, and more.

Examples of these classes of instruments are required at all stages of the design process to make the critical measurements needed to detect and solve signal integrity related problems:

- **Impedance measurements:** A sampling oscilloscope with a TDR module is used to check the impedances of PCB traces and connectors.
- **ASIC verification:** The sampling oscilloscope and a digital timing generator (signal source) are used to verify prototype ASICs and confirm that specifications such as setup and hold timing are within tolerances.
- **Basic functional verification:** The real-time oscilloscope and high-speed SMA probing tools are used to check clock signals for aberration-free performance, and to detect and analyze signal integrity issues such as glitches in the spread-spectrum clock.
- **Operational validation, fault detection, and debugging:** A logic analyzer is used to monitor functional tests to verify that logic operations proceed correctly. Signal integrity problems are detected using a real-time oscilloscope connected to the logic analyzer via a combined hardware and software toolset.

This chapter demonstrated the importance of being alert for signal integrity problems during the design process. Glitches, anomalies, and impairments can emerge at virtually any point along the way and must be eliminated before proceeding to the next step. The combination of state-of-the-art oscilloscopes and logic analyzers with automated acquisition and analysis tools provides the solution to these challenges.

7

Replicating Real-World Signals with Signal Sources

A core theme throughout signal integrity (SI) engineering is the behavioral analysis of a digital circuit in terms of its digital and analog properties, and notably how the behavior of a digital circuit is determined. Traditionally an electronic design engineer would use hardware to prototype his or her ideas and validate the design via a variety of test and measurement instruments. Today initial concepts and designs are invariably simulated where SI engineers naturally think in terms of computer simulation models. Moreover, computer simulations generate output signals and data that are presented in a variety of formats in response to an array of simulated inputs. However, modern high-performance embedded systems retain the need for prototyping, especially to meet the needs of stringent compliance tests and measurements. The fundamental method used to determine the real-time operation and characteristics of a device under test (DUT) is to externally control and observe the circuit. This chapter explores the instrumentation used to stimulate and control a DUT.

An acquisition instrument, usually an oscilloscope or logic analyzer, is probably the first thing that comes to mind when an SI engineer thinks about making electronic measurements and observing circuit behavior. Consequently, the majority of SI test and instrumentation documentation is given to the understanding of real-time measurement instruments and how they acquire data and signals, where there are a myriad of SI measurements and numerous signal analysis techniques. However, these tools can make a measurement only when they can acquire a signal of some kind, and there are many instances in which the DUT has no such signal of its own and requires an externally provided stimulus. This chapter is about the other half of the story—the signal source. This is exactly what its name implies—an externally provided signal that is used as a real-time stimulus for electronic measurements. Typically a signal source is used to govern real-time measurements that often form the core processes in SI debug and circuit validation.

7.1 OBSERVING AND CONTROLLING CIRCUIT BEHAVIOR

A simulated circuit model intrinsically produces an output in response to virtual excitations. A real-time measurement depends on some form of stimulus, whether it's generated by the DUT itself or is externally provided by a signal source, as shown in Figure 7-1.

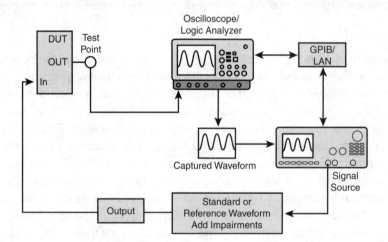

Figure 7-1   The excitation and observation of a device under test (DUT).

A generated signal may be a true bipolar alternating signal, with peaks oscillating above and below a ground reference point, or it may vary over a range of DC offset voltages, either positive or negative. It may be a sinusoidal wave or some other analog function, a digital pulse, a binary pattern, or a purely arbitrary wave shape. A generated signal source can provide "ideal" waveforms, or it may add known, repeatable amounts and types of distortion, or errors, to the signal it delivers. This characteristic is one of the principal functions of a signal source, since it is often impossible to create predictable distortion exactly when and where it's needed using only the circuit or DUT itself. The response of the DUT in the presence of these distorted signals can reveal the DUT's ability to handle stresses or fulfill regulatory compliance testing.

Today most signal sources are based on digital technology, and many fulfill both analog and digital signal generation requirements. For example, arbitrary waveform generators (AWGs) and function generators are aimed primarily at analog and mixed-signal applications. These instruments use digital sampling techniques to build and modify waveforms of almost any imaginable shape. Typically these generators have from one to four outputs. In some AWGs the main sampled analog outputs are supplemented by separate marker outputs, to aid triggering of external instruments, and synchronous digital outputs that present sample-by-sample data in digital form. Although modern signal sources often fulfill both analog and digital requirements, the most efficient signal sources are optimized for a particular digital or analog application, such as a digital waveform generator.

Digital waveform generators are sources of logic signals. They generally encompass two classes of instruments:

- Pulse generators drive a stream of square waves or pulses from a small number of outputs, usually at very high frequencies. These tools are most commonly used to exercise high-speed digital equipment.
- Pattern generators, also known as data generators or data timing generators, typically provide eight, sixteen, or even more synchronized digital pulse streams, such as a stimulus signal for a parallel computer bus or digital telecommunication system.

7.2   EXCITATION AND CONTROL

Exciting and controlling a circuit for test purposes is an exact science. Specialist texts document specific applications, such as the radio frequency (RF) characterization of a printed circuit board (PCB) trace. However, in terms of maintaining signal integrity in a mainstream digital system, or prototype device, the excitation or control typically falls into one of three basic categories: circuit verification, device characterization, and stress/margin testing.

7.2.1   Circuit Verification: Analyzing Digital Modulation

Consider the case where wireless equipment designers are developing new transmitter and receiver hardware. We can assume that they will need to simulate baseband I and Q signals both with and without impairments to verify conformance with emerging and proprietary wireless standards. Some high-performance arbitrary waveform generators can provide the needed low-distortion, high-resolution signals at rates of up to 1 gigabit per second (Gbps) with two independent channels—one for the I phase and one for the Q phase.

7.2.2   Device Characterization: Testing Data Converters

Newly developed digital-to-analog converters (DACs) and analog-to-digital converters (ADCs) must be exhaustively tested to determine their limits of linearity, monotonicity, and distortion whereby their transfer characteristics are derived. A state-of-the-art AWG can generate simultaneous in-phase analog and digital signals to drive such devices at speeds of up to 1 Gbps. Signal sources often use standard, user-created, or captured waveforms, adding impairments where necessary for special device characterization and performance measurements.

7.2.3   System Stress/Margin Testing: Stressing Communication Receivers

Engineers working with serial data stream architectures, which are commonly used in digital communications buses and disc drive amplifiers, need to stress their devices with impairments, particularly jitter and timing violations. Advanced signal sources can save the SI engineer untold hours of calculation by providing efficient, built-in jitter editing and generation tools. These instruments can shift critical signal edges as little as 300 Femto seconds.

A typical example of stress testing is in the implementation of a high-speed serial bus. A PCI Express high-speed serial bus device includes a transmitter (TX) section and a receiver (RX) section. PCI Express has a number of compliance tests whereby a particular specification recommends certain tests and equipment for design validation and manufacturing. Receiver testing requires a stimulus source to drive the DUT configured in a loopback mode. The proper tool for the job is a pattern generator (also called a data timing generator), which can generate defined multiple-Gbps test patterns. These test patterns are called training sequences and are currently defined as TS1 and TS2 for PCI Express. When the receiver section of the DUT recognizes a training sequence, the transmitter section responds by sending out a similar sequence. This transmitted sequence can be observed and analyzed with an oscilloscope and/or logic analyzer. It is common practice to alter the training sequence such that the DUT's performance can be characterized under various stressful conditions. Figure 7-2 shows a PCI Express signal with added pre-emphasis. Specific stresses often include amplitude level variations, differential skew variances, or added noise and jitter, which typically can be observed on an oscilloscope as eye diagram changes.

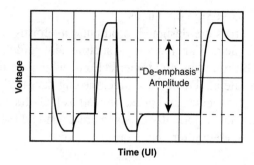

Figure 7-2   A PCI-Express test signal with added pre-emphasis.

7.3 SIGNAL-GENERATION TECHNIQUES

There are several ways a signal source can create pure waveforms and replicate real-world signals or stressed reference waveforms of the type shown in Figure 7-2. The chosen waveform and method of signal generation depend on the information available about the DUT and its input requirements, such as whether there is a need to add distortion or error signals, and other variables. Modern high-performance signal sources offer at least three ways to develop waveforms:

- Create "pure" signals for circuit stimulus and general testing.
- Replicate a real-world signal that was captured with an oscilloscope or logic analyzer.
- Generate ideal or stressed reference signals for industry-standard tests or compliance testing where the test signal has specific tolerances.

7.3.1 Arbitrary Waveform Generators and Function Generators

Arbitrary waveform generators and function generators generally are divided into mixed signal sources that primarily are analog signal sources, and logic sources that typically are pulse or pattern generators. These signal sources normally encompass the range of signal-producing instruments to provide the majority of waveforms required for signal integrity engineering. However, particular instruments have unique characteristics and assets that make them more or less suitable for specific signal integrity applications.

7.3.1.1   Mixed Signal Sources

Mixed signal sources are designed to output waveforms with analog characteristics. These may include analog waves such as sinusoidal, triangular, and square waves that exhibit rounding and imperfections, which are part of every real-world signal. In a versatile mixed signal source, the engineer typically can control waveform amplitude, frequency, and phase as well as DC offset and rise and fall times. Moreover, the engineer normally needs to create and add aberrations to the signal, such as overshoot, edge jitter, modulation, or noise.

7.3.1.2   Digital Signal Sources

True digital sources are meant to drive digital systems. Their outputs are binary pulse streams—a dedicated digital source cannot produce a sine or triangle wave. The features of a digital source are optimized for computer bus needs and similar applications. These features might include software tools to speed pattern development as well as hardware tools such as probes that are designed to match various logic families.

Although there appear to be two distinct types of instruments, today almost all high-performance signal sources, from function generators to arbitrary sources to pattern generators, are based on digital architectures. Digital signal source architectures typically incorporate programmable functions that give modern instruments flexibility and exceptional accuracy. Historically, the task of producing diverse waveforms was achieved by separate, dedicated signal sources, from ultra-pure audio sine wave generators to multi-GHz RF signal generators. Although many traditional dedicated commercial solutions still exist, the SI engineer may need to custom-design or modify a signal source for his or her particular project. It can be very difficult to design an instrumentation-quality signal generator. Also, the time spent designing ancillary test equipment can become a costly distraction from the project itself.

Fortunately, digital sampling technology and signal processing techniques have brought us signal generation functionality that provides most of today's circuit excitation needs with just one type of instrument—the arbitrary signal generator. Arbitrary signal generators can be classified into arbitrary function generators (AFGs) and arbitrary waveform generators (AWGs). Although these two types of signal sources are similar, the following sections highlight their differences.

7.4   ARBITRARY FUNCTION GENERATORS

The AFG serves a wide range of stimulus needs. It is the principal signal source architecture in the industry today. Typically, this instrument offers fewer waveform variations than its AWG equivalent, but with excellent stability and a fast response to frequency changes. If the device under test requires the classic sine or square waveforms, to name a couple, and the ability to switch almost instantly between two frequencies, the AFG is the right tool. An additional virtue is the AFG's low cost, which makes it very attractive for applications that do not require AWG versatility.

The AFG shares many features with the AWG, although the AFG is by design a more specialized instrument. The AFG has unique strengths. It produces stable waveforms in standard shapes—in particular, the all-important sine and square waveforms—with both accuracy and agility. Agility is the ability to change quickly and cleanly from one frequency to another. Most AFGs offer some subset of the familiar wave shapes, such as sine, square, triangle, sweep, pulse, ramp, and modulation. Although most AWGs can provide the same waveforms as an AFG, and customized signals, generally they are more expensive. Moreover, modern AFGs are designed with improved phase, frequency, and amplitude control of the output signal. Furthermore, many of today's AFGs offer a way to modulate the signal from internal or external sources, which is essential for some types of standards or compliance testing. In the past, AFGs created their output signals using analog oscillators and signal conditioning. More recent AFGs rely on Direct Digital Synthesis (DDS) techniques to determine the rate at which samples are clocked out of their memory.

Most digital waveform generators use waveform data stored in random-access memory (RAM). As the RAM address is incremented, the values are output sequentially to a DAC, which reconstructs the waveform as a series of voltage steps that are subsequently filtered before being passed to the output amplifier and attenuators. The frequency of the output waveform is determined by the rate at which the RAM addresses are changed.

The first arbitrary generators used a system in which the RAM address was incremented by a simple counter. The waveform's rate of replay was set by varying the clock frequency into the counter. This system was later refined so that individual sections of the RAM could be addressed and clocked at will. However, the basic principle of accessing successive RAM locations at a rate defined by a variable-frequency clock remained. This is called variable clock architecture. More recently, direct digital synthesis, a technique originally designed for generating variable-frequency sine waves, has become a widely used technique for arbitrary waveform generation.

7.4.1 Direct Digital Synthesis

DDS is a key building block found in most modern digital function generators. DDS is a technique used to generate signals directly in the time domain. In a DDS system, signal amplitude characteristics are stored as digital values in memory. Typically the waveform values are loaded into RAM. The values are recalled using a clocked digital design that allows for their discrete placement in time. Such a process gives precise control over amplitude, frequency, and phase, resulting in the ability to produce any arbitrary waveform.

DDS has a number of advantages over analog waveform generation techniques. Principally, unlike an indirect frequency synthesizer, a DDS system normally does not use a phase locked loop (PLL) that can limit an instrument's performance because of the loop settling time. In a DDS generator, the RAM address increment rate is determined by a fixed clock frequency and a digital block composed of a phase increment register and a phase accumulator. Figure 7-3 shows the DDS principle, where a sinusoidal wave is produced by a counterclockwise rotating phasor. The number of discrete points around the circumference of the circle shown in Figure 7-3 is n, which is the number of stored signal values. The rate at which the phasor outputs each discrete point is f, which is a constant number of output bits per second. An important point to note is that the number of discrete points output per second is constant. The phasor's rotational speed determines the output frequency. Therefore, the sample rate f is fixed, and the function generator must either add or subtract points n to produce a required output frequency.

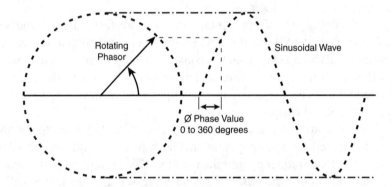

Figure 7-3 A sinusoidal wave produced by a counterclockwise rotating phasor.

In Figure 7-4, the speed of the rotating phasor is adjusted so that it skips or misses a set number of points each time it is rotated. This causes the sinusoid's frequency to change. In this case, the phasor is twice the speed shown in Figure 7-3, and the output frequency doubles, because every other point is skipped. If a lower output frequency is required, the phase accumulator circuit must slow the phasor and output additional points for each step of the rotating phasor. Therefore, for a lower frequency, the generator replicates a number of the waveform points.

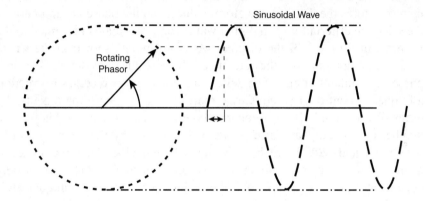

Figure 7-4   The same sinusoidal wave shown in Figure 7-3. However, the phasor is now rotating at twice the speed and skipping points, giving a doubling of frequency.

A simple formula, Equation 7-1, relates the frequency of the generated sinusoid F_{wave} to the number of discrete points around the circumference of the circle n; the rate at which the phasor outputs each discrete point f, which is constant; and the number of points skipped each time the phasor rotates S:

$$F_{wave} = \frac{S \times f}{n}$$

Equation 7-1

For example, if n is 4,096, f is 1 megabit per second (Mbps), and S is 1, F_{wave} is

$$256Hz = \frac{1 \times 1,048,576}{4096}$$

Therefore, if we change S to 2 and skip every other point, as shown in Figure 7-4, the frequency of the sinusoid doubles:

$$512Hz \;=\; \frac{2 \times 1,048,576}{4096}$$

Actual DDS technology synthesizes waveforms by using a single clock frequency to produce any frequency, albeit within the instrument's range. Figure 7-5 simplifies the DDS-based AFG architecture. In straightforward terms, the waveform memory stores the discrete waveform values, and the phase accumulator provides the selection criteria for each digital waveform value that is presented to the DAC. In terms of Figure 7-3, the discrete waveform points shown in the sinusoid are held in the memory, and the rate at which the values are output depends on the phase accumulator circuit. The delta (Δ) phase register receives instructions from a frequency controller (Freq. Ctrl.), representing the phase increments. Again, in terms of Equation 7-1, the frequency controller is S, the clock f is fixed, and the number of discrete waveform points is determined by the memory word size w, where n equals 2^w. In reality, during a single clock, the phase accumulator addresses and outputs a number of waveform values from the waveform memory, dependent on the frequency controller value. Two distinct techniques are used to control the output waveform frequency. Values can be skipped or, in the case of a high-performance instrument, the waveform values are output at a faster rate. Either way, the DAC converts digital values to an analog wave, and they are smoothed by an analog output stage. Any alias, unwanted frequencies, or spurious frequencies that could result from the DDS process are removed.

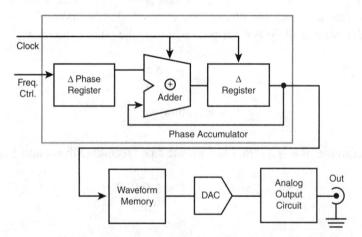

Figure 7-5   A simplified block diagram of a DDS-based AFG architecture.

In a modern high-performance AFG, the phase resolution may be as small as one part in 2^{30}—that is, approximately $1/1,000,000,000$. Put simply, the memory outputs $1,000,000,000$ waveform points in a single 360-degree waveform cycle. The output of the phase accumulator serves as the address for the waveform memory portion of the AFG. The instrument's operation is almost the same as that of the AWG, with the notable exception that the waveform memory typically contains just a few basic signals, such as sine and square waves. The analog output circuit basically is a fixed-frequency, low-pass filter that ensures that only the programmed frequency of interest and no clock artifacts leave the AFG output. Also, the filter minimizes the quantization effects of the DAC.

To understand how the phase accumulator creates a frequency, imagine that the frequency controller shown in Figure 7-5 sends a value of 1 to the 30-bit Δ phase register. The phase accumulator's Δ output register advances by $360 \div 2^{30}$ in each cycle, because 360 degrees represents a full cycle of the instrument's output waveform.

Therefore, a Δ phase register value of 1 produces the lowest-frequency waveform in the range of the AFG, requiring the full 2Δ increments to create one cycle. The circuit remains at this frequency until a new value is loaded into the Δ phase register. Values greater than 1 cause the AFG to advance through the 360 degrees more quickly, producing a higher output frequency by skipping some samples, thereby sampling the memory contents. However, the only thing that changes is the rate at which the phasor rotates in Figure 7-4, which is the phase value supplied by the frequency controller in Figure 7-5. The main clock frequency does not need to change at all. Moreover, the phase accumulator allows a waveform to commence from any point in the waveform cycle.

For example, assume that it is necessary to produce a sine wave that begins at the peak of the positive-going part of the cycle, as shown in Figure 7-6. Basic mathematics tells us that this peak occurs at 90 degrees. For simplicity, assume that our sinusoidal wave is constructed from 4,096 discrete points. Therefore, 4,096 phase increments $= 360°$ and $90° = 360° \div 4$; then, $90° = 4,096 \div 4$. When the phase accumulator receives a value equivalent to $4,096 \div 4$, it prompts the waveform memory to start from a location containing the positive peak voltage of the sine wave. Consequently, the phase accumulator can control the point in the cycle or phase at which a waveform starts and the waveform's frequency.

An AFG normally has several standard waveforms stored in a preprogrammed part of its memory. In general, sinusoidal and square waves are the most widely used for many test applications. Normally they are held in read-only memory (ROM). Arbitrary waveforms are held in a user-programmable part of the memory, and wave shapes can be defined with the same flexibility as conventional AWGs. However, the DDS architecture in an AFG does not support memory segmentation and waveform sequencing. These advanced capabilities are reserved for high-performance AWGs and are described in the next section.

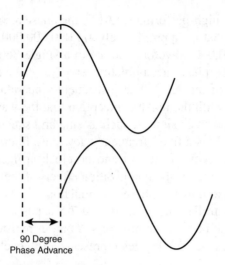

Figure 7-6   Sinusoids that begin at the peak of the positive-going part of the cycle are in effect advanced in phase by 90 degrees.

Consider the basic practical DDS example shown in Figure 7-7. RAM is loaded with the amplitude values of the individual points of one cycle, 360 degrees, of a desired waveform, such as a sinusoid. In a DDS system a number of vertical amplitude points define a waveform cycle, as shown earlier in Figure 7-4. In this example, one complete cycle of the selected waveform is stored in RAM as 4,096 12-bit amplitude values. The DDS system's 12-bit 4,096-word waveform store allows a maximum frequency of 40 MHz, with a frequency resolution of 0.0001 Hz. Put simply, the action of the phase accumulator is to increment the RAM address and sequentially output waveform values to the DAC, which reconstructs the waveform as a series of voltage steps. Sine waves and other wave shapes, such as triangles, are subsequently filtered (as shown later in Figure 7-9) to "connect the dots" and smooth the voltage steps at the DAC output.

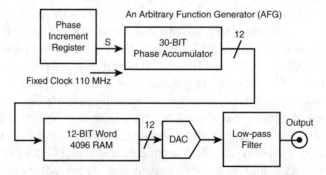

Figure 7-7   The simplified architecture of a 12-bit DDS AFG signal source.

The frequency of the output waveform is determined by the rate at which the RAM addresses are changed. In this example, the waveform generation process can be considered in five basic steps:

- The RAM is loaded with 4,096 amplitude values for the individual points of one cycle, 360°, of the waveform. Each sequential address change corresponds to a phase increment of the waveform of 360° ÷ 4,096. In this case we get an output value for approximately every 0.1 of a degree. Traditionally a counter would have been used to generate the sequential RAM addresses, but today a phase accumulator circuit is used to determine the rate at which the RAM contents are read.

- On each clock cycle, the phase increment, which has been loaded into the phase increment register, is added to the current result in the phase accumulator to produce a RAM address. In this example the 12 most significant bits of the phase accumulator drive the RAM address lines. The DDS in this case uses only the 12 most significant bits of the phase accumulator, because the RAM is limited to 4,096 12-bit words. Often the RAM is limited to an optimal memory size for the generator.

- The output waveform frequency is now determined by the phase increments at each clock. If each phase increment is the same size, the output frequency is constant. If the phase accumulator changes the size of the phase increments, the output frequency changes, but the point at which the signal starts is constant. In other words, a change of frequency should not change a signal's phase.

  This example has a requirement for a frequency resolution of 0.0001 Hz. To give a frequency resolution of 0.0001 Hz, the smallest phase increment of the 40-bit accumulator needs to be $2^{40} \times 10^{-4}$, which is 109.951 MHz. This gives a fixed clock frequency (f_{CLK}) of 110 MHz. Put another way, we now have $f_{CLK} \div 2^{40} = 0.0001$ Hz.

- If the 4,096 RAM address increments on every clock, the waveform frequency is determined by $f_{CLK} \div 4,096$—that is, 110 MHz ÷ 4,096 (110 MHz ÷ 4096) and a waveform point is output to the DAC every 9 nanoseconds. Therefore, 4,096 × 9 nanoseconds gives a waveform period of approximately 0.37 microseconds, which is a frequency of approximately 27 kHz. To output frequencies below 27 kHz, there must be a longer waveform period than 0.37 microseconds and more waveform points, where repeated addresses need to be output for each clock cycle. Similarly, at waveform frequencies above 27 kHz, there must be a shorter waveform period than 0.37 microseconds, and fewer waveform points could be output, which will cause some addresses to be skipped. This gives the effect of sampling the stored waveform. Typically different points are sampled on successive cycles of the waveform.

- The minimum number of sampled points required to accurately reproduce a wave shape determines the maximum useful output frequency, where fmax = f_{CLK} ÷ number of points. For sinusoids, a suitably designed filter allows the waveform to be reproduced accurately toward what is called the Nyquist limit, which is just two points per waveform cycle. However, limitations in the performance of a practical filter lower the limit to, say, 40 MHz for a 110 MHz clock. For other waveforms, the required number of points depends on the waveform's complexity and the precision with which the waveform must be reproduced.

The DDS principle is particularly well suited to the generation of sinusoidal waveforms. The output filter can create a low-distortion sinusoid from a very small number of points per cycle, approaching the Nyquist limit of two. For non-sinusoidal waveforms such as pulses, triangles, and ramps, sharp transitions in the signal cannot be passed through an output filter unless the signal frequency is notably low. Consequently, for a 100 MHz clock generator, sinusoids may be available up to 40 MHz or more, but triangles and ramps may be limited to 500 kHz or less.

The DDS architecture provides exceptional frequency agility, making it easy to program both frequency and phase changes on-the-fly. This is useful when you're testing frequency modulated devices, such as radio and satellite system components. Moreover, modern low-cost function generators typically provide sufficient frequency ranges to test frequency shift keying (FSK) and frequency hopping for the validation of telephony and wireless network technologies. However, the DDS arbitrary generator can only faithfully reproduce an arbitrary waveform at the repetition rate, or submultiples, of the DDS clock frequency divided by the number of waveform points. At other frequencies waveform points normally are either omitted or duplicated an uneven number of times. For repetitive waveforms this can result in waveform jitter, because the starting address can change at the beginning of each cycle. This is an issue only at high repetition rates, where sampled waveforms may have parts of their waveform skipped over. Jitter effects become relatively insignificant at low waveform repetition rates.

DDS arbitrary function generators produce the most common test signals used in laboratories, repair facilities, and design departments around the world. Moreover, the AFG delivers excellent frequency agility and offers a very cost-effective way to get the job done. Nonetheless, the AFG cannot match the AWG's ability to create virtually any type of waveform with high precision and accuracy.

7.5   THE ARBITRARY WAVEFORM GENERATOR

The AWG normally can be relied on to produce the required waveform. This is true whether the SI engineer needs a data stream shaped by a precise Lorentzian pulse to characterize a disk drive, or a complex modulated radio frequency (RF) signal to test a complex telephone handset. The engineer can use a variety of AWG methods to create a required wave shape, ranging from the application of mathematical formulae to the drawing of the waveform. Fundamentally, an AWG is a sophisticated playback system that delivers waveforms based on stored digital data, where the data describes constantly changing voltage levels that represent an AC signal. The AWG is a tool whose block diagram, shown in Figure 7-8, is deceptively simple. To put the signal source in familiar terms, the AWG is conceptually similar in operation to a CD player that reads stored music or digital disc data. However, whereas a CD player uses a disc for its digital storage, an AWG contains an integral integrated circuit waveform memory. Nonetheless, both the CD player and the AWG replay a digitally stored analog signal in real time.

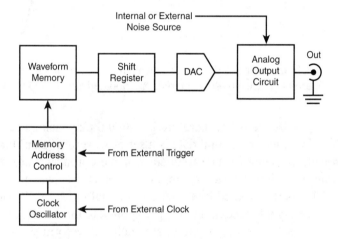

Figure 7-8   The simplified block diagram of an AWG.

To understand the AWG, first you must grasp the broad concepts of digital sampling. Digital sampling is exactly what its name implies. In essence, it designs a signal using digital samples, or data points, that represent a series of voltage measurements. The digital samples may be determined by capturing a real-world waveform with an instrument such as an oscilloscope that actually measures and records the amplitude and time of each sample point. The engineer could also use a graphical waveform technique or a mathematical technique. The left part of Figure 7-9 shows a series of sampled points. All the points are sampled at uniform time intervals, even though the curve makes their spacing appear to vary. In an AWG the sampled values are stored in binary form in a fast RAM. Reading the stored information from memory and passing the data through a DAC enables the stored signal to be reconstructed, as shown on the right side of Figure 7-9. It should be noted that the output of the DAC requires interface circuitry to filter the signal and effectively join the signal points to create a clean, continuous waveform shape.

Figure 7-9   On the left are sampled, discrete, vertical amplitude points for a single-cycle sinusoid. On the right are the corresponding filtered, or smoothed, sinusoidal waveform points.

Conceptually the AWG therefore is a digital playback machine described by Figure 7-8, in which the digital data is a set of waveform amplitude points stored in the waveform memory. The frequency of the output waveform is determined by the rate at which the stored amplitude points are read from the waveform memory. This is determined by either the internal clock oscillator or an external supplied clock signal. An external trigger applied to the waveform address control can be used to set the time, or phase, at which a waveform starts. The shift register is purely used to enable the waveform data to be presented to the DAC in the correct digital format. Figure 7-8 also shows an internal or external noise source that can be used to add a specified amount of distortion to the output signal to enable the signal to be applied in a controlled DUT stress or compliance test.

7.5.1   AWG Parameters, Characteristics, and Performance Standards

AWG parameters define the characteristics of the instrument. Put simply, the following set of AWG parameters describe the performance standards of a particular signal source.

7.5.1.1 Memory Depth or Record Length

Memory depth, or record length, goes hand in hand with clock frequency. Memory depth determines the maximum number of samples that can be stored. Each waveform sample point occupies a memory location. Each memory location holds an amplitude sample of the output signal and equates to a point in time where the time between samples is determined by the current clock frequency. For example, if the clock is running at 100 MHz, the stored samples are separated by 10 nanoseconds.

Memory depth plays an important role in the output signal fidelity, especially at high frequencies, where the memory depth determines how many sample points are used to define the output waveform. Moreover, in the case of complex waveforms, memory depth is essential for the accurate and faithful reproduction of multifaceted signals. Incorporating a large depth of memory in an AWG has a number of benefits. For example, different signals can be combined in a sequence and flexibly joined with infinite loops, event jumps, and patterns to generate complex wave shapes. Also, more waveform detail can be stored, allowing complex waveforms to include such detail as high-frequency information in their pulse edges and transients. Normally it is difficult to interpolate fast transitions for the faithful reproduction of a complex signal. A large depth of memory makes available waveform memory capacity for the storage of transitions and fluctuations rather than simply more cycles of the signal. Typically high-performance mixed signal sources offer memory depth and high sample rates. Modern high-performance AWGs can store and reproduce complex waveforms such as pseudorandom bit streams. Similarly, these fast sources with deep memory can generate very brief digital pulses and transients that have precise signal characteristics.

7.5.1.2 Clock Frequency or Sample Rate

The sample rate affects the frequency and fidelity of the main output signal. Sample rates denote the maximum clock frequency used to read the output waveform points. Sample rates generally are specified in terms of the number of samples per second, which typically is mega- or giga-samples per second and signifies the maximum rate at which an instrument can operate. As stated earlier, we must remember that the theoretical minimum number of sampling points for an AWG to define a waveform is determined by the Nyquist sampling theorem. It states that the sampling frequency, or clock rate, must be more than twice that of the highest spectral frequency component of the generated signal to ensure accurate signal reproduction. To generate a 1 MHz sinusoidal signal, it is necessary to produce two sample points per cycle. Therefore, the AWG must provide at least 2 mega-samples per second. Although the Nyquist theorem is clearly defined for sinusoidal waveforms, the theorem is imprecise when applied to the definition of the minimum number of samples for the faithful reproduction of complex waveforms. Theoretically, nonsinusoidal signals must have a sample rate at least twice the highest frequency found in the waveform. Empirically it has been found that complex waveforms must have considerably more than two sample points per waveform cycle for their faithful reproduction. Nonetheless, with sufficient memory depth and an adequate sample rate, the AWG signal source can reproduce complex, high-frequency wave shapes.

Calculating the frequency of the waveform that an AWG signal source can produce is a matter of solving a few simple equations. As a simple example, consider an instrument with a single sinusoidal waveform cycle stored in its memory. Assume that the instrument has 100 mega-samples per second clock frequency and a memory depth, or record length, of 4,000 samples. The generated waveform, Foutput, is as follows:

Foutput = clock frequency ÷ memory depth
Foutput = 100,000,000 ÷ 4,000
Foutput = 25,000 Hz (or 25 kHz)

This concept is illustrated in Figure 7-10.

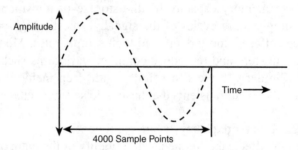

Figure 7-10   A 25 kHz waveform generated with 4,000 sample points.

At the stated clock frequency, the samples are about 10 nanoseconds apart. Of course, this is the time resolution, which is the waveform's horizontal axis. Be sure not to confuse this with the amplitude resolution, which is the vertical axis.

Carrying this process a step further, assume that the waveform memory contains not one, but four cycles of the sinusoidal waveform:

Foutput = (clock frequency ÷ memory depth) × (cycles in memory)
Foutput = (100,000,000 ÷ 4,000) × (4)
Foutput = 25,000 Hz × 4
Foutput = 100,000 Hz (or 100 kHz)

The new frequency is 100 kHz. Figure 7-11 depicts the new waveform.

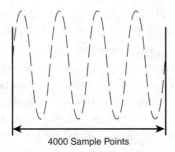

4000 Sample Points

Figure 7-11   The new 100 kHz waveform that is generated with four stored cycles and 4,000 sample points.

In this instance, the time resolution of each waveform cycle is lower than that of the single-waveform example shown in Figure 7-10. In fact, it is exactly four times lower. Each sample now represents 40 nanoseconds in time. The increased time interval comes at the cost of some horizontal, or time, resolution. In essence, the number of sample points is dictated by the complexity of the required waveform. Given the Nyquist sampling theorem for a sinusoid, which states that two samples are required per cycle, theoretically we could store 2,000 sinusoidal cycles in the waveform memory and obtain an output frequency, Foutput, as follows:

Foutput = (clock frequency ÷ memory depth) × (cycles in memory)
Foutput = (100,000,000 ÷ 4,000) × (2,000)
Foutput = 25,000 Hz × 2,000
Foutput = 50,000,000 Hz (or 50 MHz)

7.5.1.3   Timing or Horizontal Resolution

Horizontal resolution expresses the smallest time increment that can be used to create waveforms. Typically the timing figure is simply the reciprocal sampling frequency, T = 1/F, where T is the timing resolution in seconds and F is the sampling frequency. For example, by this definition, the timing resolution of a signal source whose maximum clock rate is 100 MHz would be 10 nanoseconds. In other words, the features of the output waveform from this mixed signal source are defined by a series of steps 10 nanoseconds apart. Some instruments offer tools that significantly extend the effective timing resolution of an output waveform. Although these tools do not increase the instrument's base resolution, the tools simply apply changes to the reproduced waveform to replicate the effect of increased samples by interpolation techniques and moving edges typically by increments in the picosecond range.

A region shift function shifts a specified edge of a waveform either right or left, toward or away from the programmed center value. If the specified amount of the shift is less than the sampling interval, the original waveform is resampled using data interpolation to derive the shifted values. Region shift makes it possible to create simulated jitter conditions and other tiny edge placement changes that typically exceed the instrument's resolution. Again, considering the example of a signal source with a 100 MHz clock, it would be meaningless to shift a stimulus edge in 10-nanosecond increments to simulate the effects of jitter. Real-world jitter operates in the low picoseconds range. Region shift makes it possible to move the edge by a few picoseconds with each step, which is a much closer approximation to a real-world jitter phenomenon. Figure 7-12 shows a region shift where an edge has been incrementally shifted in time to introduce jitter into a waveform, allowing the AWG to be used for a standard stress test.

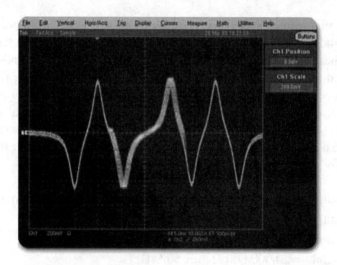

Figure 7-12   A highlighted section of waveform cycle showing an added region shift.

7.5.1.4   Signal Amplitude or Vertical Resolution

In the case of mixed signal sources, vertical resolution relates to the binary word size, in bits, of an AWG DAC, where more bits equals higher resolution. The DAC's vertical resolution defines the amplitude accuracy and distortion of a generated waveform. A DAC with inadequate resolution contributes to quantization errors and may result in a substandard output waveform. Ideally, the higher the bit rate, the better the instrument. However, in many cases a high-frequency AWG has limited vertical resolution—8 or 10 bits, compared to a general-purpose signal source that offers 12 or 14 bits. For example, an AWG with 10-bit resolution provides 1,024 sample levels spread across the instrument's full voltage range. In this instance, if the 10-bit AWG has a total voltage range of 2 volts peak-to-peak, each sample represents a step of approximately 2 mV, which is the smallest voltage increment the instrument could deliver without additional attenuators. This assumes that the AWG is not constrained by other factors in its architecture, such as limited output amplifier gain or a signal offset voltage. Figure 7-13 illustrates the quantization effects of using a DAC that has a limited number of bits. The top trace shows a low vertical resolution DAC, and the bottom trace shows a DAC with a higher number of bits, thereby giving better vertical resolution. However, there is always a trade-off between the vertical resolution, which defines the amplitude accuracy of the reproduced waveform, and the frequency of the reproduced waveform. A high-frequency DAC is technically limited in the number of bits it can convert at any one time. For example, a 14-bit flash converter would require 16,384 high-performance matched comparators, compared to 1,024 comparators for a 10-bit DAC. Also, often overlooked, the higher sample rate DAC has other benefits. It spreads quantization noise over a proportionally wider bandwidth, thus reducing the quantization noise spectral density.

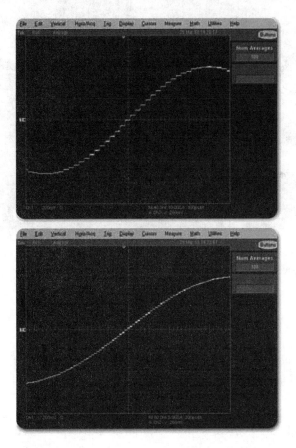

Figure 7-13   The top trace shows a low vertical resolution DAC. The bottom trace shows a DAC with a higher number of bits and better high vertical resolution.

7.5.1.5 Bandwidth

The bandwidth of an AWG is an analog term that is independent of the sample rate in describing the instrument's performance. The analog bandwidth of a signal source generally is determined by the instrument's output circuitry, where the bandwidth must be sufficient to handle the maximum frequency that the AWG sample rate supports. In other words, there must be enough bandwidth to pass the highest frequencies and transition times that can be clocked out of the memory without degrading the output signal characteristics. In Figure 7-14, an oscilloscope display reveals the importance of adequate bandwidth. The uppermost trace shows the uncompromised rise time of a high-bandwidth signal source. The lower traces show the degrading effects that result from reduced output circuit bandwidth.

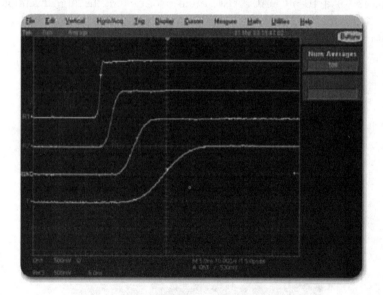

Figure 7-14   The top trace shows the uncompromised rise time of a high-bandwidth signal source. The lower traces show the degrading effects that result from a reduced AWG output circuit bandwidth.

7.5.1.6 Output Circuitry and Channels

Many applications require more than one output channel from a signal source. For example, testing an automotive antilock brake system requires—assuming that the vehicle has four wheels—four stimulus signals. Moreover, biophysical research applications typically call for multiple signal sources where the instruments simulate electrical signals similar to those produced by the human body.

And complex IQ-modulated telecommunications equipment requires a separate signal for each of the two phases. In answer to these needs, a variety of AWG output channel configurations has emerged. Some AWGs can deliver up to four independent channels of full-bandwidth analog stimulus signals. Other instruments may offer up to two analog outputs, supplemented by up to sixteen high-speed digital outputs for mixed-signal testing, where this class of instrument can provide analog signals and digital data via an integrated instrument.

Some AWGs include separate digital outputs that fall into two categories: marker outputs and parallel data outputs. Marker outputs provide a binary signal that is synchronized with the main analog output signal of a signal source. In general, markers allow the SI engineer to output a pulse or pulses to be synchronized with a specific waveform memory location or sample point. Marker pulses can be used to synchronize the digital sections of a DUT at the same time the DUT is receiving an analog stimulus from the mixed-signal source. Equally important, markers can trigger acquisition instruments that are external to the DUT, such as an oscilloscope or logic analyzer. Marker outputs typically are driven from memory that is independent of the main waveform memory. Figure 7-15 shows a mixed signal AWG output with a marker pulse for the purposes of external instrument signal acquisition synchronization.

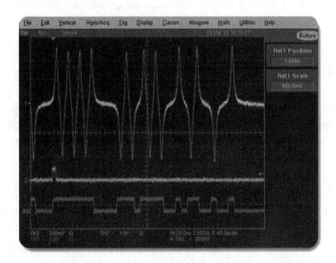

Figure 7-15   An oscillogram of a mixed signal AWG output with a marker pulse.

Parallel digital outputs generally take digital data from the same memory as that supplying the analog waveform samples. When a particular waveform sample value is present on the analog output, its digital equivalent is available on the parallel digital output. For example, this digital information is ready-made for use as comparison data when testing ADCs, among other things. Alternatively, the digital outputs can be programmed independently of the analog output.

Once the basic waveform is defined, other operations, such as filtering and sequencing, can be applied to modify or extend the basic waveform. Filtering allows the engineer to remove selected bands of frequency content from the generated signal. For example, when testing an ADC, it is necessary to ensure that the analog input signal, which comes from the signal source, is free of frequencies higher than half the ADC clock frequency. This prevents unwanted aliasing distortion in the ADC output, which would compromise the test results. Aliasing is simply the intrusion of distorted conversion by-products into the frequency range of interest. A DUT that is putting out an aliased signal cannot produce meaningful measurements. One reliable way to eliminate these frequencies is to apply a steep low-pass filter to the waveform, allowing frequencies below a specified point to pass through and drastically attenuating unwanted frequencies above the filter's cutoff point. Filters can also be used to reshape waveforms such as square and triangle waves. Sometimes it's simpler to modify an existing waveform in this way than to create a new one. Traditionally, it was necessary to use a signal generator and an external filter to create various waveforms. Fortunately, many of today's high-performance signal sources can digitally create almost any desired waveform. They have built-in filters that can be controlled to remove unwanted artifacts from the DAC process.

7.5.1.7   Sequencing

Often, it is necessary to create long waveform files to fully exercise a DUT. Where portions of the waveforms are repeated, a waveform sequencing function can save the engineer tedious, memory-intensive waveform programming.

Sequencing allows the engineer to store a vast number of "virtual" waveform cycles in the AWG memory. The waveform sequencer borrows instructions from the computer world, where loops, jumps, and so forth allow the instrument output to be programmed. The program instructions, which reside in sequence memory separate from the waveform memory, cause specified segments of the waveform memory to repeat. Programmable repeat counters, branching on external events, and other control mechanisms determine the number of operational cycles and the order in which cycles occur. With a sequence controller, the engineer can generate waveforms of almost unlimited length.

As a simple example of sequencing, imagine that a 4,000-point memory is loaded with a clean pulse that takes up half the memory, 2,000 points, and a distorted pulse that uses the remaining 2,000 points. If we were limited to basic repetition of the memory content, the signal source would always repeat the two pulses, in order, until it was told to stop. But waveform sequencing allows any cycle order. Suppose the engineer wanted the distorted pulse to appear twice in succession after every 511 cycles of a clean pulse train. The engineer could write a sequence that repeats the clean pulse 511 times and then jumps to the distorted pulse, repeats it twice, and goes back to the beginning to loop through the steps again—and again, as shown in Figure 7-16.

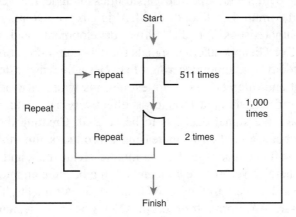

Figure 7-16   A waveform sequence that repeats a clean pulse 511 times and then jumps to a distorted pulse, repeats it twice, and goes back to the beginning to loop through the steps 1,000 times.

Loop repetitions can be set to infinite or to a designated value or can be controlled via an externally input event. Considering the trade-off between the number of stored waveform cycles and the resulting timing resolution, sequencing provides much-improved flexibility without compromising the resolution of individual waveforms.

However, it should be noted that any sequenced waveform segment must continue from the same amplitude point as the segment preceding it. If the last segment in a sample of a sine wave had a value of 1.2 volts, the starting value of the next segment in the sequence must also be 1.2 volts. Otherwise, an undesirable glitch can occur when the DAC attempts to abruptly change output at the voltage discontinuity. Although this example is quite basic, it represents the kind of capability that is needed to detect irregular pattern-dependent errors. One example of a pattern-dependent error is intersymbol interference in communications circuits. Intersymbol interference can occur when a signal state in one cycle influences the signal state in the subsequent or following cycle, which can distort the following cycle or even change its value. With waveform sequencing, it is possible to run long-term stress tests, extending to days or even weeks, with the signal source providing a programmed sequence of premeditated signal aberrations as the test stimulus.

In addition to programmable sequencing, many AWGs have built-in waveform development resources, whereby some mixed-signal sources offer optional application-specific signal creation functions. In effect, these are specialized waveform editors for engineers working with, say, high-speed disk drive or communications equipment. Specific capabilities include the generation of Partial Response Maximum Likelihood (PRML) signals with Nonlinear Transition Shift (NLTS) characteristics for disk drive development, and the generation of 100BASE-T or Gigabit Ethernet signals for data communication applications.

With today's faster engineering life cycles achieving faster time to market, it is important and often necessary to test designs with real-world signals or simulate real-world signal characteristics as effectively as possible. To generate real-world signals, the signal data has to be created. Creating these waveforms has historically been a challenge, increasing a product's time to market. However, modern PC software packages let you import, create, edit, and send waveforms to an AWG, where waveform creation and editing become an intuitive and relatively straightforward process for both AWG and AFG instruments. The PC applications let you capture waveforms from oscilloscopes or create a signal from a standard waveform library. Waveforms can be resampled to match the timing resolution of the destined signal generator. The engineer typically can define waveforms freely based on a standard menu of waveforms or point drawing tools or via a numeric data table entry. Once the engineer has created a waveform, anomalies can easily be added with mathematical functions or waveform editing tools. PC waveform editing tools allow waveform segments or a complete waveform to be easily shifted in the time or amplitude axis. This enables real-world signal generation to be achieved with relative ease. Figure 7-17 shows an industry-standard waveform editing suite that is hosted on a PC.

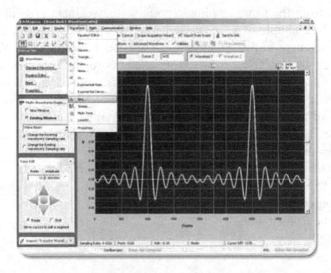

Figure 7-17 PC-based waveform editing.

7.6 LOGIC SIGNAL SOURCES

Logic signal sources are more specialized tools designed for specific digital test requirements. They meet the special stimulus needs of digital devices that often require long, continuous streams of binary data that generally need specific information content and timing characteristics. Logic sources fall into two classes of instruments:

- Pulse pattern generators
- Data timing generators

7.6.1 Pulse Pattern Generator (PPG)

Pulse pattern generators drive a stream of square waves or pulses from a small number of outputs, usually at very high frequencies. Unless the stream is modulated, usually no information content or data is expressed in the pulses. However, the high frequency and fast rise time capabilities of an advanced pulse generator make it an ideal tool to test high-speed digital equipment.

7.6.2 Data Timing Generator (DTG)

Whereas the AWG and AFG are designed primarily to produce waveforms with analog shapes and characteristics, a data timing generator is designed to generate volumes of binary information. The data timing generator, also called the pattern generator or data generator, produces the streams of binary 1s and 0s needed to test computer buses, microprocessors, integrated circuit (IC) devices, and other digital elements. In the design department, the data timing generator is an indispensable stimulus source for almost every class of digital device. In broader terms, the DTG is useful for functional testing, debugging new designs, and performing failure analysis on existing designs. The DTG is also well known for its ability to support the timing and amplitude margin characterization of a DUT. A DTG can be used early in the product development cycle to substitute for system components that are not yet available. For example, it might be programmed to send interrupts and data to a newly developed bus circuit when the processor that would normally provide the signals is absent form the hardware. Similarly, the DTG might provide addresses to a memory bus, or even the digital equivalent of a sinusoidal wave to a DAC under test. With its extraordinarily long patterns and its ability to implant occasional errors in the data stream, the DTG can support long-term reliability tests to ensure compliance with military or aerospace standards.

A typical device characterization application is where a DTG is made to respond to external events from the DUT to provide a particular digital pattern test for interoperability or compliance testing. The DTG is equally at home testing semiconductor devices, hard disk drive circuits, DVDs, image sensors, and digital display drivers/controllers. Put simply, the DTG is a universal signal source for the development, evaluation, and test of most digital devices and systems.

Like the AWG and AFG, the DTG architecture contains an address generator, waveform (or pattern) memory, and a shift register, as shown in Figure 7-18. However, the DAC is absent from the pattern generator's architecture. The DAC is not necessary, because the pattern generator does not need to trace the constantly shifting levels of an analog waveform. Most DTGs provide a way to program the logic 1 and 0 output voltage values and incorporate analog output circuits to enable edge parameters and other waveform parameters to be set.

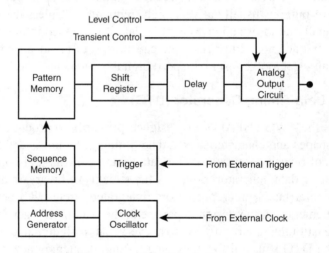

Figure 7-18 A simplified block diagram of a DTG.

The DTG has some digital features designed to support jitter and timing tests. A special delay circuit is responsible for implementing the small changes in edge positioning that these applications require. The delay circuit can deliver tiny changes, on the order of picoseconds, in edge placement. Some state-of-the-art DTGs provide simple front-panel controls that allow the engineer to move all edges or selected edges in increments of 0.2 picoseconds within a range of 5 nanoseconds or more. These small timing changes model the classic jitter phenomenon in which the placement of a pulse edge in time moves erratically about a nominal center point. The engineer tests jitter tolerance by changing and observing the effects of edge timing in relation to the clock. In today's best DTGs, it's possible to apply this jitter throughout the pattern, or on isolated pulses via a masking function that pinpoints specific edges. Figure 7-19 shows the oscilloscope capture of a pattern generator's output signal, with the addition of edge effects. The inset illustration provides a simplified and enlarged view of the same events. Other features give the modern DTG even more flexibility for critical jitter testing. Some instruments have an external analog modulation input that controls both the amount of edge displacement in picoseconds and the rate at which the displacement occurs. With so many jitter variables at the SI engineer's disposal, it's possible to subject a DUT to a wide range of real-world stresses.

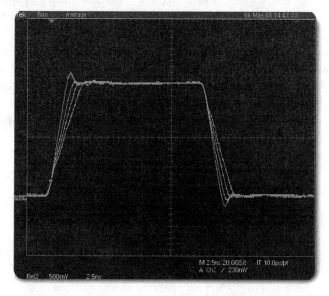

Figure 7-19   An oscilloscope capture of a pattern generator's output signal showing added edge effects.

The delay circuit plays a second, equally important, role in testing timing problems such as setup-and-hold violations. Most clocked devices require the data signal to be present for a few nanoseconds before the clock pulse appears (setup time) and to remain valid for a few nanoseconds (hold time) after the clock edge. The delay circuit makes it easy to implement this set of conditions. Just as the DTG can move a signal edge a few picoseconds at a time, it can move that edge in hundreds of picoseconds or hundreds of nanoseconds. This is exactly what is needed to evaluate setup-and-hold time. The test involves moving the input data signal's leading and trailing edges a fraction of a nanosecond at a time while holding the clock edge steady. The resulting DUT output signal is acquired by an oscilloscope or logic analyzer. When the DUT begins putting out valid data consistent with the input condition, the location of the leading data edge is the setup time. This approach can also be used to detect metastable conditions in which the DUT output is unpredictable.

Because a DTG is dedicated to digital testing applications, it contains unique features that neither an AWG nor AFG can provide, such as a sophisticated sequencer, multiple outputs, various pattern data sources, and distinctive display.

No internal memory is deep enough to store the many millions of pattern words, or test vectors, required for a thorough digital device test. Consequently, pattern generators are equipped with sophisticated sequencers. The pattern generator must provide very long and complex patterns, and typically it needs to respond to external events. Although a DTG pattern memory capacity is large in a similar way to the AWG, the DTG needs to loop on short pattern segments to produce lengthy data streams. The DTG sequencer normally offers many levels of loop nesting and branch conditions that can be controlled using normal programming conventions to produce address, data, clock, and control signals for most digital devices. A DTG sequencer has the unique ability to expand pattern length indefinitely. For example, Figure 7-20 shows a flow diagram for a DTG sequence that allows a few short instructions and pattern segments to unfold into millions of lines of stimulus data.

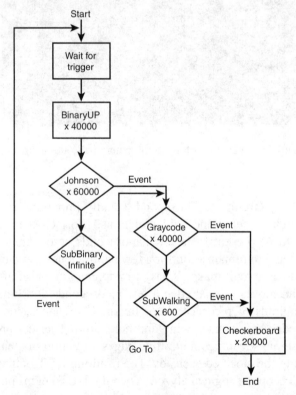

Figure 7-20   A flow diagram for programming a DTG sequence.

Multiple outputs also make the DTG ideal for digital design applications. Where the AWG or AFG may have two or four outputs, the pattern generator may have eight, thirty-two, or even hundreds of output channels to support the numerous parallel buses found in a typical digital device. Because a complex digital pattern would be impossibly tedious and error-prone if entered by hand, the modern DTG accepts data from logic analyzers, oscilloscopes, simulators, and spreadsheets. Moreover, digital data usually is available from various simulation software packages that are used in the design and verification processes of a digital design. Figure 7-21 shows a multichannel DTG output sequence.

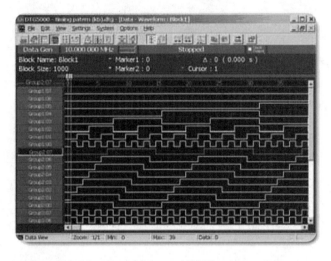

Figure 7-21   A multichannel DTG output sequence.

CONCLUSION

Many SI engineers look at tasks such as troubleshooting and design verification as purely "measurement" challenges and tend to view their oscilloscope or logic analyzer as the whole solution to the problem. However, these acquisition instruments are only half the story; the partner in their work is the stimulus instrument—the signal source. SI engineers need to both control and observe their device or system by working with a signal source and oscilloscope, or logic analyzer, where they can explore the limits and performance of their design. Numerous specialist toolsets exist for driving and measuring particular devices or systems, such as transmission line or microwave testing. However, AWG and similar stimulus instruments can drive a typical embedded system or device with complex real-world signals and quantifiable waveform aberrations for margin testing and system characterization. Moreover, introducing deliberate stresses with a signal source and measuring the results with precision acquisition instruments are almost universal prerequisites for statutory compliance and interoperability testing.

8

Signal Analysis and Compliance

The proliferation of new bus standards and the need for interoperability has focused new attention on analysis and testing to achieve compliance with these standards. This chapter examines how analyzing signal integrity is a major consideration for compliance testing. It describes the various standards, with particular emphasis on high-speed serial buses. It also shows why, at frequencies of more than 2.5 GHz, real-time test and measurement is the only way to achieve compliance. Again, the practical use of oscilloscopes, logic analyzers, and signal sources is emphasized, along with techniques such as eye diagrams and statistical analysis. This chapter also looks at new developments in signal-path analysis and scatter parameter measurements, which become increasingly important at these higher frequencies.

A well-worn adage tells us that "the only constant is change." Clearly this is true of serial data technology and the standards that define it. Serial architectures and protocols advance from one generation to the next, gaining performance as they go. Standards committees and working groups meet to debate evolutionary changes. And specifications—including compliance measurement specifications—evolve as new tools, technologies, and methods emerge.

In the computing and communications industries, a degree of standardization at the system, subsystem, and component levels is the foundation that technology builds on. Examples of standardization range from low-voltage differential signaling to the PCI Express serial bus designed to replace current PCI technology. Standards pervade semiconductor architectures, network protocols, and software components. And for every standard there must be some means of certification—some way to prove that new products are in compliance with the standard.

How can engineers stay ahead of compliance requirements that will inevitably change? How can they be sure that their products will comply with performance standards—many intimately linked with signal-integrity considerations—that are still being defined? It is a process of making assumptions, building on experience, taking some risks, and "leading the target."

Compliance is not without cost. Interpreting the standards, purchasing measurement equipment, making the necessary measurements, and documenting the whole process are activities that cost money, time, and resources.

Parallel data transmission architectures have traditionally dominated digital systems design, but this situation is rapidly changing. Parallel constructs are nearing their practical bandwidth limits, and there is a tremendous need for an order-of-magnitude increase in data throughput. Serial data architecture is virtually certain to prevail as data rates continue to escalate, since it provides some key benefits over parallel data transfer:

- Higher bandwidth per pin
- Scalable performance through the use of aggregated lanes
- Lower overhead and latency (maximizing payload efficiency)

The advance of serial transmission technology means that design and validation engineers will have to adapt to system-level testing and debugging on serial technologies.

8.1 STANDARDS FRAMEWORK

The key to the success of any emerging standard is a systematic plan to address specifications, testing, and compliance requirements. Typically, such plans are developed by industry working groups. Companies with an interest in supporting a new proposal work cooperatively to develop and stabilize standards and define qualification criteria. Interoperability does not occur in a vacuum, and the only way for a company to keep pace with emerging standards is to participate in the committees and working groups that develop them. Standards typically include measurement requirements and, in many cases, specific test equipment recommendations as well. Tests usually are performed at "plugfests" where prospective vendors gather to evaluate their products for interoperability and compliance with the standard.

A plugfest typically is sponsored by an industry working group on behalf of a particular standard such as Serial ATA or PCI Express. The event brings together a community of people who have an interest in the standard: product developers working with the new technology, working group members, industry watchers and media editors, and vendors of the tools and instrumentation used to measure compliance and performance.

Figure 8-1 shows a simplified product development cycle, from concept to announcement. The plugfest is the turning point in this continuum. It can be the difference between achieving time-to-market goals and literally going back to the drawing board.

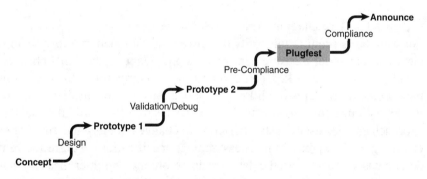

Figure 8-1   The path to compliance.

As the plugfest name implies, various devices are connected to confirm their interoperability. But plugging a transmitting device into a receiving component, for example, and confirming that the arrangement works is not enough. Measurements must verify compliance with voltage, timing, and impedance requirements. Bit error rates must be assessed, failures pinpointed, and, importantly, disparities resolved. The ideal of the plugfest is to put every candidate product on the same footing with regard to test conditions, methods, and equipment.

Measurement industry leaders participate in standards groups with the intent of smoothing their customers' path toward plugfest success. Test equipment vendors, with their understanding of measurement solutions and their cross-disciplinary work with multiple serial standards, act as a working group's advocates for best practices in compliance testing. Measurement vendors that invest time and resources in committee work can use their experience to guide solutions that dovetail with evolving standards. In the case of instruments such as oscilloscopes or bit error rate testers (BERTs), this implies keeping pace with advancing performance demands. More importantly, it means delivering software acquisition and analysis tools that anticipate the changing measurement needs that come with changes in the standard.

Increasingly, serial analysis relies on measurement automation, DSP, and complex mathematics to overcome challenges such as de-embedding fixture effects and maximizing the observability of inaccessible signals.

8.1.1   Measurement Innovations and Compliance Testing

Some of today's most difficult serial measurement challenges range from producing eye diagrams quickly and effectively to capturing impedance characteristics to inform SPICE models. Consider the following examples.

8.1.1.1   Eye Diagram Analysis Software

Software tools running integrally on proven measurement platforms can add efficiency to the serial measurement process and adapt as standards change. Consequently, software is a key battleground in the competition to solve compliance measurement problems of all kinds. Applications are now available that use plug-ins to optimize the broader toolset to the needs of a particular standard. This approach enables users to stay in step with changes in the standards. For example, emerging PCI Express Gen2 measurement specifications require the removal of de-emphasis effects from the data stream before making jitter measurements. This requirement currently is not incorporated in all standards, although it may get ported to other serial bus standards eventually. This situation is tailor-made for a plug-in solution that can be adapted to meet the new specification without affecting the measurement application as a whole.

8.1.1.2   Transmission Models in the Development of Compliant Serial Devices

In a perfect world, aberration-free differential serial data signals would travel through noiseless transmission channels and arrive intact at the receiver. In the real world, however, things are very different. Signals can get badly degraded by high-frequency losses, crosstalk, and other effects. Increasingly, serial measurement procedures are taking this into account. One innovative approach for serial data network analysis (SDNA) applications uses time domain reflectometry (TDR)-based S-parameter measurement tools, which are emerging as an efficient, cost-effective solution. These tools offer the performance to support uncompromised SDNA measurements with ample dynamic range for serial applications.

Moreover, these platforms have gained a host of software tools to speed and simplify SDNA work. SDNA requires time and frequency-domain characterization of the interconnect link using TDR measurements. SDNA uses the data to support SPICE modeling, eye diagram analysis, and a wide range of impedance parameters. Tools are now available that can run on a sampling oscilloscope, alongside eye diagram and jitter analysis applications, to provide a comprehensive toolset for multigigabit differential serial data link characterization and compliance verification.

8.1.1.3   Common-Mode Jitter De-embedding

Serial FB-DIMM (fully buffered double inline memory module) signals are notorious for their tendency to accumulate noise and crosstalk, with jitter as the consequence. Until now it has been very difficult to isolate the data jitter components in the midst of these degraded signals. A new technique, best described as "common-mode jitter de-embedding," offers an effective solution for FB-DIMM jitter testing using a real-time (RT) oscilloscope. FB-DIMM architecture maintains a reference clock channel that is separate from the data channel. Both these paths are influenced by the same board losses, and both exhibit essentially the same amount of noise and crosstalk. But the data channel also has some jitter contributed by the transmitter. This is obscured by the noise and signal degradation but nevertheless has the effect of increasing the channel's bit error rate. The common-mode jitter de-embedding technique acquires both channels and finds the differences. Since both channels carry the same noise and crosstalk, the difference between the clock channel and the data channel is the jitter value. Common-mode jitter de-embedding has been proposed to the appropriate working groups and is currently being evaluated.

8.1.1.4   Virtual Test Points to Reveal Hidden Signals

Observability is a challenge when measuring serial receiver performance. The receiver input in a serial device is an almost meaningless access point for viewing signals. This is because the serial receiver itself processes the input signal through a built-in decision feedback equalization (DFE) filter designed to offset the degradation that occurs during transmission through cables, PCB traces, and connectors. The signal that goes into the active portion of the receiver—where eye diagrams and other characteristics must be evaluated—is encapsulated inside the device and is thus inaccessible. Some oscilloscopes now include built-in finite impulse response (FIR) filters to mimic the effect of the receiver's DFE filter. The user can load the same coefficients into the oscilloscope that were used to design the filter in the device under test. With the filter applied, the oscilloscope user can probe the input pin and view the signal as it would be seen if the device could be probed internally. This virtual test point reveals the receiver's post-filter signal, even though the physical test point is a pin on the device package.

8.1.1.5 Simplifying Receiver Jitter Measurements

Figure 8-2 shows a traditional jitter tolerance measurement setup using a pattern generator with external modulators for receiver testing. The device under test (DUT) must be driven first with a setup sequence (built-in self-test frame information structure [BIST FIS]), followed without interruption by a jitter-laden data signal. Clearly this is a complex arrangement, and some compromises in signal quality occur as well. The solution is to use an arbitrary waveform generator (AWG) using direct digital synthesis to provide sufficient bandwidth to minimize complexity, as shown in Figure 8-3 With this approach, jitter in any form can be merged into the test signal itself, and the effects of both random and deterministic jitter can be modeled. Moreover, the AWG can incorporate the BIST-FIS instructions as part of the data, eliminating the power combiner and its effects on signal fidelity. Many standards that are moving into their second generation, such as 3 Gbps SATA, require maximum control of the rising edge of test signals to accurately stress the receiver. These signal sources must also be able to accurately simulate jitter profiles in terms of Tj, Sj, and Rj. Although mathematically Nyquist says we only need to sample at twice the highest frequency components, in reality accurate reconstruction of signals requires more than this. AWGs that can generate six samples per unit interval and create long waveform lengths are required for this task.

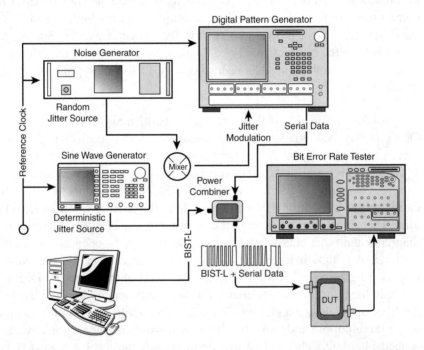

Figure 8-2   Modulated digital waveforms are the basis of this complex jitter tolerance measurement setup.

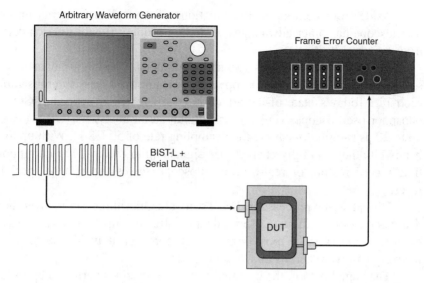

Figure 8-3   AWG-based setup for receiver jitter tolerance measurements.

8.2   HIGH-PERFORMANCE TOOLS FOR COMPLIANCE MEASUREMENTS

At this point, it should be clear that serial compliance measurements require the resources of high-performance measurement instruments. Five classes of instrumentation are most often used for PCI Express validation and compliance testing, as well as debug tasks.

8.2.1   Oscilloscopes

The tool of choice for measuring the electrical portion of the active physical layer is a high-performance oscilloscope. State-of-the-art digital oscilloscopes can capture waveforms in the multigigabit range with very good signal integrity. Once a clock is recovered from a serial bit stream, the oscilloscope can create an eye diagram from the waveform. Eye diagrams created by oscilloscopes provide a good view of signal characteristics. An extensive library of built-in measurements is available for immediate quantitative evaluation of both the eye diagrams and the acquired waveform.

Further processing with dedicated tools yields the time interval error (TIE). The TIE can be analyzed to separate random and deterministic jitter. This analysis can then also provide an estimate for total jitter at the 10^{-12} bit error rate (BER).

Aside from combining the functions of several instruments into one, the oscilloscope has other advantages, including probing flexibility, rich display, and triggering capabilities.

Real-time oscilloscopes capture continuous, contiguous data records. For serial link analysis, the oscilloscope needs to capture at least several samples from each bit. Today's state-of-the-art digital storage oscilloscopes (DSOs), digital phosphor oscilloscopes (DPOs), and digital serial analyzers (DSAs) can sample up to 20 ps sample-to-sample (at a sampling rate of 50 GS/s). With up to 20 GHz bandwidth they can fully characterize signals up to 8 Gbps. The continuous, real-time record feature of real-time oscilloscopes gives them these special advantages.

Digital signal processing (DSP) software algorithms can recover the embedded clock from the digitized serial data bit stream. This (software-based) method of clock recovery is the most flexible one; moreover, it avoids the need for clock-recovery hardware and its unavoidable jitter.

The signal path in the best real-time oscilloscopes performs up to (or nearly up to) the instrument's bandwidth. Therefore, the extensive triggering capabilities of the hardware triggers in these real-time oscilloscopes can be used to trigger on data or events of interest. They often capture events so rare that they are invisible through other means.

Real-time capture of the data allows analysis with the fewest constraints. The whole bandwidth of the instrument is captured so that, for example, the jitter spectral information is complete and unaliased for all jitter types. Even completely random data can be captured, stored, and decoded.

Finally, the completeness of the data capture is invaluable in debugging unexpected or unpredictable events.

Equivalent-time (ET) sampling oscilloscopes, also known as sampling oscilloscopes or communication (signal) analyzers, are ideal for analyzing signal integrity at very high speeds. With a maximum bandwidth of more than 70 GHz, they can analyze optical and electrical signals from below 1 Gbps to 40 Gbps and beyond. Because sampling oscilloscopes do not attempt to capture at real-time speeds, they can perform more precise signal capture. As a result, they offer higher digitizer resolution and superior noise performance. Both of these features are advantageous for acquiring the low-voltage signals common in serial links today. Similarly, because the samples are captured in only "equivalent time" sequence, they can be spaced within femtoseconds of one another, removing any concerns about sample-to-sample spacing or interpolation. On the other hand, jitter analysis is more complex and sometimes constrains the signal measured—to a repetitive pattern, for example.

Because of their high bandwidth, sampling oscilloscopes also offer TDR and S-parameter measurement capability, thereby eliminating the need for a separate vector network analyzer for performing the S-parameter measurements on serial data devices.

Sampling oscilloscopes can also be equipped with clock recovery; in this case, the function is provided by hardware.

Some real-time oscilloscopes do offer both real-time and equivalent-time eye reading techniques using hardware clock recovery, each of which has its advantages.

8.2.2   Signal Generators

Good high-speed engineering practices include exercising designs under real-world conditions. The right tool for mimicking these conditions as closely as possible is a programmable signal generator. Generating test signals at today's data rates requires high-speed data timing generators (DTGs) and arbitrary waveform generators (AWGs). Without these instruments there would be no way to test and validate new physical layer designs.

Many signal sources can replay signals that have been captured with an oscilloscope. These signals can act as a reference signal or may be modified to stress the device under test.

Data timing generators are especially useful for generating multiple streams of channels of parallel data—up to 96 channels in today's most advanced instruments. These tools currently can deliver 3.3 Gbps data rates. At the same time, these advanced instruments provide a host of signal manipulation features, including independent level, rise/fall, and jitter controls.

High-speed digital signals inevitably have analog attributes. Arbitrary waveform generators can provide stimulus signals with analog content (usually deliberate impairments) on a bus channel. Capable of delivering any type or shape of waveform, AWGs are universally applied in design and manufacturing. Current AWGs have sample rates of up to 20 GS/s with 5.8 GHz data rate bandwidth.

8.2.3   Logic Analyzers

The preferred tool for measuring the logical subblock of the physical layer and the data link and transaction layers of a high-speed serial bus is the logic analyzer (LA). Unlike both RT and ET oscilloscopes, logic analyzers provide protocol disassembly of all layers of the link with packet level, symbol level, and link event triggering across all lanes of the link. The purpose of the logic analyzer is to simplify acquisition and analysis of the purely digital aspects of both serial and parallel transmission. To carry out the serial bus debug mission, the LA must deliver features consistent with the needs of high-speed buses: physical layer acquisition, deep memory, flexible triggering, and synchronization with other system buses.

And, like the oscilloscope, it must offer low-impact probing tools that provide direct insight into physical layer signals.

8.3   VALIDATION AND COMPLIANCE MEASUREMENTS

Serial bus specifications typically include amplitude, timing, jitter, and eye diagram measurements for the transmitter and receiver.

Insertion loss, return loss, and frequency-domain crosstalk are also normally specified. In the PCI Express standard the CEM and Cabling Specifications define jitter and eye diagram margins at the system level compliance points and represent the tests performed at plugfests. Automated measurement and analysis tools are commonly used to speed the selection and application of these tests.

All serial standards include amplitude, timing, jitter, and eye diagram measurements within their compliance testing specifications. Of course, standards differ from one to the next, and not all measurements are required for compliance with every standard. Following is an overview of some key compliance measurements:

- Amplitude tests ensure that the signal has enough amplitude tolerance to do its job under worst-case conditions.
- Differential voltage tests confirm that a specific minimum differential voltage will arrive at the receiver under worst-case media conditions (maximum loss). This ensures proper data transfer.
- Common-mode voltage measurements detect any common-mode imbalances and noise. They also help locate crosstalk and noise effects that may be coupling into one side of the differential pair and not the other.
- Waveform eye height tests characterize the data eye opening in the amplitude domain. The height is measured at the 0.5 unit interval (UI) point, where the UI timing reference is defined by the recovered clock.
- Timing measurements detect aberrations and signal degradation that arise from distributed capacitance, crosstalk, and other causes that affect signal integrity.
- Unit-interval and bit-rate measurements look at variations in the embedded clock frequency over a large number of consecutive cycles.
- Rise/fall time measurements confirm that the transition times are within acceptable limits. Edges that are too fast can cause electromagnetic interference (EMI) issues, and those that are too slow can cause data errors.

- Waveform eye width measurements can verify the signal's general "health" if accompanied by statistical details about the number of edges used in the measurement.
- Jitter measurements to determine factors such as random jitter, deterministic jitter, and total jitter (at a specific bit error rate) are required in many serial bus specifications.

This is a partial list of compliance measurements. Full compliance tests may include 50 individual parameters or more, although many of these are extracted from the basic eye diagram acquisition.

8.4 UNDERSTANDING SERIAL ARCHITECTURES

To better understand how these measurement tools are used for the signal-integrity testing needed to ensure compliance with serial bus standards, it is useful to examine serial architecture in more detail. Although each serial protocol has its own unique qualities, most of them share some fundamental characteristics, particularly in the areas of bandwidth, signaling, protocol architecture, and data formatting.

Figure 8-4 illustrates the components of a sample system. This implementation, with minor variations, is found in most high-speed serial buses in use today. This figure shows a simplified Serial ATA (SATA I or SATA II) installation. The host in this case is a PC motherboard.

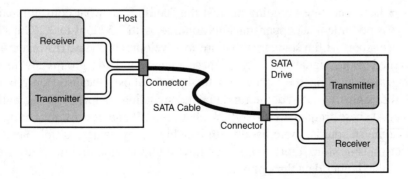

Figure 8-4 A SATA transceiver system. Each element—transmitter, link, and receiver—must be tested for compliance to ensure interoperability.

Viewing the block diagram from left to right, a short PCB trace carries the signals from the host transmitter to a connector. A compliant cable links the data to the receiver, which is situated within a hard disk device. Of course, data goes in both directions, so each side of the link incorporates both transmitters and receivers. Note that the transmission and receiving elements exchange differential signals, requiring two signal-carrying conductors per channel.

Some buses differ in detail from this model. FB-DIMM, for example, normally does not send its signals through a cable on the way to their destination. FB-DIMM memories usually are installed in connectors mounted directly on the motherboard along with the CPU and chipsets that access them, so the path consists of PCB traces and the FB-DIMM connector. But the fundamental measurement needs remain the same, as is the case with PCI Express, XAUI, Fibre Channel, and many other buses. Note that virtually every existing serial bus implementation is migrating toward second- or even third-generation standards. Inevitably this means higher data rates and even more stringent requirements for signal behavior such as noise and jitter.

Serial data signaling is achieved by embedding the clock and data in the same transmission medium. The serdes (serializer/deserializer) found in all links is responsible for serializing the data when transmitted and deserializing it when received. By embedding the clock in the data bit stream, fewer pins are used to achieve the same data transfer rate. In addition, many traditional testing concerns, such as setup and hold time and clock-data jitter, are avoided. However, the clock and data still need to be recovered by a receiver in a serial data serdes. This means that excessive jitter can cause the receiver not to recover the clock correctly and cause bit errors.

Data rates are bumping against the "ceiling" of what conventional wisdom says is possible in a copper transmission line, with 2.5 Gbps to 3.2 Gbps being the "sweet spot" of the standards that are in development today. However, as more is learned about signal integrity at these rates, standards groups are in the early stages of pushing these rates up to 4 to 6 Gbps. Because non-return-to-zero (NRZ) signaling is used, the fundamental signaling frequency is one half the data rate. At these frequencies, rise times between 50 and 100 ps are not uncommon. In some standards, more bandwidth is achieved by using multiple lanes of serial data transmission. Table 8-1 shows how multiple lane architectures are used to facilitate higher data throughput.

Table 8-1   Serial Bus Speeds

Serial Bus Standard	Bit Rate	Maximum Raw Data Rate
PCI Express	2.5 Gbps	Up to 80 Gbps (32X)
Infiniband	2.5 Gbps	Up to 30 Gbps (12X)
10 GbE XAUI	3.125 Gbps	12.5 Gbps (4X)
Fibre Channel 2X	2.125 Gbps	2.125 Gbps (1X)
Serial ATA II	3.0 Gbps	3.0 Gbps (1X)

8-bit/10-bit encoding is another common denominator of these specifications. This technique involves data being transferred over a single differential pair, with the clock embedded in the data bit stream. As a 10-bit set of data arrives at the pins of the receiver, it is deserialized by the receiver device. The result of this function is an 8-bit data word that can be used as data. Signal rate, the number of lanes deployed, and the encoding provide the maximum data transfer rate for a serial standard.

8.4.1   Differential Signaling

Differential signaling is not new. However, it is turning out to be the best solution for transmitting the sensitive high-speed signals used in today's serial protocols. Fast serial buses often rely on very low-voltage signals simply because it takes less time to change states over the span of a few hundred millivolts, for example, than it takes to make a full 1 V transition. This has led to the widespread adoption of environments such as low-voltage differential signaling (LVDS) and has created a need for a transmission medium that protects signals from noise and losses.

The differential transmission approach is the answer to this challenge.

Differential transmission delivers the signal along two "balanced" paths. Figure 8-5 contrasts differential transmission with the single-ended technique.

In the differential system, the two sides of the path are exact complements of one another: a signal level of +1 V on one side is mirrored by its –1 V complement on the other. The two signals "swing" in opposite directions at all times. If an unwanted external noise component couples into the line, it does not share this complementary characteristic; instead, it tends to induce the same signal into both lines. This is known as a common-mode signal. It is automatically rejected at the receiving end of the circuit, theoretically leaving only the valid signal. The extent of this rejection, or attenuation, of the unwanted signal is called the common-mode rejection ratio.

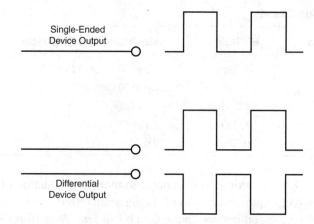

Figure 8-5   Single-ended and differential signals. The differential environment uses two complementary paths.

Differential signaling has long been the mode of choice for long "box-to-box" links. But with the advent of high-speed serial architectures it has become the best way to ensure reliable data transmission over any distance—even at the chip-to-chip level.

The advantages of differential signaling in high-speed data transmission include the following:

- Common-mode noise rejection
- Increased noise immunity
- Reduced crosstalk
- Reduced ground noise
- Reduced EMI

Differential signaling has many advantages, which has led to its increased use in high-speed data applications. However, it does lead to increased layout complexity, and it requires balanced signals and interconnections. Routing two traces for every signal clearly results in more complexity than routing only one trace per signal. In addition, the common use of point-to-point connections rather than a shared bus structure results in separate paths for transmit and receive signals, effectively doubling the number of signal pairs required in a high-speed serial link. Since the two traces of a differential signal generally are routed together as a coupled pair, this also adds to the complexity of the routing task. The routing of the traces in a differential signal must also be matched carefully in length and provide a symmetrical interconnection path and matched termination.

Gigabit data rates have a number of high-frequency effects that need to be considered with differential signal interconnects.

As signaling frequency increases, electrical signals suffer increased attenuation due to skin effect and dielectric losses. In the case of cable interconnects this attenuation is usually characterized as insertion loss and pulse dispersion. For circuit-board interconnects, these same frequency dependent loss effects are seen, although the dielectric loss tends to dominate at multigigabit data rates—particularly for the commonly used FR4 circuit board material.

Because of these loss effects, the trace length of differential signals should be minimized for high-speed data signals. Several of the high-speed serial data standards allow the use of pre-emphasis or equalization to try to compensate for these frequency-dependent loss effects. Pre-emphasis (sometimes also called de-emphasis) is an increase in the transmitter drive signal in the first bit period following a change in signal polarity. The transmitter drive is then reduced in bit periods where there is no signal polarity change.

Many of the new serial data communication standards implement the physical layer interconnect with serdes devices or circuits. A serdes provides the interface between a byte-wide data stream and a high-speed serial data stream and may contain encoding, synchronizing, and clock recovery circuitry. The high-speed serial output of some serdes devices is implemented with current-mode logic (CML). CML originally was a variant on a bipolar emitter coupled logic (ECL) output structure, where the output emitter followers were removed. CML can also be implemented, however, with MOSFET transistors, with the transistors configured as a steered current source switch. A CML output structure is a switched current source output topology that is usually loaded with an internal 50-ohm pull-up resistor on each line of the output pair. These CML output pull-up resistors serve as a source termination or back termination for the transmission-line interconnect. The CML transmission line interconnect should, of course, also be terminated at the receiver end of the line. This use of both source and load terminations on the transmission-line interconnect provides an ideal signal path for high-speed serial data signals and is in common use, especially for multigigabit data signals.

8.4.2 "Stack" Architecture

Within most serial standards, the architecture is viewed as a "stack" of layers, as shown in Figure 8-6.

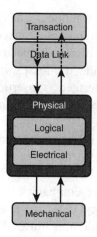

Figure 8-6 Layers of a serial data link.

Layers communicate with one another while buffering one another's operations from adjacent levels. The stack includes the physical layer in which electronic signals pass through transmission media, a logical layer in which these signals are interpreted as meaningful data, a transaction layer, and more. Each layer has its own applicable standards and compliance procedures.

Figure 8-7 shows the physical layer (PHY) partitioning of a serial data link.

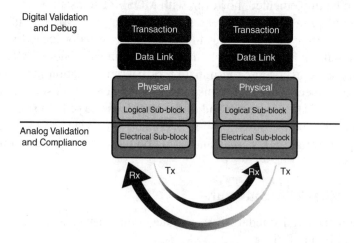

Figure 8-7 A layered model of the physical layer.

The PHY isolates the higher layers of the protocol stack and encompasses two layers: logical and electrical. The electrical section of the PHY handles the high-speed serial packet exchange and power management mechanisms. The logical layer of the PHY handles reset, initialization, encoding, and decoding. Both the electrical and logical subblocks also may incorporate standards-specific features.

Each of the two blocks of the PHY has unique test requirements. Analog waveform characteristics such as jitter are a priority when making electrical interface measurements. In the logical layer, digital packets must be interpreted, embedded clocks extracted, and so forth.

Figure 8-8 shows a typical physical layer link interface of a serial bus implementation.

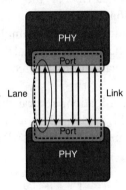

Figure 8-8   Serial PHY link implementation.

This layout is typical of a 4X PCI Express link and others. Each lane of the link consists of two differential channels: one transmit and one receive.

8.4.3   Packetized Data

Serial bus data is packetized, and the packets contain contributions from several layers. Figure 8-9 shows an example of this concept:

- The logical subblock of the PHY adds framing to signal the beginning and end of each packet and is responsible for symbol transmission and alignment.
- The data link layer provides error checking and retry services. Packets include ACK (acknowledge), power management information, and more.
- The transaction layer handles initialization, instruction generation and processing, and flow control.

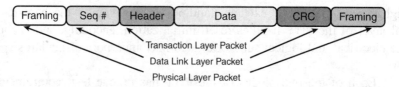

Figure 8-9   The layers of a serial data packet.

8.4.4   Characteristics of the Physical Layer

The physical layer is the carrier of the packetized differential signals just described. From the functional perspective, the physical layer mechanics—circuit board traces, connectors, and cables—are simply a path for data expressed in the form of binary signals. These binary signals are the subject of physical-layer measurements for debug, validation, and compliance tests, typically using an oscilloscope.

In a copper PHY specification, several different bus configurations can be defined. Figures 8-10, 8-11, and 8-12 show three typical mechanical environments for a serial physical media-dependent (PMD) interface.

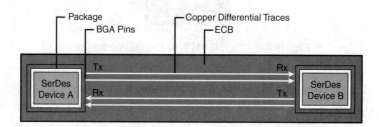

Figure 8-10   Chip-to-chip lane.

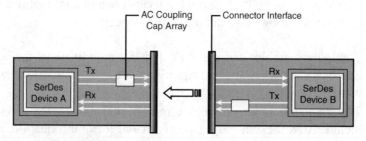

Figure 8-11   The connector part of the lane.

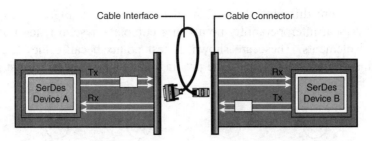

Figure 8-12   The cable connection part of the lane.

Every individual lane of the PHY consists of transmit and receive differential pairs. The transmit pair and its transmission medium are often called a channel. Within a channel, signals may traverse three basic types of copper paths: chip-to-chip, card-to-card, or card-to-cable:

- **Chip-to-chip:** A chip-to-chip lane normally resides on an etched circuit board (ECB), which serves as the transmission medium. A typical application is a PCI Express bus on a motherboard, where large amounts of data need to be transmitted from one device to another.

- **Card-to-card:** When a connector is part of the lane, as in an edge card application, the specification may require AC coupling capacitors on the transmit or receive side of the link. This eliminates potential common-mode bias mismatches between transmit and receive devices. Low-cost links may employ long runs on FR4 board and inexpensive connectors, both of which can add jitter, crosstalk, and potential imbalances due to layout.

- **Card-to-cable:** Introducing a cable connection into the lane adds yet another source of loss and jitter. Cable-connected serial links are prevalent in server applications such as InfiniBand, in storage applications such as Serial ATA and Fibre Channel, and in peripheral applications such as 1394B.

A cabled system such as that shown in Figure 8-12 includes contributions from the maximum potential number of vendors. Compliance and interoperability testing are key to the success of any product in which many independently developed elements must function together flawlessly. This is especially true in the early stages of a standard's life cycle, when details are still in flux and interpretations may vary.

All standards address the important issue of transmit and receive loss budgets. They also define compliance test points at which system-level testing must be performed. Essentially, compliance points are those at which system components (usually from different vendors) need to interoperate. Figure 8-13 summarizes some typical interoperability points in a complete system made up of interconnected elements. These are shown as test points because they are specifically nominated in the standard as the probing attachment points for test instrumentation.

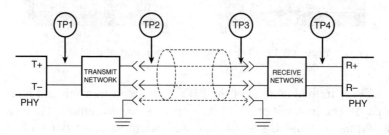

Figure 8-13   Common interoperability test points in a serial system.

8.5   PHYSICAL LAYER COMPLIANCE TESTING

While today's serial architectures operate in the digital realm, much of the compliance testing task consists of analog measurements. Why? Isn't this all about digital data? The answer is a qualified yes. There are important exceptions. Digital signals exist in a world of distributed capacitance, noise, power-supply variations, crosstalk, and other imperfections. Each of these phenomena detracts from "ideal" digital signals, sometimes to the point of compromising their ability to carry data. As a result, serial bus specifications set limits on signal distortions and degradation.

The device under test must meet these limits to be considered "in compliance." Test conditions and test points are explained in detail in the standards, sometimes to the extent of recommending specific test equipment, including the particular make and model. Figure 8-14 summarizes a typical compliance test process—in this case applicable to the InfiniBand and PCI Express standards but very similar to those for the other leading standards.

8.5.1   Common Compliance Measurements

Serial standards normally include amplitude, timing, jitter, and eye diagram measurements within their compliance testing specifications. Automated measurement/analysis tools are commonly used to speed the selection and application of these tests. Amplitude and timing measurements are covered in Chapter 6, so this discussion concentrates on eye diagram and jitter measurements.

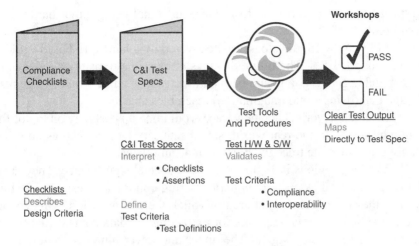

Figure 8-14   The compliance procedure.

8.5.2   Eye Diagrams: The Cornerstone of Compliance Measurements

The eye diagram has become the definitive tool for validation and compliance testing of digitally transmitted signals. It is a display, typically viewed on an oscilloscope, that quickly reveals impairments in a high-speed digital signal (see Figure 8-15).

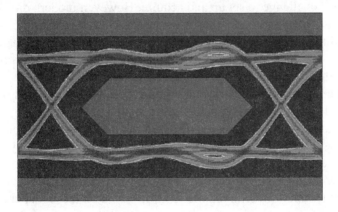

Figure 8-15   An eye diagram. The blue areas are the mask violation zones.

The eye diagram is constructed by overlaying the waveform traces from many successive unit intervals (UIs). Eye diagrams display serial data with respect to a clock recovered from the data signal using either hardware or software tools.

In Figure 8-15, the clock was recovered by a hardware-based "golden phase locked loop (PLL)." The figure displays all possible transitions (edges), positive-going and negative-going, and both data states in a single window. The result is an image that (using some imagination) resembles an eye.

In an ideal world, each new trace would line up perfectly on top of those that came before it. In the real world, signal integrity factors such as noise and jitter cause the composite trace to "blur" as it accumulates.

The blue regions in Figure 8-15 have special significance. They are the violation zones used as mask boundaries during compliance testing. The blue polygon in the center defines the area in which the eye is widest. This encompasses the range of safe decision points for extracting the data content (the binary state) from the demodulated signal. The upper and lower blue bars define the signal's amplitude limits.

If a signal peak penetrates the upper bar, for instance, it is considered a "mask hit" that presumably will cause the compliance test to fail (although some standards may tolerate a small number of mask hits). More commonly, noise, distortion, transients, or jitter cause the traces to thicken. The eye opening "shrinks," touching the inner blue polygon. This too is a compliance failure, because it reveals an intrusion into the area reserved for evaluating the state of the data bit.

The eye diagram's compelling advantage is that it enables a quick visual assessment of the signal's quality. The information-rich display looks as if it might be a challenge to set up and acquire. However, modern digital oscilloscopes can be equipped with tools that expedite the complex clock-recovery, triggering, and scaling processes automatically, before performing quantitative measurements on the data. With these software and hardware facilities, the eye diagram measurement has become a single-button operation.

The key benefit of the eye diagram is that, in one captured screen, all possible signal transitions of the signal are displayed: positive-going, negative-going, leading, and trailing. This single display provides information about the eye opening, noise, jitter, rise and fall times, and amplitude. The display can be used for qualitative analysis, and the embedded statistical database of the oscilloscope can be used to make quantitative measurements.

Rather than extracting numeric information on the signal characteristics, the two-dimensional shape can easily be compared to a group of violation zones called a mask. Figure 8-16 shows a mask as outlined in a typical standards document.

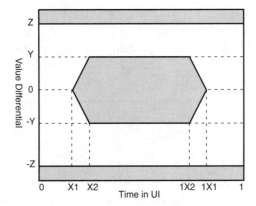

Figure 8-16   Eye pattern mask.

Masks are defined by and for each standards group. Comparing the shape of an eye to a mask is a quick, efficient way to ensure that the transmitter source signal will not cause excessive degradation of the receiver performance. Looking again at the eye diagram shown in Figure 8-15, you can see that the boundaries of this image, which passes the compliance test, are within the ranges expressed in Figure 8-16.

8.5.3   Jitter Measurements

Jitter measurements have been a topic of extensive discussion in industry working groups in recent years. Jitter is so important, in fact, that specialized analysis tools exist to help designers penetrate this difficult problem.

Time-interval error (TIE) is the basis for many jitter measurements. TIE is the difference between the recovered clock (the jitter timing reference) and the actual waveform edge. Performing histogram and spectrum analysis on a TIE waveform provides the basis for advanced jitter measurements.

Jitter measurement has few consistent measurement methods, aside from eye diagrams. In Fibre Channel, InfiniBand, and XAUI, methods have been developed by the T11.2 jitter working group, which defines total jitter as "the sum of random and deterministic jitter components." The relationship between total jitter (TJ), jitter eye opening (at 10^{-12} BER), and unit interval is as follows:

$$\text{total jitter} + \text{jitter eye opening} = 1 \text{ unit interval}$$

Total jitter is defined by establishing a jitter timing reference defined by a "golden PLL" model for clock recovery from the serial bit stream. The "golden PLL" loop bandwidth is defined as $f_c/1667$, where f_c is the bit rate. From the jitter timing reference, a cumulative distribution function (CDF), also known as a "bathtub curve" (see Figure 8-17), is established, showing where the eye opening at 10^{-12} BER occurs.

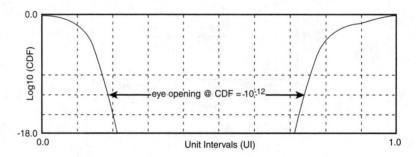

Figure 8-17   "Bathtub" curve, or cumulative distribution function.

Traditionally these measurements have been cumbersome to make with a BERT scanner or time-interval analyzer (TIA) instrumentation. But now automated software toolsets are available for performing these measurements with a real-time oscilloscope and for separating random and deterministic jitter.

Standards such as Serial ATA and PCI Express agree that total jitter is the sum of random and deterministic components. However, various models for clock recovery exist. These standards call for the clock to be recovered from a set number of consecutive bits in the serial bit stream. These jitter methods are used because different forms of clock recovery such as oversampling or phase interpolation can be implemented.

Figures 8-18 and 8-19 show two different jitter measurement methods used on the same 2.5 Gbps signal.

Figure 8-18 shows the total jitter and deterministic jitter (as specified in InfiniBand). Total jitter is determined using the bathtub curve method. In this case, the jitter eye opening is 0.583 UI, and the total jitter is 0.417 UI at 10^{-12} BER. Figure 8-18 shows the "250-cycle jitter" test specified in PCI Express. The median to maximum outlier is determined by the median of the TIE histogram to the maximum outlier on the histogram—in this case, 44.87 ps measured over any 250 consecutive bits.

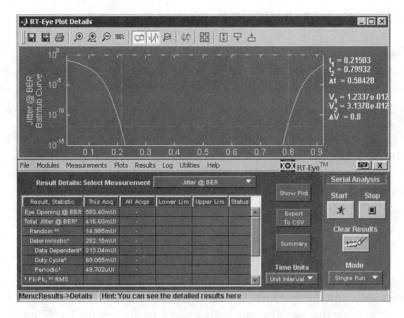

Figure 8-18   Total jitter and deterministic jitter measurement determined using InfiniBand compliance methodology.

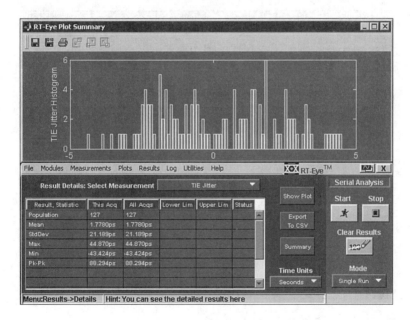

Figure 8-19   The results of a 250-cycle jitter test performed by following PCI Express compliance methodology.

8.6 MEASUREMENTS ON OPTICAL SIGNALS

Many serial data standards such as InfiniBand, Fibre Channel, and 1394B have both copper and optical PMD variants. Examples of commonly used optical measurement are average optical power (AOP) and extinction ratio (ER). Somewhat less known is optical modulation amplitude (OMA), which is growing in popularity because of its use in the Fibre Channel and 10 Gigabit Ethernet (IEEE 802.3ae) standards—by far the fastest-growing segments of the very high-speed optical market.

Extinction ratio (ER) is a very common optical measurement; it is defined as follows:

$$ER = log_{10}\left(\frac{high}{low}\right)$$

The high and low have changed somewhat over the years, but today they are the mean values in the aperture, which for the NRZ eye is a region horizontally centered between the eye crossing, with a width of 20% of the UI. Figure 8-20 shows the ER measurement on an oscilloscope screen.

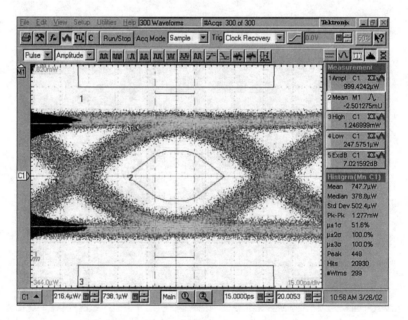

Figure 8-20   Extinction ratio measurement.

The basic issue of ER measurement is that the low level greatly influences the measurement; typically it is necessary to run "dark level compensation" to obtain a reliable result.

8.6.1   Optical Modulation Amplitude (OMA)

The performance of an optical link is related to the transmitted power of the high and low signal levels, which therefore need to be measured. Instead of measuring these parameters directly, telecommunication standards have typically specified extinction ratio and average optical power.

The 10 GbE (Gigabit Ethernet) specification uses OMA instead of the ER/AOP pair. This change does not remove the references to ER, however, so both OMA and ER have to be measured. AOP remains only in non-signaling-related specifications (such as safety).

The equation to obtain OMA is simply

$$OMA = high - low \text{ (in watts)}$$

or, in its more common usage in dBm units:

$$OMA_{dBm} = 10\,log\left(\frac{high - low}{1mW}\right)$$

Referring to Figure 8-21, note that the n (in the n*UI expressions at the top of the figure) differs from standard to standard. For example, for 10 Gigabit Ethernet it is allowed to be between 4 and 11.

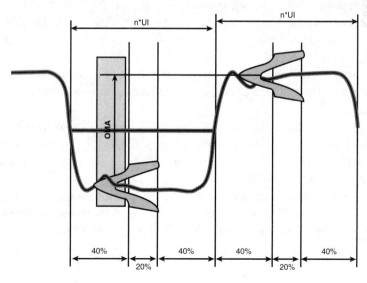

Figure 8-21   Measuring OMA on a square wave as prescribed in the 10 GbE standard.

For the sake of comparable results, the lowest allowed value (such as n = 4 for 10 GbE) is recommended.

Practical measurement is shown in Figure 8-22.

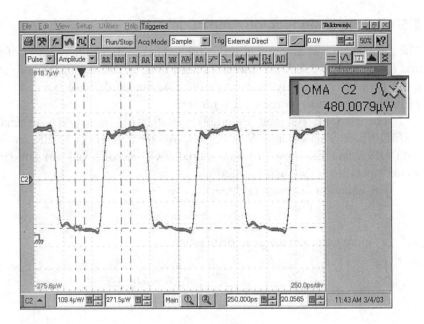

Figure 8-22   Practical measurement of OMA.

The measurement on a square wave is specified because it is perceived to be more repeatable than a measurement on the eye diagram. However, it cannot be performed on live data or on a pseudo-random binary sequence (PRBS). For this reason, another definition is sometimes used based on the measurement of OMA on an eye diagram.

Figure 8-23 shows an example of this measurement, which illustrates the intersymbol interference (ISI) that is relatively common in today's low-cost transmitters.

In some standards, this eye-diagram-based method is the only one used for OMA. However, the 10 GbE standard only allows this method for approximating OMA and insists on the square-wave method as the only binding technique for conformance verification.

ER and AOP will remain the preferred way to specify longer-haul links. OMA is being used more frequently because of the growing number of shorter-reach standards such as 10 GbE and 10GBFC (Gigabit Fibre Channel).

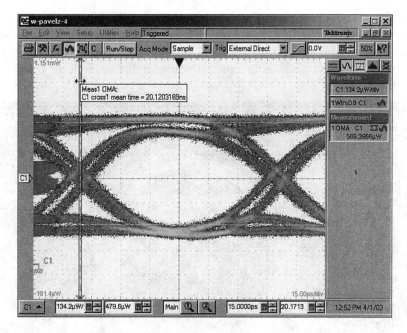

Figure 8-23   Eye diagram measurement of OMA.

8.7   COMPLIANCE MEASUREMENT CONSIDERATIONS: ANALYSIS

Jitter tests and eye diagram measurements are among the most complex procedures that an oscilloscope can perform. Both require statistical analysis of vast amounts of data, and both are key elements of serial compliance testing.

Eye diagrams have been the province of sampling oscilloscopes for many years. The sampling methodology provides accurate measurements with a very low jitter noise factor. However, the requirement for a stable, uninterrupted signal makes the sampling instrument impractical when performing industry-standard jitter measurements. Such techniques require continuous (contiguous) data to perform the analysis—something that is not provided by sampling oscilloscopes.

More recently, some real-time oscilloscopes have incorporated clock-recovery circuits based on a PLL-based clock-recovery circuit. They perform eye diagram tests using random equivalent-time sampling. This technique is similar to that used in sampling oscilloscopes and requires repetitive trigger events to build an eye diagram. As a result, it is susceptible to trigger jitter.

Fortunately, some real-time oscilloscopes offer another means of eye rendering. They rely on the single-trigger nature of real-time acquisition, capturing a contiguous series of complete waveform cycles as they occur after the trigger event. The embedded clock is recovered in software after the acquisition, and the actual waveform edges are "redrawn" using the recovered clock as the reference. Figure 8-24 shows an eye diagram derived with this methodology.

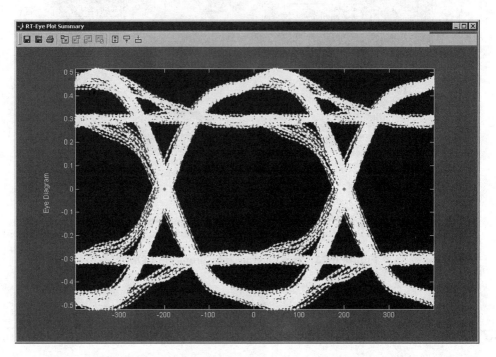

Figure 8-24   A real-time eye diagram displayed with post-processing techniques.

This approach has several advantages. It provides a jitter noise factor as low as 700 fs on certain real-time oscilloscope models, making it the equal of sampling oscilloscopes in terms of eye diagram precision. Moreover, because the clock is recovered using software-based digital signal processing, clock recovery is not limited to a single algorithm. The software-based package provides an interface for changing the method by which the clock is recovered—a valuable asset in a world of constantly evolving standards. Real-time clock recovery and eye rendering also provide a means for separating transition bits from nontransition bits and performing separate mask testing operations on each type of bit, as is required for PCI Express. Figure 8-25 shows the separation of transition and nontransition bits from the eye diagram shown in Figure 8-24.

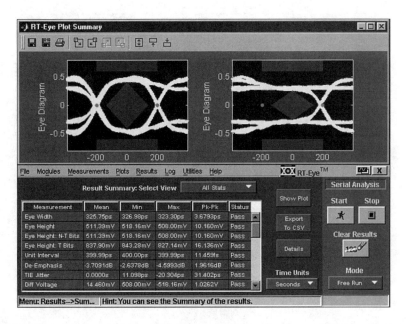

Figure 8-25   The serial compliance summary screen showing the eye pattern mandated by the standard as well as quantitative results in tabular form.

It is often necessary to analyze the data more deeply in the context of the compliance measurements. For example, an eye diagram test may fail because the embedded clock signal has too much modulation. The eye diagram shows only violations to the mask, but a time-interval error waveform trend or frequency spectrum can reveal further clues. Other built-in oscilloscope tools such as cursor measurements and zoom controls can aid the in-depth analysis.

8.7.1   Automating the Analysis Process

Eye diagrams, and jitter measurements in particular, produce a volume of data that would be difficult to manage without the help of automated analysis tools. Figure 8-26 shows the user interface of an integrated analysis package running on a real-time oscilloscope.

Key measurements such as rise and fall time can be set up with just one on-screen "button." The acquisition window presents the raw serial waveform, but Figure 8-27 brings out the true benefit of the analysis package. In Figure 8-27, the automated tools work on the waveform from Figure 8-25 to produce an information-rich jitter trend analysis display.

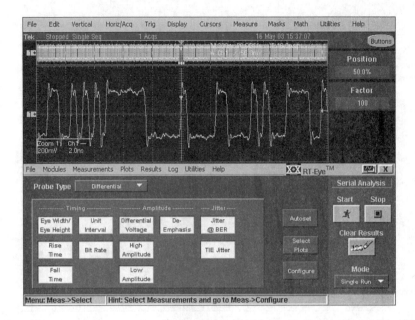

Figure 8-26   The user interface window of a serial analysis tool running on a real-time oscilloscope.

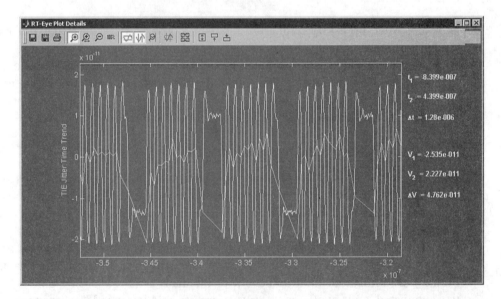

Figure 8-27   Jitter trend analysis. The real-time waveform is shown in blue and the jitter trend in red.

8.8   TESTING THE SERIAL LINK

Having looked at the basic concepts of compliance testing as applied to the physical layer, it is now appropriate to look at key elements of the serial link and their key measurement requirements. The examples presented in this section refer to SATA testing, but the principles apply to any high-speed serial bus.

8.8.1   Transmitter Testing

In many ways, transmitter testing is the simplest step in the whole process. Signals measured at the transmitter output are as clean and free of aberrations as they ever will be. Even so, these signals are not perfect. Optimizing them is in the best interests of every transmitter design. And they must live up to compliance requirements as well.

Transmitter measurements address general PHY conformance specifications including (among others) unit-interval and spread-spectrum modulation frequency. Transmitted signal requirements include differential output voltage, rise/fall time, differential skew, both total jitter and deterministic jitter at the connector over a span of 5 UI and 250 UI, and other measurements. Finally, out-of-band (OOB) signal measurements have their own unique test requirements.

8.8.1.1   Bandwidth and Rise Time

As a rule of thumb, the oscilloscope should have enough bandwidth to acquire the fifth harmonic of the signal's highest frequency. For SATA II, the data rate is 3.0 Gbps, which equates to an underlying clock frequency of 1.5 GHz. Thus, the oscilloscope should have a bandwidth of at least 7.5 GHz for SATA II measurements. Transmitter measurements may be performed with either a real-time oscilloscope or an equivalent-time sampling oscilloscope equipped with the appropriate software tools.

Although the fifth harmonic is a rule of thumb, an oscilloscope's rise time performance is the critical factor in second- and third-generation standards. Accurate measurement of the rising and falling edge of signals is extremely important. Edges that are too slow may cause bit errors, and edges that are too fast can contain very high frequency components that can cause interoperability issues or EMI.

The displayed rise time on an oscilloscope is equal to the square root of the sum of the squares:

$$risetime\ display = \sqrt{\{(risetime\ of\ signal)^2 + (risetime\ of\ oscilloscope)^2\}}$$

The smaller the rise time of the oscilloscope in this equation, the closer the displayed rise time is to the signal's rise time. To measure rise time signals accurately, the oscilloscope's rise time specification must be about 3 to 5 times that of the signal.

In reality, although the fifth harmonic rule leads us to believe that an oscilloscope with a 7.5 GHz bandwidth will do for SATA, in fact the SATA-IO recommendation is for at least a 10.5 GHz oscilloscope because of rise time considerations. This is similar for other standards such as USB2.0 and PCIe.

8.8.2 Link Testing Measurements

The transmission link, although it is a passive path, is critical to the success of the system design. Its characteristics impact the signals passing through it, to the extent that a noncompliant cable or connector can make the signals unintelligible at the receiving end. PCB traces, pads, vias, connectors, and cables all contribute to the environment that the signal encounters. In addition, the growing use of multilane serial implementations means more potential for crosstalk and interaction within the conductors that make up the link. Transmission link properties such as impedance (and its variations over the length of the path) and S-parameters must be thoroughly characterized.

The range of link measurements is extensive. The list includes (but is not limited to) mated connector impedance, insertion loss, rise time, intrapair skew, and more. In addition, there is growing urgency to predict the link's contribution to the bit error rate.

As described earlier, most serial buses use differential transmission techniques to minimize external interference with their low-amplitude data signals. Thus, every signal path is made up of two conductors carrying the signal and its complement. To accurately characterize the behavior of such a path, true differential TDR is required.

The TDR must send simultaneous complementary pulses and acquire both responses at once. The alternative is to send pairs of positive-going pulses sequentially and then use calculations to correct the polarity and alignment before displaying and analyzing the signal. However, this latter method cannot be used to make measurements on powered links and may not comply with some standards.

Testing the bit error rate of serial elements has long been a time-consuming task—on the order of hours for a single link component. Designers are eager for solutions that can reduce test time without degrading the accuracy of the BER results.

8.8.3 Receiver Testing

Some aspects of receiver testing remain the least resolved issue of the whole measurement process. When a signal leaves the transmitter, (normally) it is clean and sharp. By the time it reaches the receiver, it may be so degraded that the eye opening is barely distinguishable. This is an inevitable consequence of the increasing data rates and edge speeds discussed earlier, as fast edges and narrow pulses react to distributed capacitance and physical impediments such as PCB vias. De-emphasis and equalization techniques can reduce the signal deterioration, but in any event, measurements at the receiver input must deal with a very compromised signal.

Several well-established receiver test procedures rely on applying a signal with known characteristics to the input. These characteristics may range from a near-perfect stream of square wave pulses to variants containing de-emphasis filtering, jitter, noise, and other distortions.

The receiver's output should respond in a predictable manner. The receiver's response is the measure of its tolerance for signal imperfections in the end-user application.

Among the specific receiver tests are amplitude sensitivity, timing, and jitter tolerance. Some receiver measurements require PRBS patterns and may require special loopback settings within the DUT itself.

Standards and procedures for receiver testing are still being defined. Until such time as the necessary definitions are resolved, it may be difficult to achieve repeatable results with every equipment configuration. However, SATA has approved three methods of carrying out receiver testing from Synthesis, Agilent, and Tektronix.

Receiver testing requires an input signal source with bandwidth matched to that of the DUT. Typically this is a pulse, data, or pattern generator with at least two outputs to provide the complementary sides of the differential signal.

Applying jitter and/or de-emphasis to the signal adds substantial complexity to the test apparatus when a digital source is used. External analog function generators, power combiners, and associated fixturing may be required to modify the generator's output before it is submitted to the DUT.

The receiver output may be acquired by either a real-time or equivalent-time oscilloscope; usually the selection is made based on the data rates involved. The very highest frequencies, above a fundamental of 5 GHz (10 Gbps), require an ET instrument, which trades off on behalf of bandwidth the ability to acquire one-time (single-shot) events.

The challenge is to assess all possible states of the binary information and confirm that waveform characteristics such as rise time and amplitude are within acceptable margins.

Increasingly, designers are relying on serial analysis application packages (see Figure 8-28) to set up, acquire, and analyze the eye diagrams that summarize these areas of DUT performance.

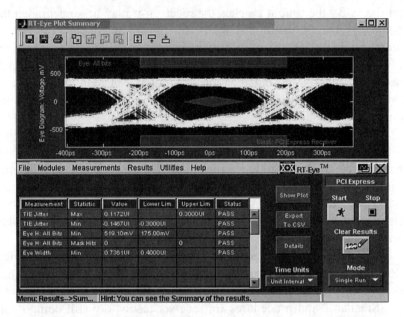

Figure 8-28   The results page from an oscilloscope-based serial analysis software application. The toolset encompasses quantitative measurements, pass/fail tests, masks, eye diagrams, and statistics.

8.8.3.1   Acquiring and Interpreting Data

System testing takes us from the top of the OSI physical layer to the data link layer. In a sense, the "measurements" that occur here are not really measurements at all. Although they are acquisitions of electronic data, the waveform characteristics are a settled issue by this time. What matters in system testing is the content that those electronic signals express. This is the information that all the elements working together are meant to deliver. It is necessary to decode serial buses and view actual transactions. These must be correlated with activities occurring on other buses or within the chipset and CPU.

8.8.3.2   Bus Support

For system testing, it is essential to support a broad range of buses. In this context, "support" means not only the software tools to decode and interpret the data, but also hardware probing and interconnect devices and interfaces to expand serial data to its readable parallel form. These interfaces must be able to deal with issues such as multiple lanes, embedded clocks, and, of course, differential transmission.

8.8.3.3 Acquisition and Storage Capacity

Ultimately the support interfaces that feed an instrument—typically a logic analyzer—must have enough channels to accommodate word lengths of 32 bits or more on multiple buses—potentially hundreds of channels. These word lengths must be accompanied by massive memory depth to capture thousands or millions of operational cycles.

8.8.4 Transmitter Testing

Transmitter physical layer tests fall into three categories, each containing several measurements as prescribed in SATA standard 7.2.1 and/or in the SATA Unified Test Document 2.12.1:

- General requirements (PHY 1 to 4)
- Transmitted signal requirements (TSG 1 to 12)
- Out-of-band requirements (OOB 1 to 7)

For the sake of simplicity, the following discussion focuses on the host transmitter, with references to the hard disk drive (HDD) transmitter where appropriate. Either transmitter must be put into a self-test state initiated by commands from a controller or surrogate hardware. Both make similar demands on the test equipment and procedures, although some hosts may lack the necessary utilities to put the host into BIST mode, as explained later.

Serial data transmission is all about speed—the data rate of a serial bus or device in gigabits per second. From the perspective of measurements with an oscilloscope, data rates relate directly to instrumentation bandwidth.

Either real-time (RT) or equivalent-time (ET) oscilloscope platforms may qualify for measurements on the transmitter element. No explicit requirement in the SATA specification prefers one over the other for evaluating signal characteristics.

Table 8-2 summarizes present and foreseen SATA data rates. The "future needs" entry does not specifically apply to SATA, but the XFI example is included to demonstrate that bandwidth requirements can be expected to continue escalating.

Both SATA Gen I (SATA I), with its 1.5 Gbps data rate, and SATA II, running at 3.0 Gbps, are compatible with the capabilities of real-time oscilloscopes. The target bandwidth for such tools is five times the maximum bus frequency—

enough to encompass the signal's fifth harmonic. In the case of SATA II, this multiplies out to 7.5 GHz. Yet another SATA generation is in development, with an expected data rate of 6 Gbps. This rate is also within reach of today's top real-time oscilloscopes.

Table 8-2   SATA Test Requirements

	Data Rate	Maximum Bus Frequency	Fifth Harmonic	Platform
Serial ATA	1.5 Gbps	0.75 GHz	3.75 GHz	RT
Serial ATA Gen II	3.0 Gbps	1.5 GHz	7.5 GHz	RT
Serial ATA Gen III	6.0 Gbps	3.0 GHz	15 GHz	RT or ET
Future needs (such as XFI)	10 Gbps	5 GHz	25 GHz	ET

The ET platform is likely to remain the only solution for the very fastest technologies at any given time. Sometimes an ET instrument is an expedient solution even though its 70 GHz bandwidth far exceeds what is needed for most serial measurements. An ET oscilloscope is the tool of choice for required TDR-based impedance measurements on the link elements in the system—PCB traces, connectors, and cables—and therefore may be available for other tests as well. The recent emergence of advanced jitter measurement software (addressing both temporal jitter and "vertical" noise-related jitter) gives the ET oscilloscope another toolset for transmitter tests. The ET platform offers the lowest available instrument noise and jitter and is always a candidate when high-speed serial measurements are needed.

However, the ET oscilloscope requires a repeating signal that may not always be available. It is less suited to troubleshooting work and for measurements in which single-shot acquisitions are the order of the day.

In general the RT oscilloscope is more commonly used for measurements such as the SATA transmitter compliance test series. This section focuses on using a real-time instrument for the job.

Two proven real-time oscilloscope architectures are available: digital storage oscilloscopes (DSOs) and digital phosphor oscilloscopes. Both tools offer models with ample bandwidth, record, and timing accuracy to meet the needs of serial bus technologies such as SATA I and SATA II.

Many of these instruments can run onboard serial measurement software applications that expedite everything from initial setup to eye diagram display to statistical analysis.

Various industry committees have concluded that oscilloscope bandwidth must be sufficient to acquire the fifth harmonic of the maximum bus frequency. This is simply because "square waves" of the type that make up digital data streams theoretically consist of a fundamental and many odd harmonics. An oscilloscope's bandwidth sets limits on how many harmonics it can capture—in effect, how accurately it can acquire the square wave. The SATA-IO group recommends an oscilloscope bandwidth of at least 10 GHz bandwidth for SATA II measurements.

Similarly, rise time is a concern that is inextricable from bandwidth when selecting an oscilloscope for SATA measurements.

Assuming 5% time measurement accuracy as a nominal goal, an oscilloscope should have a bandwidth of almost 5 GHz to reproduce a 100 ps signal edge. To capture a 30 ps rise time with the same degree of accuracy, the oscilloscope needs a 16 GHz bandwidth.

Not to be overlooked is the need for record length in the RT oscilloscope. High sample rates such as 40 GS/s consume vast amounts of record length. To capture a complete 223-1 PRBS pattern, as required by many serial standards, the instrument's memory must store 1.4 ms worth of data. Other deep-memory applications are similarly demanding, including spread-spectrum clock modulation analysis and low-frequency noise tests—each requiring a large population of samples.

Influences other than bandwidth often guide the platform choice for SATA measurements. RT oscilloscopes are ubiquitous in engineering and research labs, so it is often feasible to use existing equipment for the job. When these general-purpose instruments are equipped with serial measurement software tools and accessories such as high-speed differential probes, they become powerful application-specific analysis platforms.

Both RT and ET platforms have unique strengths for digital serial analysis. Real-time oscilloscopes offer versatility and efficiency for troubleshooting, and they adapt readily to digital serial analysis tasks. Equivalent-time instruments are simply unsurpassed in bandwidth, and they contribute vanishingly low jitter and noise to the measurement. In most laboratories and design departments involved with serial bus systems, both platforms can be put to efficient use.

8.9 PROBES AND PROBING

Serial components rely on fast, low-amplitude signals to deliver the high data rates that today's systems demand. Differential signaling protects these signals from intrusive noise and crosstalk. However, from the designer's perspective, differential signaling means getting two complementary (mirror-image) signals from the device under test to the oscilloscope's acquisition system for every data channel.

There are two methods of connection for making these differential measurements. The first uses two channels of the oscilloscope, each acquiring one side of the signal's differential pair. The resulting waveforms are summed using the instrument's built-in mathematics functions. This approach calls for very careful deskewing of the two channels and precisely matched cables to connect them to the DUT.

The other approach uses a dedicated differential probe (see Figure 8-29).

Figure 8-29   An SMA differential oscilloscope probe suitable for high-speed digital serial
analysis applications.

A single probe (with two inputs and one output to the oscilloscope) captures
both sides of the differential signal pair. The probe cables are terminated with
SMA connectors that attach via a fixture to the DUT's SATA connector. This integrated differential probe simplifies connections between the instrumentation and
the DUT and minimizes skew discrepancies. It also consumes just one oscilloscope input, a useful benefit when several differential channels need to be
observed simultaneously.

During troubleshooting, it is not always feasible to connect to the DUT
through SMA connectors. For this kind of work, a movable differential probe is
the right choice. Here too, the probe brings both sides of the differential signal
pair through one oscilloscope input.

8.9.1   Fixturing

Compliance tests of all kinds require fixturing between the measurement instruments and the DUT. The purpose of this is twofold.

First, a well-designed fixture preserves the integrity of the signals under observation. In most high-speed measurements, precise fixturing is the only way to ensure accurate, repeatable results and to prevent aberrations and rise time degradation effects from simply overwhelming the signal.

Second, fixturing ensures the same baseline signal environment for the DUT, irrespective of who is doing the testing. A fixture is a medium with correct impedances, matched path lengths, mating connectors, and mechanical support for connections. The fixture itself must comply with specifications. This provides a consistent, controlled-impedance interface and minimizes variables that can affect measurements.

Fixtures for SATA and other serial buses (HDMI, FB-DIMM, and more) are commercially available. Fixturing specialty vendors and instrumentation makers cooperate to ensure compatibility among their respective tools. Figure 8-30 depicts a set of SATA fixtures: near-end, far-end, and live probing types.

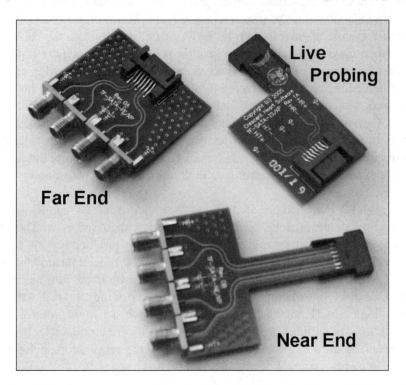

Figure 8-30   SATA fixtures are necessary for clean, reliable acquisition of high-speed signals.

8.10   SOFTWARE TOOLS

The specified measurements for SATA PHY and TSG (Transmitted Signal Requirement) tests include complete statistical analysis applied to captured SATA waveforms containing many different test patterns. These measurements are complex procedures both in terms of setting up the oscilloscope's operation and interpreting pass/fail or quantitative results. A waveform view expresses massive amounts of data.

Jitter, for example, usually encompasses so many edge timing variations that it is impossible to distinguish any one edge with the unaided eye. Furthermore, what matters in jitter analysis is not so much the individual edge placement as the statistical summary of all edge placements.

Tasks like these are tailor-made for computer-aided automation and analysis. When the demands of compliance testing are considered, application-specific software tools are indispensable.

As discussed elsewhere, many modern RT oscilloscopes can be equipped with both eye diagram analysis software and jitter measurement/analysis software. These oscilloscopes are integrated with computer platforms that can run such applications on the instrument itself. Individual plug-ins adapt the application to specific standards such as SATA or PCI Express. In the case of the real-time eye diagram tools, the application post-processes acquired signal data to produce a detailed eye view plus actual measurements expressed in numeric form. The embedded clock is extracted by the software. Masks can be imposed on the eye and "hits" automatically calculated. Histograms are also automatically generated. Certain ET oscilloscopes take jitter measurement to the next level by adding tools to acquire, separate, and analyze the vertical (amplitude-related) components of jitter. As explained earlier, the ET oscilloscope offers the highest bandwidth of any oscilloscope platform. With this comes the lowest noise floor and the lowest internal jitter.

The ET methodology is different from that of the RT application packages. Clock extraction is done with hardware rather than software. The application builds a database of samples that can be used to produce reports ranging from numeric values to three-dimensional curves that correlate to actual bit error rates.

A third useful application toolset is a hardware/software pairing that provides flexible control options for addressing and driving the DUT with the pattern data. Figure 8-31 shows a user screen for such a pattern control application, which resides within a PC that manages a hardware emulator tool. The application can be used to place the host device in the required state for compliance measurements, protocol debugging, and related tests.

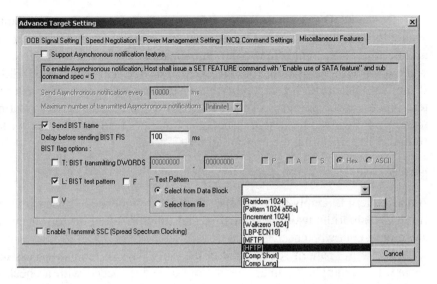

Figure 8-31   A pattern control tool that can be used to support SATA compliance testing. The package manages standard patterns such as HFTP, MFTP, lone bit, and others.

8.10.1   Following a Pattern

The SATA specification requires that hosts and drives support BIST FIS modes. Several modes are defined for BIST FIS, but the commonly used ones are BIST-T (pattern generation) and BIST-L (loopback).

SATA compliance tests require specific data patterns that, once initiated, are sustained by the device's BIST elements. Test results are meaningless unless these defined patterns are used to make the prescribed tests.

Some systems include features that allow the host itself to initiate the BIST patterns. Others require external hardware devices to program the host port to the BIST mode. In truth, no single BIST initiation method is guaranteed to work on all host devices. Different methods of placing DUTs into BIST mode have different advantages and disadvantages. Vendor-specific software or digital testing tools are often the best and easiest way to get DUTs into the correct mode. However, these require a disconnection step, which can cause about 20% of products to go into a "sleep" mode. AWGs with Direct Digital Synthesis (DDS) can always put a properly working product into BIST mode and do not require the product to be disconnected. On the other hand, they require some manual setting up. Procedures have been developed that allow an arbitrary waveform generator to send a series of SATA commands to the host to place it in the BIST-L (loopback) mode.

On the HDD side, most drives contain dedicated standards-compliant BIST-T circuits that can be activated by external system initialization hardware running a pattern control application. After initialization, the patterns are free-running without further intervention from the external device. This is the BIST FIS mode. If the drive lacks BIST-T capability, provisions are made to accept external pattern inputs for the purposes of the PHY tests. Note that the drive's rotating disk itself is not involved in these transactions. It neither reads data to nor writes data from the platters during the transmitter tests.

8.10.2   What Is a Pattern?

A data pattern is simply a predictable, predefined sequence of binary information. It may be algorithmically generated in real time or stored bit-for-bit in memory and clocked out when needed.

The term "pattern" dates back to the early days of random-access memory testing. For example, an image read from a device loaded with a checkerboard pattern actually looked like a checkerboard of alternating 1s and 0s. The term remains in use, and the context has expanded to include almost any standardized binary sequence, serial or parallel, used as digital stimulus data.

The SATA transmitter test specification is unusual in that it prescribes several different patterns over the course of its PHY and TSG tests (see Figure 8-32). Most other standards specify only one pattern for the purpose. The SATA patterns are as follows:

- HFTP: High-frequency test pattern
- MFTP: Mid-frequency test pattern
- LFTP: Low-frequency test pattern
- LBP: Lone-bit pattern

The HFTP pattern delivers a change of state (an edge) in every cycle, with MFTP and LFTP providing successively lower frequencies of change. The LBP pattern provides a 1 at periodic intervals, preceded by 4 0 bits and followed by 3 0 bits.

The transmitter physical layer test series includes measurements in three categories:

- PHY measurements

  The PHY tests encompass the SATA specification's general signal requirements and evaluate the underlying timing characteristics of the SATA signal environment—essentially cycle length (unit interval) and clock performance, including spread-spectrum clocking (SSC) behavior.

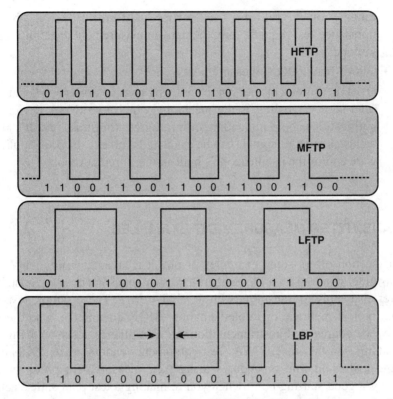

Figure 8-32   Industry-standard test patterns for SATA compliance. The arrows on the bottom
trace bracket the "lone bit" in the LBP pattern.

SSC is designed to minimize radiated emissions from devices such as PC
motherboards and SATA drives. Even a few picoseconds of modulation can
substantially reduce emission problems. Ranges and tolerances are critical,
since the clock affects every transaction on the SATA bus. The modulation
frequency of the SSC must be in the range of 30 to 33 kHz. The modulation
deviation is specified to be no more than 0.035% from the nominal frequency and is "down-spread" for a total allowable range of 0 to –0.5%.

A full-featured jitter measurement application running on a real-time or
equivalent-time oscilloscope can automate the acquisition and analysis of
these entities.

• Transmitter signal measurements

Transmitter signal measurements examine the "analog" behavior of SATA
signals: voltages, skew, and edge times. The series also includes jitter measurements spanning various sample populations. The tests typically are performed with the aid of either an eye diagram tool or a jitter application

running on a real-time oscilloscope platform. However, the amplitude imbalance test requires only the native measurement capability of the oscilloscope.

• Out-of-band (OOB) measurements

OOB measurements pertain to signaling parameters as opposed to data characteristics. Signaling is the means by which two SATA devices exchange notifications about impending or in-progress transmissions. It's also called a handshake. OOB signals can be evaluated with eye diagram application software and/or the oscilloscope's built-in measurement tools.

8.11 TRANSMITTER MEASUREMENT EXAMPLES

This section offers some examples of physical layer measurements on a SATA transmitter residing on a PC motherboard, which also acts as the host. The procedures also apply to transmitters on the SATA drive. These sample tests are drawn from typical methods of implementation (MOI) documents, which explain specific procedures and instrumentation for compliance tests on particular serial standards—in this case, SATA. Test equipment vendors create MOIs after developing and evaluating procedures using their measurement products. A MOI is always a good starting point when you're making a test plan and selecting equipment for serial devices.

8.11.1 Measurement Configuration

The general measurement configuration for PHY and TSG measurements is shown in Figure 8-33. The RT oscilloscope receives signals from the DUT via a differential SMA probe connected to the test fixture. Two 50-ohm cables conduct the complementary halves of the signal pair.

This figure shows a true differential probe on the oscilloscope, but it is also acceptable to use a pseudo-differential arrangement with two separate probes feeding two channels of the instrument. If the pseudo-differential approach is used, the oscilloscope's mathematical functions can easily calculate a differential measurement.

A typical results page using eye diagram application software is shown in Figure 8-34.

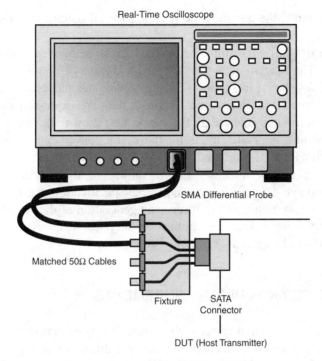

Figure 8-33 The configuration for PHY and TSG measurements.

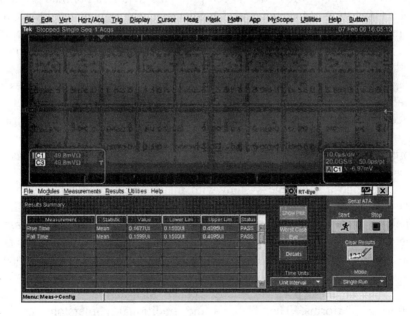

Figure 8-34 Results of the transmit rise/fall time measurement.

Of course, the test must be repeated for all pattern variants and applicable data rates.

The rise/fall edge time for SATA I must be between 100 ps and 273 ps, inclusive. The rise/fall edge time for SATA II must be between 67 ps and 137 ps, inclusive. As shown in Figure 8-34, the device has passed. Note that the results are not expressed in picoseconds; instead, they are expressed as a decimal value relative to a full unit interval. In the application shown, the time units can be set to seconds or unit interval. This is for the user's convenience. Some specifications are written with a UI limit, and others are written with a seconds limit.

This example illustrates the versatility of the eye diagram application. Because no eye diagram is required in the specification, the user can select only the tests that are required and forego the generation of the actual eye diagram, as demonstrated here. But the application includes features that automate other important—and required—serial measurements.

8.12 IMPEDANCE AND LINK MEASUREMENTS

The full scope of digital serial analysis necessarily encompasses both active components and "passive" elements including cables, connectors, and PCB traces. Understanding the characteristics of the link between the transmitter and the receiver is critical. Most high-speed protocols require a nominal 50-ohm impedance (100 ohms differential, plus a tolerance) throughout the link. Compliance with this specification is extremely important not only to meet the letter of the law but also to ensure the least possible degradation of the signals traveling along the path.

Modern design tools implement the applicable impedance rules for high-speed protocols, but physics, circuit-board materials, and human error can introduce unforeseen departures. As a result, many developers have learned that a rigorous process of verifying impedance characteristics can help them detect and correct problems early. Design choices can be reconsidered, if necessary, before quantity orders are placed with a vendor.

Imagine a full-scale system with many—perhaps hundreds—of channels, like the transmitter/receiver pairing under discussion. The connectors involved are likely to make up a significant part of the materials costs involved in manufacturing the system. The cabling must be uniform in length and performance, and it too is costly compared to simple PCB traces. Characterizing the impedance of these elements is the key to cost-effective design in this area of the system.

For the most part, impedance-related measurements rely on one of two differing platforms. The TDR is the traditional cornerstone of impedance measurement technology. This oscilloscope-based tool offers the precision to resolve tiny circuit features and deliver accurate readings spanning the entire path of the device under test.

Serial data network analysis (SDNA) methodology is an alternative approach that was originally embodied in a tool known as the vector network analyzer, developed for RF applications.

SDNA is a frequency-domain toolset that derives S-parameters and uses the data to support a wide range of impedance parameters as well as SPICE modeling, bit-error-rate prediction, and more. Serial data network analysis is no longer the exclusive province of the vector network analyzer platform. Today's ET oscilloscopes offer the performance to support uncompromised SDNA measurements with ample dynamic range for serial applications. They have gained a host of software tools to speed and simplify SDNA work.

8.12.1 Impedance Measurement Basics

The tool of choice for measuring impedances is a sampling oscilloscope equipped with a TDR module. The TDR module permits the signal transmission environment to be analyzed in the time domain, just as the signal integrity of live signals is analyzed in the time domain. Time-domain reflectometry measures the reflections that result from a signal traveling through a transmission environment. The TDR instrument sends a fast step pulse through the device under test—in this case, a cable and connectors—and displays the reflections from the observed transmission environment.

The TDR display is a voltage waveform that includes the incident step and the reflections from the transmission medium. The reflections increase or decrease the step amplitude depending on whether the nature of the discontinuity is more inductive or capacitive, respectively.

A reflection from an impedance discontinuity has a rise time equal to or (more likely) slower than that of the incident step.

The physical spacing of any two discontinuities in the circuit determines how closely their reflections will be positioned relative to one another on the TDR waveform. Two neighboring discontinuities may be indistinguishable to the measurement instrument if the distance between them amounts to less than half the system rise time.

The quality of the incident step pulse is critical, especially when measuring short traces. In addition to its fast rise time, the step must be accurate in terms of amplitude and free from aberrations.

8.12.2   True Differential TDR Measurements

SATA and most other high-speed serial standards rely on differential transmission techniques using complementary signals. Two conductors carry time-aligned mirror images of the signal. Though more complex than the single-ended approach, differential transmission is less vulnerable to external influences such as crosstalk and induced noise and generates less of the same.

These differential paths require differential TDR measurements.

The incident pulse must be sent down both sides of the differential pair and the reflections measured. This can be done in two ways:

- In the "virtual" or "calculated" differential TDR method, the TDR sends two positive-going incident steps in serial fashion, one following the other, into the two sides of the differential pair. The TDR corrects both the polarity and alignment before the measurement waveform is displayed on the oscilloscope screen.
- In the true differential TDR method, the TDR sends true complementary signals that are accurately and correctly aligned in time. The DUT receives a differential stimulus signal more like those it will encounter in its end-user application, potentially producing better insights into the device's real-world response. The TDR system does not need to manipulate the displayed step placements.

The calculated differential TDR method can provide valid results in many practical situations. However, the more modern true differential TDR method has gained broad acceptance among designers for several important reasons.

Some protocol standards stipulate that measurements be performed on powered devices. Certain SATA drive Rx/Tx measurements, for example, require this. If the DUT requires powered measurements using differential transmission, the serially launched approach will not suffice.

The calculations used to synthesize a differential display from nondifferential signals will produce errors if the signal contains nonlinearities. Such errors are extremely difficult to detect.

8.12.3   Setting up and Making Connections

The validity of TDR measurements depends explicitly on knowing the characteristics of the instrument itself and the path leading to and from the DUT. For this reason, it is essential to precalibrate both before proceeding with impedance measurements.

For the sampling oscilloscope itself, the instrument's internal clock output can be used to provide a reference signal. This is fed to one channel and stored for comparison against all other channels that will be used. Each channel's delay is measured against the stored reference, and appropriate channel deskew (time offset) values are entered.

The TDR pulses traversing the cables connecting the instrument to the DUT must arrive at the ends of the cables at precisely the same time. The calibration process that ensures this is similar to that of the previous step. Delay differences are measured and deskew values recorded to equalize the response of all cables.

These processes should match all the TDR channels in use within a 2 ps spread.

8.12.4.   Basic TDR Impedance Measurement

The TDR is connected to SATA cable through a pair of SMA fixtures, one at each end of the cable. This fixture is simply a conduit for the step and return signals. It adapts the oscilloscope's SMA connectors to the standard SATA connectors mounted on the cable. The incident step will see the effect of the fixture connectors as well as those on the cable itself. If no termination is present at the far end of the cable, the TDR trace drives toward an infinite impedance.

The TDR display tells a detailed story about the impedance variations in the signal path. Figure 8-35 is a typical TDR screen showing an impedance measurement.

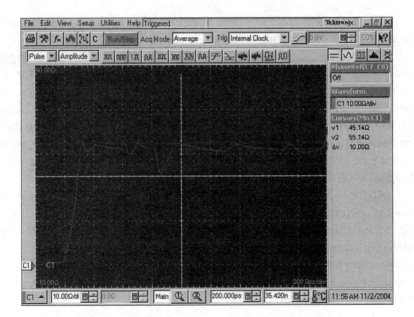

Figure 8-35   The TDR impedance measurement screen.

Figure 8-36 is a simplified diagram of the result.

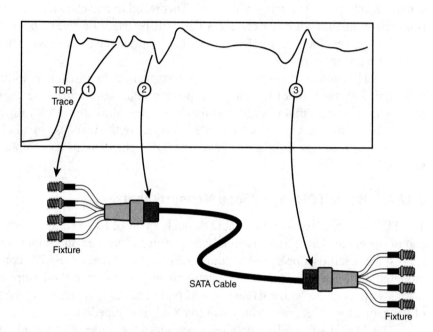

Figure 8-36   The TDR impedance measurement produces a trace on which amplitude expresses variations in impedance.

The horizontal scale has been compressed to depict events occurring over a trace whose length depends on the cable length.

To see all these events in sufficient detail on the TDR itself, it might be necessary to expand the horizontal (time) scale and scroll through the waveform record. The impedance display reads from left to right. The leftmost events are those physically closest to the step generator—the signal's origin. Again, these reflections are produced by the elements along the signal path when they receive the energy from the incident step. A perfect path with perfect terminations would produce no reflections. But no such signal path exists in the real world. It is the designer's job to understand and account for impedance variations.

At the top of the initial step transition in Figure 8-36, a momentary overshoot is sourced within the step generator itself. Event 1 denotes the step's crossing at the fixture's input connector. Events 2 and 3 are classical TDR responses to relatively large disruptions in the impedance profile. These deviations may be within acceptable tolerances, but the disparity between the cable's low-cost, consumer-grade connectors and the instrumentation-quality SMA connectors on the fixture is pronounced.

A new class of software tools is transforming the process of TDR and time domain transmission (TDT) impedance measurement. These tools provide results that can be used to predict signal losses, jitter, crosstalk, reflections and ringing, digital bit errors, and eye diagram degradation, and to ensure reliable system operation. Moreover, the data can inform the SPICE models used to refine the evolving system design. The impact of this new technology is profound. It dramatically accelerates the evaluation of gigabit interconnect links and devices, delivering interconnect analysis results in minutes instead of days.

The tools convert acquired TDR/TDT measurements into the frequency domain and extract S-parameters from the data. S-parameters describe the electrical functions of linear networks in the presence of certain small stimulus signals at known frequencies. The S stands for "scattering," which denotes the effect of impedance variations on signals traversing a transmission line.

The bandwidth of an S-parameter measurement system stems from the rise time of the TDR/TDT platform. Simply stated, a high-bandwidth TDR/TDT system can deliver high-bandwidth S-parameter analysis.

S-parameter computation procedures follow established industry standards such as JEDEC or IBIS where applicable. Models may include skin-effect and dielectric loss, insertion and return loss, eye diagram degradation, and frequency-dependent RLGC parameters. The models can analyze the effect of equalization and pre-emphasis on cable assembly performance. Eye mask testing accounts for the effects of crosstalk on the eye diagram.

An impedance deconvolution algorithm takes care of multiple reflections in the TDR impedance measurement and produces a true impedance profile. In effect, the new software tools not only improve the TDR oscilloscope impedance measurement accuracy but also increase the resolution.

Using a TDR system with 12 ps incident rise time, 15 ps reflected rise time, and 60 dB dynamic range, it is possible to resolve distances (and features) as small as 1 mm. Enhanced TDR modules can deliver S-parameter measurements up to 65 GHz with up to 70 dB dynamic range. In both cases, the performance is more than sufficient for digital design applications, where 1% (–40 dB) of crosstalk is usually allowed to be ignored. Electrical compliance masks typically specify measurements in the –10 to –30 dB range, so the 60 dB provided by the S-parameter measurements offers ample headroom above the specification.

A modern oscilloscope-based SDNA tool performs both phase and magnitude S-parameter measurements and requires less setup and calibration time than a vector network analyzer. At the same time, it supports multiple differential channels—a valuable asset when dealing with multilane serial buses.

8.12.5   SATA Impedance Measurements Using S-Parameters

Impedance measurements for the SATA interface normally are conducted separately on the cable and on the conductors and connectors that bring the SATA signals from the transmitter or receiver integrated circuit's internal bonding pads to the PCB edge connector. Both paths require essentially the same impedance measurement steps using the same tools. For the purposes of this discussion, we will focus on the cable, with its more standardized configuration and features.

The cable evaluation for SATA encompasses a broad range of individual impedance-related measurements:

- Mated connector impedance
- Cable absolute differential impedance
- Cable pair matching
- Common-mode impedance
- Differential rise time
- Intrapair skew
- Insertion loss
- Differential-to-differential crosstalk (NEXT)
- Intersymbol interference (ISI)

All of these tests are described in the Serial ATA specification (Rev 2.5) for internal cable and connectors.

It is beyond the scope of this chapter to discuss every measurement in detail. However, a closer look at some examples will illustrate why S-parameter techniques are gaining favor with designers. These examples are not meant to be comprehensive procedures, but they summarize the basic steps that allow the S-parameter tools to produce fast, accurate results.

8.12.6   Cable Absolute Differential Impedance

According to the specification, the cable absolute differential impedance measurement must be performed on each differential pair of a cable assembly, and on both ends of the cable assembly. The impedance for the DUT transmitter port must be between 90 and 110 ohms.

The general test configuration for a measurement using the SDNA technique with a TDR is shown in Figure 8-37.

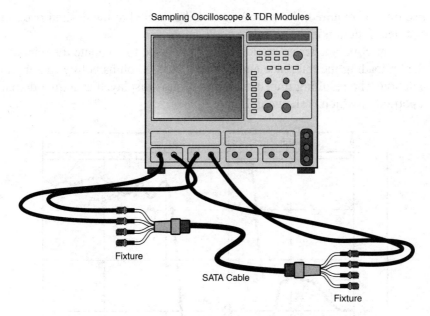

Figure 8-37   The measurement system setup for acquiring S-parameter information to evaluate a SATA cable.

This applies to several measurements in the series, not just the cable absolute differential impedance test. Differential pairs that are not used in a given test should be protected with 50-ohm terminations (not shown). The cable's host transmitter port—the one being measured in the procedure—is attached to the fixture on the left.

The test setup follows the calibration and deskew steps described earlier. In addition, it is necessary to set the oscilloscope's acquisition window to acquire the first 500 ps of cable response following any vestige of the connector response, and then disconnect the DUT and acquire the open reference waveform as the math difference of the step signals. The S-parameter software tool helps with this step.

Note also that the acquisition of S-parameters with a TDR/TDT instrument (TDR/transmitter) requires that the DUT's reflections settle to their steady DC level. Following the approximate rule of thumb, the width of the acquisition window should be set to four or five times the nominal time delay of the DUT.

To make the actual differential impedance measurement, the DUT is connected, and the TDR_{dd} (the mathematical difference of the TDR channels) is acquired for one side of the differential pair. The step impulses should be between 55 ps minimum and 70 ps maximum and as close to 70 ps as possible (20 to 80%)

rise time. If required, the waveforms can be filtered to the desired rise time using a "filter" function in the software application.

The S-parameter extraction tool can be used to compute the impedance profile by loading the reference and setting Z_o to 100 ohms before running the computation. The resulting impedance waveform is displayed as a time-domain view, a portion of which is shown in Figure 8-38.

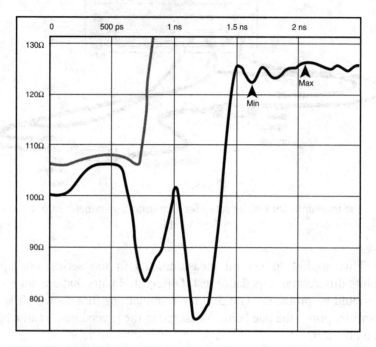

Figure 8-38   The impedance profile determined by the S-parameter extraction tool.

The oscilloscope's cursors are used to measure the minimum and maximum values, and the preceding measurement steps are repeated for all differential pairs. In Figure 8-38, arrows have been added to indicate the minimum and maximum values.

8.12.7   Intersymbol Interference

The preceding measurement example is among the simplest cable measurements in the series. The ISI test is among the most complex in terms of its analytical results, but the S-parameter toolset makes short work of this problem as well.

The setup procedure for the test is identical to that of the differential impedance test, as is the first step in the test itself.

The DUT is connected and the TDT_{dd} (the mathematical difference of the TDT channels) is acquired for one side of the differential pair.

The reference and TDT_{dd} waveforms become the basis for eye diagram calculations. A lossy line modeling tool (a component of full-featured S-parameter applications) computes the eye diagram.

For the ISI tests, the maximum frequency is set to 4500 MHz, and the display options are set to conform to the SATA eye mask specifications. Typically, this step is aided by a simple graphical tool that simplifies the "drawing" step used to create the mask outline, as shown in Figure 8-39.

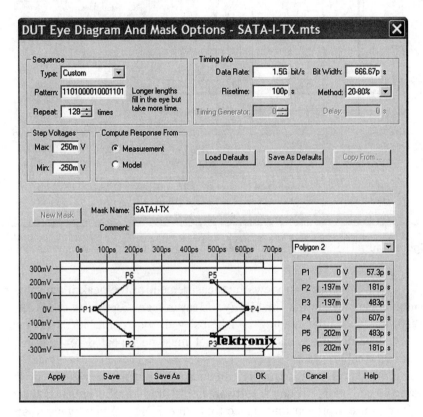

Figure 8-39 A graphical tool used to define mask outlines for the ISI eye diagram.

The resulting eye diagram is shown in Figure 8-40.

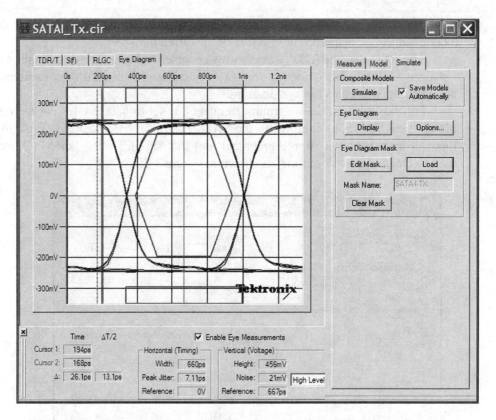

Figure 8-40   The eye diagram acquired and analyzed with the S-parameter toolset.

With this image in view, the measurement of the peak jitter value proceeds automatically when the Enable Eye Measurements option is checked. In Figure 8-40, the peak jitter reading is 4.12 ps, well within the 50 ps tolerance of the SATA specification. This measurement must be repeated for every differential pair in the cable.

Impedance measurements can locate layout problems, potential crosstalk sources, and more. Today's software tools dramatically enhance the designer's ability to use a proven impedance tool—the TDR—in new ways to analyze and solve problems ranging from signal distortion to ISI. Impedance measurements can be performed in far less time than the older methods of "brute-force" bit-error counting. They are the key to producing compliant designs that are interoperable with other system elements.

Understanding a system's underlying impedance characteristics is a key part of digital serial analysis.

8.13 RECEIVER TESTING BRINGS UNIQUE CHALLENGES

One of the most challenging aspects of serial bus design, and one in which signal integrity is the key parameter, is receiver testing. Receiver tests bring the system evaluation to its final step, and it is here that digital serial analysis is put to its most challenging test.

At the block diagram level, nothing indicates that receiver measurements might be any more difficult than the preceding testing steps. But, looking at Figure 8-41, it is easy to see why receiver testing remains a challenge for serial bus experts and novices alike.

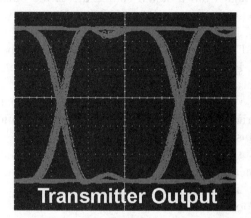 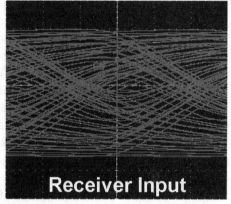

Figure 8-41   Serial signals are badly degraded by their passage through the link between transmitter and receiver.

The quality of the signal at the receiver input is badly degraded by its journey through the link. The receiver functionality of interest is encapsulated within an integrated circuit, inaccessible to probing. Without tools of the type discussed in the link measurements section, it is difficult to define the BER inherent in the signal going into the receiver. Essentially the only visible, measurable receiver signals are regenerated loopback streams. Were the errors in these streams contained in the input signal, or did they originate in the receiver itself? These are the questions that confront designers of a system's receiver elements. Standards groups and industry committees are working hard to develop specifications and effective procedures for receiver testing.

Like the transmitter tests described earlier, receiver tests evaluate data channel functions and OOB signaling functions separately. Many of the measurements in both categories are similar to those of the transmitter tests. A real-time oscilloscope remains the most common solution for the task. However, the receiver data channel tests require a stimulus signal source. Configuring and managing the stimulus is a challenging process apart from the complexities of acquisition and analysis.

The receiver's job is to acquire the data stream sent over the link and to condition and retime it for distribution to subsequent elements in the system. As explained earlier, both the host and destination devices include receiver components. Because we have defined our "system" as being a channel extending from the transmitter in the host device through the receiver in a SATA HDD, we will consider the receiver at the HDD end to be the destination and the subject of our receiver tests. The same tests apply to the receiver on the host.

During receiver (Rx) testing, data from a signal source drives the device under test, while an oscilloscope monitors the output to ensure that the device will work at nominal and stressed levels. Rx specifications include three basic signal characteristics that must be verified: amplitude (including de-emphasis where applicable), timing (including skew, PLL bandwidth, and more), and jitter tolerance. In addition, out-of-band tests evaluate the device's response to signaling.

Because the Rx section of the device must be interoperable with many device types connected through various interconnects, its specification is often more challenging than that of the transmitter (Tx) section.

In an earlier section, we discussed the performance requirements for the oscilloscopes used in transmitter measurements. It was also noted that certain tests can be performed by either real-time or equivalent-time instruments.

The same guidelines apply to tools used for receiver measurements.

Bandwidth, rise time, and accuracy demands are similar. Some of the same analysis software is equally applicable to either transmitter or receiver measurements. Jitter tests, for example, require exactly the same capability when applied to the transmitter or receiver.

From the system standpoint, it is important to remember that the transmitter and receiver measurement should be performed with the same platform: either an RT oscilloscope and appropriate integrated analysis tools, or an ET oscilloscope with its dedicated software tools.

8.13.1   Looking Inside the Receiver

Serial receivers, like their transmitter counterparts, employ serdes components internally. This configuration enables loopback testing and other techniques. To guarantee interoperability, the Rx section has to be tested over a broad range of conditions, many of which are designed to exercise the device's internal clock data recovery (CDR) and deserializer components.

Incoming data reaching the receiver contains an embedded clock that must be recovered (extracted) under a wide range of amplitude and jitter conditions.

This is the job of the CDR element, whose operation depends on a PLL that multiplies the embedded clock upward to produce a data clock signal.

Similarly, the de-serializer must be able to handle specified amplitude, jitter, and skew variations. Manufacturers developing these serial components typically provide a way to insert test data (stimulus) streams that exercise the component in the manner prescribed by the applicable standard.

8.13.2   Rx Amplitude Sensitivity Measurements

SATA receiver testing (as well as standards such as PCI Express, Fibre Channel, and others) calls for a programmable signal source that can produce data patterns. The generator must be able to generate packets or PRBS data that can be varied in amplitude.

In general, two types of instruments can provide this functionality: a digital data/pattern generator (DG) or an AWG.

8.13.3   Rx Timing Measurements

These measurements deal with inserting delays or skewing the data between the differential pairs and data lanes. This ensures that the receiver can tolerate variations in circuit board traces and cable and connector components. Other tests require inserting specific patterns or training sequences to verify that the receiver can tolerate differences in edge transition or rise and fall rates.

8.13.4   Rx Jitter Tolerance Measurements

To guarantee interoperability with other system elements, the CDR and serdes must be able to tolerate a defined amount of jitter.

The receiver jitter tolerance is a measure of how well the CDR within the receiver can tolerate jitter in its various forms. Two questions pertaining to CDR behavior are important.

First, how much horizontal eye closure can the CDR tolerate with its recovered bit clock strobe optimally placed in the eye? This result reflects how well the CDR centers its recovered bit strobe in the data eye. It also reveals the setup and hold times for the CDR's input PLL circuit.

Second, how much does the CDR's recovered bit clock wander as it attempts to track jitter within or below its passband frequency? The result is very much influenced by the jitter spectral components present in the serial data and the amount of system noise that is coupled to the CDR bandpass filters.

Note that any jitter tolerance property can be affected by other signal characteristics, such as amplitude and rise time.

In effect, jitter tolerance is a BER measurement. A bit sequence with a known amount of jitter is applied to the input, and the receiver's resulting error ratio is measured.

This procedure requires an error-detecting instrument in addition to a pattern source and jitter generator. Note that jitter tolerance measurements generally require long test times, since they must record literally trillions of cycles to ensure 10^{-12} BER performance. A DPO with high capture rate and jitter analysis tools can reduce test times for these jitter tolerance tests.

8.13.5   Out-of-Band Tests

OOB tests confirm the receiver's response to signaling commands. OOB signaling is a simple mechanism, based on the timing of a series of low-frequency pulses, that allows the host and the SATA HDD to communicate with each other to indicate that each is alive and active. OOB signals encompass the following:

- COMRESET, a signal that originates from the host controller and forces a hard reset in the connected HDD. The COMRESET signal must consist of at least six data bursts.
- COMINIT, a signal that originates from the HDD for the purpose of communication initialization. The COMINIT signal is electrically identical to the COMRESET signal.
- COMWAKE, a signal that may originate from the host controller or the HDD. It is made up of six bursts of data separated by an idle bus condition. The signal must contain no less than six data bursts.

OOB tests include the following:

- Signal detection threshold
- UI during OOB signaling
- COMINIT/reset and COMWAKE transmit burst length
- COMINIT/reset transmit gap length
- COMWAKE transmit gap length
- COMWAKE gap detection windows
- COMINIT gap detection windows

Fundamentally the tests answer two questions about the HDD device under test:

- Does the device respond to signaling commands intended for it?
- Does the device reject signaling commands not intended for it?

OOB tests require both the receiver and transmitter on the DUT to be operating. Both are involved in the test. The test series uses a very basic methodology: send a normal "compliant" pattern and detect the response, and then send a pattern with out-of-specification timing to verify that the DUT does not respond. The test specification is written accordingly.

8.13.6 Receiver Amplitude Sensitivity

The receiver amplitude sensitivity test is common to many serial standards, including SATA I and SATA II. The test confirms that the receiver meets interoperability requirements even when it experiences attenuated or higher-amplitude voltage swings. The test procedure can be summarized as follows:

- Using a data timing generator or an AWG, set the data pattern appropriately, and adjust the data packet amplitude to the nominal value.
- Decrease or increase the amplitude until the unit fails to respond correctly.
- Verify that the amplitude is outside specification when the failures occur.

8.13.7 Receiver Timing Skew Measurements

The receiver must tolerate a certain amount of timing skew (misalignment) within respective data channels and differential pairs. Each specific standard defines a limit required for this skew tolerance.

The test starts with the clock and data pairs set to zero skew and then increases the skew until the device displays an error. The maximum skew setting that still provides error-free operation is defined as the tolerated skew; this result is compared against the published limit. If the skew tolerance is greater than the specified value, the device is considered compliant with the standard. A data generator with a differential timing offset capability and differential outputs is the right stimulus tool for this task.

8.13.8 Receiver PLL Bandwidth Measurements

One common means of determining the PLL bandwidth is to use a jitter transfer function. This test characterizes the jitter amplitude response of the DUT as a function of jitter frequency. The test feeds the DUT a modulated reference clock signal (preferably with the modulation derived from a Gaussian noise source) and measures the jitter at the input and output of the DUT. The ratio of output jitter to input jitter, plotted in the frequency domain, forms the transfer function plot and illustrates the loop bandwidth.

The PLL bandwidth test typically takes one of two stimulus approaches:

- A digital data generator equipped with an appropriate jitter feature. Available modular solutions provide these tools. An external Gaussian noise generator modulates the reference clock signal from the data generator.
- An AWG with digital marker outputs, loaded with a PRBS pattern.

In both these applications, an oscilloscope running a jitter measurement software application can monitor the input and output of the PLL device and plot the results.

8.13.9   Receiver De-emphasis Generation and Testing

Serial bit streams are often modified in real time to compensate for signal losses in the transmission medium. This is known as de-emphasis, and it must be applied when a succession of bits of the same polarity occurs. Bits following the first bit are driven at a lower voltage level than the first bit itself.

Figure 8-42 is a simplified version of de-emphasis. There are two logic voltage values, with a slightly higher voltage applied to any bit that follows a sequence of bits having one polarity (state). De-emphasis signals can be produced by either a data generator or an arbitrary waveform generator.

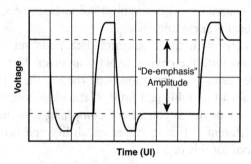

Figure 8-42   De-emphasis applied to a serial signal. The first bit following a series of bits of the opposite state is higher in amplitude than the "de-emphasized" bits that follow it.

The stimulus source must simulate de-emphasis to mimic the real-world conditions the receiver's serdes will encounter. When performed with a data timing generator, the de-emphasis tests require a pair of outputs from the instrument to drive a pair of power combiners, as shown in Figure 8-43. The power combiners mix two serial streams containing bits that coincide at specific times. This produces the needed attenuation after the full-amplitude bit preceding a sequence of same-state data.

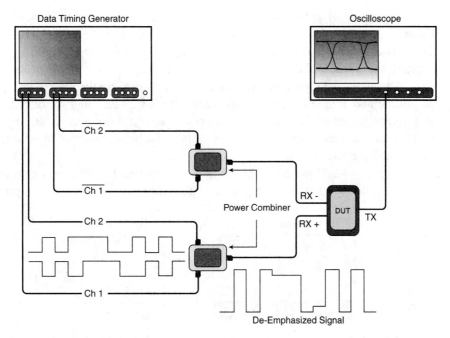

Figure 8-43 Using a data generator to create the de-emphasis effect on a serial data stream. Each side of the differential pair is made up of two constituent signals that are mixed in a power combiner.

Signal levels from the data generator must be set to counteract losses incurred in the power combiners (usually 50%) and to vary the amounts de-emphasis during stress testing.

Tools included in the oscilloscope's analysis application measure the actual de-emphasis value. This approach using the data timing generator as a stimulus source has earned wide acceptance among developers and standards committees.

The AWG-based approach is gaining favor following the advent of fast new AWG platforms. An AWG can issue a bit stream with pre-emphasis—or any other signal characteristic—already embedded in the data. Imperfections such as over-shoot or rise time degradation can be added (with or without de-emphasis) to simulate the effects of the signal's passage through a transmission line.

8.13.10 Receiver Jitter Tolerance Measurements

Jitter tolerance for receivers is defined as the ability to recover data successfully in the presence of jitter. Meeting the specification guarantees that the serdes and PLL circuits can recover the clock even when a certain amount of jitter is present.

Rigorous jitter tests are especially critical in applications such as PCI Express, in which the clock is embedded in the 8-bit/10-bit encoded data stream. There is an absolute requirement for a signal source that can supply jitter with specific amplitude and frequency characteristics.

Advanced data generators include features that supply jitter with specific amplitude and frequency modulation characteristics; these include built-in noise and jitter generators. The jitter generators can accept diverse modulation profiles such as sine, square, triangle, and noise. They can be set to apply jitter to the rising or falling edge, or both.

Figure 8-44 shows an example of a test system layout for jitter tolerance tests. Here, an external function generator provides the jitter modulation profile.

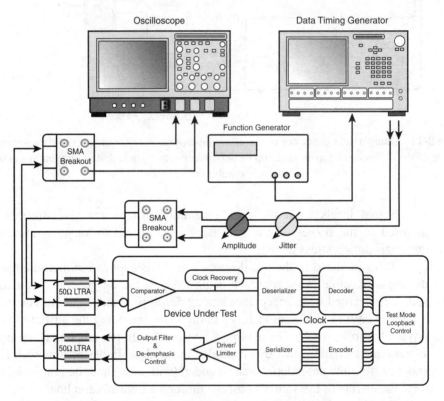

Figure 8-44   A jitter test system configuration for a serial device. The function generator provides a signal to modulate the jitter produced by the data timing generator.

A means of initiating BIST patterns with a special software tool has already been described. When testing a receiver component, it is necessary to first instruct the device to go into loopback mode and then to deliver the test pattern.

If the host is used to set the loopback mode (without the software tool), it must be disconnected before the pattern can be applied. This is a problem. Many hosts and devices exit loopback mode and return to their normal operation as soon as they are disconnected. Therefore, the signal source must provide some means of driving both the loopback commands and stimulus pattern contiguously. When using a data generator, a power splitter can be used to deliver the loopback command data followed by the "jittered" pattern, eliminating the need to disconnect. Alternatively, an arbitrary waveform generator can send a sequence containing both the commands and the jittered pattern.

8.13.11   Out-of-Band Test Example: Signal Detection Threshold

The following OOB signal detection threshold test illustrates a required compliance measurement. It also illustrates an AWG-based test methodology that contrasts with the data timing generator-based approaches presented thus far.

An external stimulus source is usually needed for the OOB tests because the host devices are not equipped to produce suitable OOB signal variations to support compliance testing.

Historically the tool of choice for OOB measurements has been either an arbitrary waveform generator or a data generator. In either case, the signal source performs the relatively simple task of delivering a purely digital signal and varying the timing, usually by means of preset routines loaded before testing begins. If an AWG is selected, the conventional approach is to use its marker outputs rather than the main analog output. This provides binary data only.

However, there is an idle state in the signal detection threshold test sequence, wherein the signal may go into a tri-state condition whose value is neither a 1 nor a 0, as shown in Figure 8-45.

Figure 8-45   OOB signals commonly incorporate idle states at amplitudes other than 1 and 0. Note that the DUT is configured in a loopback mode. This is a typical method of observing the device's response and implementing BIST protocols.

To address this with a data generator or AWG marker outputs, special measures must be taken to modify the bias level of the instrument's output driver. In some cases power splitters are used between the differential outputs.

There is a better way to solve this problem. Using an AWG's analog output—not its digital marker outputs—it is easy to generate this signal with discrete amplitude levels other than the established logic 1 and 0 voltages. This is more in keeping with what the DUT expects. Figure 8-46 shows the test instrumentation setup.

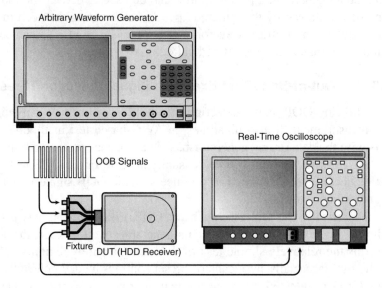

Figure 8-46   An instrumentation setup for OOB measurements. The AWG provides the necessary tri-state signal levels without external modification (signal connections only; trigger not shown).

The concept of using an AWG to embed jitter, noise, aberrations, and timing variations in both OOB and data signals for serial devices is gaining favor among designers and industry committees.

Until recently, AWG applications for serial testing were limited to lower-frequency standards and marker-based digital work, as mentioned. However, an emerging class of high-performance AWGs is making it practical to source uncompromised serial data signals from the instrument's analog outputs. These tools offer true differential analog outputs, ample bandwidth and sample rate, deep memory, and productivity tools such as internal sequencers (which enable simple generation of OOB and other patterns). The OOB signal shown in Figure 8-47 was produced by a modern AWG with 5.8 GHz bandwidth and 20 GS/s sample rate.

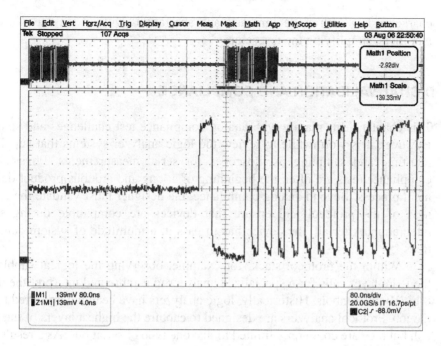

Figure 8-47 The OOB waveform delivered by the AWG includes the tri-state portion of the signal. This is produced without the aid of external attenuators or power splitters.

These high-performance AWGs eliminate the need to inject jitter and noise from external sources. They remove the requirement for combining multiple channels to create pre-emphasis. They also make it possible to transmit, as an integral part of the signal itself, artifacts such as intersymbol interference and other system anomalies. Full-featured AWGs also include powerful sequencers that can invoke loopback modes so that test patterns can be sent without the need to disconnect the DUT.

Receiver measurements call for tools that can activate loopback modes, deliver complex patterns, vary the signal amplitude and apply voltage values outside the normal binary range, and—of course—meet the rigorous bandwidth and timing requirements of today's high-speed serial buses.

Receiver testing may be the most difficult discipline in serial system evaluation, and it is not a task that responds well to shortcuts.

Signal sources play a key role in receiver measurements: they provide a stimulus for the DUT. Emerging standards typically do not mandate specific configurations to deliver those stimulus signals. Consequently, both data generators and arbitrary waveform generators have found a place in receiver testing. The data generator is a proven tool that can drive many channels of data with excellent timing accuracy. The AWG is earning a reputation for its ability to apply signal characteristics such as de-emphasis and aberrations without the use of external modifiers, simplifying the test setup.

8.14   DIGITAL VALIDATION AND COMPLIANCE

This chapter deals mainly with analog compliance test challenges and solutions. However, it is worthwhile to review the logic analyzer systems that supplement real-time and sampling oscilloscopes for serial measurements. Physical-layer compliance tests are often the beginning of a long and arduous product-development process. Beyond analog compliance, the designer must validate the correctness of the protocol and ensure that devices are compatible during system integration. After all, the device has to pass in a multitude of system configurations.

Within the protocol stack, for example, observing the logical subblock of the PHY layer calls for a specialized analyzer to acquire and interpret the data as 8-bit/10-bit symbols. Historically, logic analyzers have been the preferred tool for the job. Protocol analyzers are designed to capture the highest layers of the protocol, but they are commonly limited to just one type of serial bus. As a result, if the compliance task requires a simultaneous view of diverse buses or general-purpose signals in both serial and parallel formats, the logic analyzer is the tool of choice. Protocol analyzers have historically worked well when the bus is operating correctly, whereas logic analyzers display physical-layer aspects of the bus, including detailed insight into link training and power management states.

In a logic analyzer, captured data is displayed in the waveform window for cross-bus analysis or analog correlation, and in the state listing window, which provides protocol decode and error detection. With appropriate analysis tools, this recorded data can be automatically interpreted to help engineers confirm that higher-level programmatic instructions are being carried out correctly and discover the source of errors that have caused a failure on the serial bus link.

Until recently, acquiring serial data with a logic analyzer meant using a complex external preprocessor to interface the instrument to the device under test. But, with serial bus architectures becoming almost universal in digital systems, a more efficient solution is needed, and serial acquisition is now being integrated into the logic analyzer itself. This new capability dramatically simplifies serial acquisition and enables designers to mix serial and parallel functions as needed.

8.14.1   Data and Protocol Analysis

Physical layer compliance tests are often the climax of a long and arduous product-development process. However, other concerns must be considered before and after compliance, because the devices under test have to pass not only pulses but also valid binary data. In the discussions so far, nothing has proven the device's ability to deliver the correct information at the data and protocol levels. Debugging, troubleshooting, and design validation at these levels are as much a part of the project as is preparation for compliance tests.

For these tasks, the logic analyzer offers the capability to acquire and interpret the data as a whole, rather than as individual pulse features. In the logic analyzer's digital environment, captured data creates a timing diagram of the bus activity. Serial data is fanned out to parallel form and stored in the instrument's memory. With appropriate decoding tools, this recorded data can even be disassembled to help engineers confirm that higher-level programmatic instructions are being carried out correctly.

8.14.2   Acquiring the Signal: from Packet to Parallel

Serial signals deliver deep packets of data that are a few bits wide; parallel signals appear in the form of "words" that are 1 bit deep but many bits wide. A logic analyzer is an innately parallel architecture, with many channels (bits) accepting data simultaneously. But, thanks to special processor and bus support packages, logic analyzers can acquire and analyze serial data just as readily as parallel data. The hallmark of a state-of-the-art logic analyzer is the breadth of its processor and bus support packages.

Figure 8-48 shows how the logic analyzer display breaks apart serial data and displays it in the logical and electrical subblock formats.

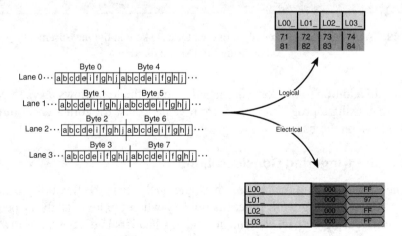

Figure 8-48   In this example, the logical view descrambles, deskews, and decodes the data and sets the bits to the correct polarity. The results are displayed as a stream of hexadecimal data.

The electrical view shows waveform data, but it is neither decoded nor unscrambled, and its polarity appears in the inverse. This view does, however, show the time alignment of every change in the data values in all four lanes.

At higher speeds, channel-to-channel skew can make it difficult to acquire valid data. The solution to this problem is a measurement tool that automates adjustment of setup/hold windows to remove skew from the measurement system, allowing valid data to be captured. Figure 8-49 shows such a tool.

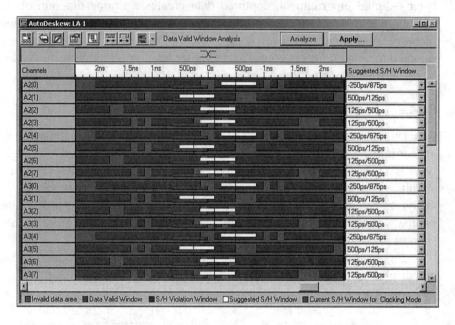

Figure 8-49   An automated deskewing tool counteracts skew in the measurement system when acquiring high-speed data.

In addition to this key automated feature, the tool allows users to verify their settings using a programmable margin value to test for corner cases of data-valid windows on the bus.

8.14.3   Triggering Considerations

One of the defining features of any logic analyzer is the flexibility of its triggering system. Debug work proceeds quickly when the logic analyzer permits real-time triggering on packet elements such as headers and payload or raw symbols. Protocol-specific trigger interfaces are now available that can quickly isolate and capture specific transactions, packets, ordered sets, link events, or link errors.

Triggering templates such as the one shown in Figure 8-50 allow the user to select from menu items and fill in a "form" to specify events of interest.

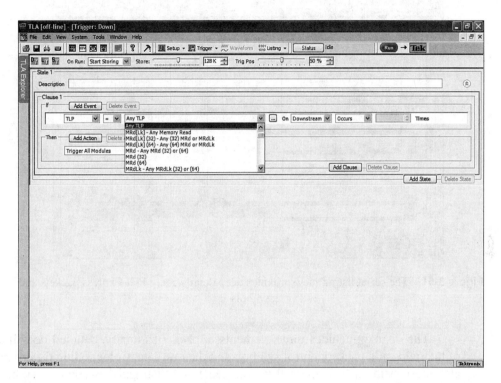

Figure 8-50   Protocol-based trigger criteria simplify the setup step.

In this instance, the fields are optimized for the unique requirements of the PCI Express serial protocol.

8.14.4   Analyzing the Results

With the acquisition completed, the stored data must be decoded into meaningful results and displayed in data windows for analysis. Software tools can help disassemble, decode, and display the captured data in a packet-style view using the logic analyzer's listing window. An X16 PCIe acquisition is shown in the listing window in Figure 8-51.

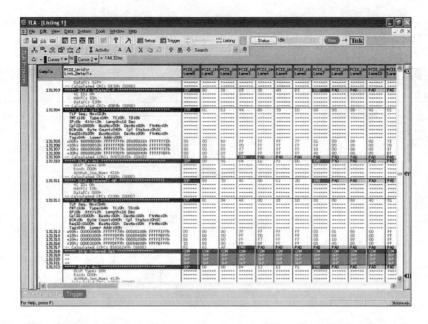

Figure 8-51   The serial listing view provides deep details about PCI Express packets and symbols.

The display includes three elements: a packet summary, detailed decoding of the fields of a packet/control symbol, and the raw data. Color coding differentiates the text in each of these three elements and distinguishes control symbols from packets in the packet/control symbol summary.

Figure 8-52 is a waveform view of serial data. The serial analyzer delivers advanced disassembly features, such as deep capture of serial buses time-correlated to all other system buses using the logic analyzer's common system timestamp.

The disassembler provides control symbol decoding and display of individual symbols for all layers of the protocol. Simultaneous decoding of both transmit and receive data ports is available, along with any other system buses being probed by the logic analyzer. Moreover, analog and digital waveforms can be time-corrected.

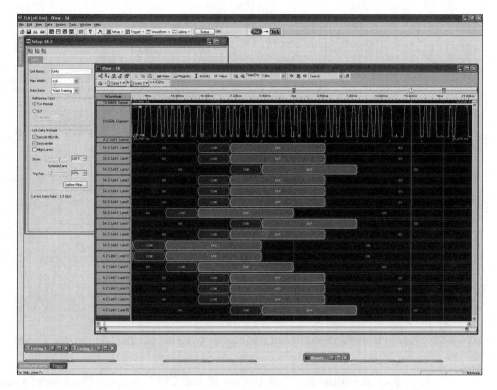

Figure 8-52   Time-aligned packet data shown as digital waveforms correlated with an analog waveform. This display is equivalent to the conventional logic analyzer timing view, but it represents decoded serial data.

8.15   MULTIBUS SYSTEMS

As mentioned, a logic analyzer equipped with a serial module is in a unique position to capture uncompromised time-correlated data from any and all buses in a system, whether serial or parallel. While it is rarely necessary to capture every bus at once, it's common to acquire a full set of PCI Express lanes along with the concurrent front-side bus transaction or the parallel output of a critical register, for example. The following example illustrates this point.

Figure 8-53 depicts a typical (but simplified) motherboard architecture. Buses that are not relevant to this discussion are shown in gray, although they too may require time-correlated acquisition and analysis at some point.

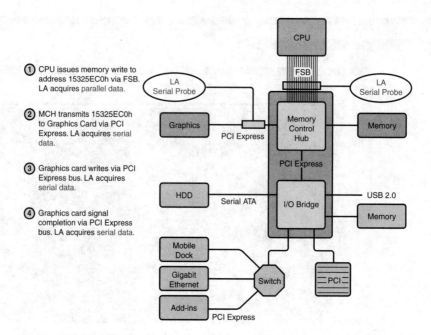

Figure 8-53 A typical motherboard block diagram showing the parallel and serial test points.

Consider the situation that occurs if the CPU issues a "memory write" instruction to the graphics card, with the destination address assumed to be 12325EC0h: a 32-bit value expressed in hexadecimal form.

The instruction exits the CPU and travels across the front-side bus (FSB) to the memory control hub (MCH). A parallel logic analyzer probe acquires the FSB. Figure 8-54 shows the result of this acquisition.

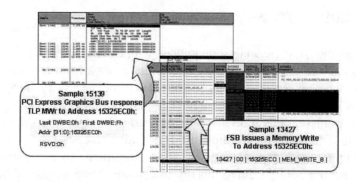

Figure 8-54 A time-correlated acquisition from serial and parallel buses on the motherboard.

The window in the foreground is the FSB trace. The MCH, receiving the instruction, tells the graphics card to write to memory address 15325EC0h. The instruction is relayed through the serial PCI Express bus connecting the MCH to the graphics card. The logic analyzer's mid-bus "connectorless" serial probe sits on a group of pads on the bus, capturing the serial traffic.

The result acquired by the logic analyzer is shown in the background window in Figure 8-54. In this example, the transaction has executed correctly: the graphics card has written to the specified address. Note the samples in which the two respective events occurred. The FSB event was recorded 1,712 samples earlier than the response from the graphics card.

The two events are time-correlated. The 1,712-sample separation is equivalent to a time interval that remains consistent across all transactions of this type. One of the defining features of any logic analyzer is the flexibility of its triggering system. When coupled with a support package designed for a particular serial standard (such as RapidIO or PCI Express), the logic analyzer delivers advanced triggering capabilities to easily isolate and capture specified transactions.

Debug work proceeds quickly when the logic analyzer permits triggering (at speed) on transactions or other packet elements such as control symbols. Some instruments even include triggering templates that allow the user to fill in a "form" to specify events of interest. The template fields can be optimized for the unique requirements of each serial protocol.

8.15.1   Analyzing the Results

A logic analyzer has to decode data into meaningful results and display it appropriately. Bus support packages include sophisticated software tools to disassemble, decode, and display the captured data in a packet-style view using its listing window, as illustrated in Figure 8-55 for an acquisition from a RapidIO bus.

The display includes three elements: a packet/control symbol summary, detailed decoding of the fields of a packet/control symbol, and the raw data. Color coding differentiates the text in each of these three elements and distinguishes control symbols from packets in the packet/control symbol summary. Color also differentiates the transaction and operation levels of request and response packets in the Details column.

When equipped with an optimized support package, a logic analyzer can deliver advanced disassembly features such as deep capture of onboard processor activity correlated to transmit and receive ports and deep synchronous display of these ports. The disassembler provides control symbol decoding and display of individual fields in the physical layer; packet decoding and display of individual fields for physical, transport, and logical protocol layers; and simultaneous decoding of both transmit and receive data ports.

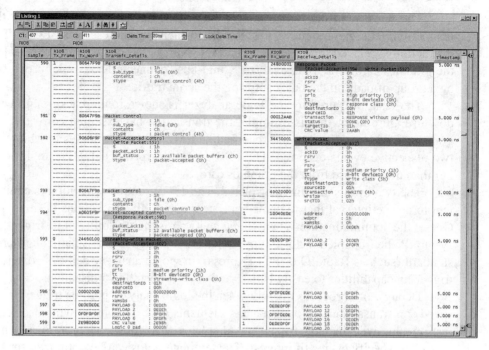

Figure 8-55   A cycle-by-cycle analysis of RapidIO bus activity.

CONCLUSION

This chapter has demonstrated both the need and the means for system-wide testing of emerging serial designs. Fast serial signals interact with their environment from beginning to end, and these interactions must be understood and controlled. Measurements are the key to this understanding.

Industry-standards bodies, working committees, and special-interest groups develop and promote both operational and test guidelines for every active serial protocol. The test documents range from general recommendations to specific step-by-step procedures for compliance testing. Here again, measurements are the cornerstone. Compliance testing is not optional. It is mandatory under industry standards, and end users expect it.

The discipline of digital serial analysis is constantly advancing to keep pace with the continuing increases in serial performance. New hardware tools answer the specific probing and acquisition needs of high-speed serial buses.

Methodologies for measurements ranging from simple amplitude assessment to complex jitter analysis are adapting to the demands of serial applications. S-parameters developed from time domain TDR measurements, for example, promise to speed up impedance analysis and bit error rate prediction. And software tools are evolving to complement the entire range of measurement requirements, delivering information-rich displays, including eye diagrams, bathtub curves, histograms, and much more.

Serial buses, components, and transmission elements are here to stay. They are destined to grow in importance as technology markets continue to demand ever-accelerating data rates. Design and validation engineers have a new and perhaps unfamiliar discipline to learn—serial compliance measurement—even as they confront aggressive development schedules and fast-changing standards.

9

PCI Express
Case Study

When this book went to press, the PCI Express™ specification was five years old, and the 5.0 Gbps upgrade had just been published. Like DDR2, its market penetration has given PCI Express a place in the portfolios of many signal integrity engineers. This makes it an ideal candidate for a case study.

Our team recently shipped a product that marries a high-performance graphics processor with 5 GB of memory on a PCI Express add-in card. Packaging density and cooling forced us to depart from the well-known design guidelines. When we find ourselves at the crossroads of a project and the team is looking to us for input on a critical decision point, we have three possible choices:

- Invoke the gods, and just say no.
- Be a team player, cross your fingers, and hope the hardware comes up okay.
- Begin the process of quantifying your margins.

Options one and two run the risk of someone smarter than you calling your bluff, but you might get away with it. Or you might find yourself in the middle of a line-down crisis in one or two years. Option three involves convincing project management to pay for the extra homework. There's no free lunch. On the positive side, once you've done the work to understand the margins, extrapolating to similar scenarios becomes a lot easier. People start showing up at your office door for advice.

The reference simulator for this chapter is Advanced Design System (ADS) from Agilent Technologies, a high-frequency SPICE signal integrity simulator.

9.1 HIGH-SPEED SERIAL INTERFACES

When source synchronous interfaces, like their predecessor, eventually succumbed to the constantly decreasing size of the transistor, system designers turned to a technique already familiar to engineers who had to transmit data over long lengths of cable: clock data recovery. If skew between clock and data becomes a problem, why not get rid of the clock? Of course, there is always a clock somewhere. Both the transmit and receive chips need a clean reference clock to run their phase locked loops (PLLs). But this clock does not travel along with the data between two chips. Rather, the transmitting chip multiplies the frequency of the reference clock and uses this faster clock to send data that is guaranteed to have no more than 5 consecutive bits in the same state. The transmitting chip rearranges the data and inserts extra bits here and there to satisfy this condition; the receiving chip must perform the inverse operation on the other side of the link. When the transmitting chip has no data to send, it transmits an idle pattern to keep the analog circuitry on the receiving chip humming along and ready for the next data packet.

Encoding is the name for this manipulation of the data, and the algorithm itself is not as important as the fact that it places lower and upper limits on the frequency of the data on the wire. Five 0s followed by five 1s translates into 250 MHz if the bit time is 400 ps (2.5 Gbps). A 1010 pattern translates into 1.25 GHz. This means that the circuit designer can customize the receiver to ignore everything outside this band. In the time domain, encoding implies a limit on the amount of intersymbol interference (ISI) caused by losses: An edge can never be more than 5 bits long before turning around and heading the other way. PCI Express also requires drivers with de-emphasis to further reduce intersymbol interference.

In the absence of a clock signal, the traditional electrical constraints of setup and hold timing at the pins of the chip no longer apply, yet there is still a flip-flop somewhere inside the receiving chip. In this example, both transmit and receive chips use the same 100 MHz reference clock and their own on-chip PLL with a 25x frequency multiplier. The receive chip also has a phase alignment circuit, designated by φ in Figure 9-1, that locates the clock at the optimum sample point with respect to the incoming encoded data stream. All the rules of metastability avoidance still apply at the flip-flop that lies on the other side of the clock data recovery logic. The flip-flop's behavior is more statistical than ever in this environment, where a jitter distribution is the metric. To ensure an acceptable error rate at this internal flip-flop, the architects of the PCI Express specification divided the jitter budget shown in Table 9-1 into four general categories for a typical application.

Figure 9-1   PCI Express link with phase interpolator clock data recovery.

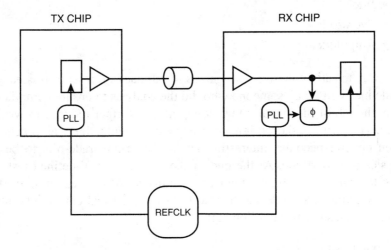

Table 9-1   2.5 Gbps Jitter Budget

Category	Total Jitter	Percentage
Transmitter (chip and package)	100	22%
Reference clock	108	24%
Media (interconnect)	90	20%
Receiver (chip and package)	160	34%
Totals	**458**	**100%**

In the world of high-speed serial interfaces, signal integrity engineers have very little to say about the transmitter and receiver—although it's still important to understand how they work. An unknown analog circuit group designs the high-speed serial macro (TX and RX) whose schematics remain proprietary. Even if a new application-specific integrated circuit (ASIC) package is associated with the silicon, the same company that designed the analog circuitry also writes a set of design rules for the package. The company is responsible for ensuring that the combination of the chip and package meets the PCI Express specification at the package pins. This makes life simpler but less interesting for us: no more driver selection, simultaneous switching output (SSO) noise analysis, package crosstalk analysis, and so on. The interconnect is all that remains.

Anyone who wants to claim compliance with the PCI Express specification must satisfy five fundamental electrical constraints:

- Unit interval variation
- Voltage-time eye mask
- Jitter
- Eye width
- Amplitude

One way to satisfy these constraints is to fall back on a set of trusted and proven guidelines written by someone who did the analysis to ensure compliance. In this case, the job of the signal integrity engineer is further reduced to writing design rules. However, if the design goes beyond the guidelines, as practical designs often do, life becomes interesting again. There are trade-offs to be made and physics to understand. At the end of the project, a well-defined test bench at a plug fest will be the final arbiter of compliance. At the beginning of the project, however, you need to decide how to translate electrical constraints into physical design rules, such as the following:

- Dielectric material
- Transmission line construction (stripline or microstrip)
- Add-in card line length
- System board line length
- Line width
- Spacing between true and complement
- Spacing between diff pairs
- Via count
- Via location
- Via diameter
- Via pad diameter
- Via antipad diameter
- AC coupling capacitor surface artwork
- AC coupling capacitor location

9.2 SENSITIVITY ANALYSIS

Like the preceding DDR2 memory discussion, the PCI Express case study examines the effects of each contribution to the channel one at a time. The goal is to give you a tool, sensitivity analysis, with which to make decisions about the assignment of design rules for each unique application. Sensitivity analysis involves observing how some end result, such as jitter, responds to poking one parameter at a time. Think of it as a practical application of multivariable calculus. It's nothing new—just something we tend to overlook in the rush to ship product.

The case study will conclude by filling out the last four columns in Table 9-2 and comparing simulation results to the corresponding PCI Express jitter specifications and loss recommendations. The PCI Express specification addresses losses at 625 MHz and 1.25 GHz, but this table shows only the 1.25 GHz point. Be aware that frequency behavior out to 5 GHz and beyond also affects deterministic jitter in the link, and resonances may not show themselves at either of these two frequencies.

Table 9-2   Sensitivity Analysis Template

	Component	Model	Incremental Jitter (ps)	Cumulative Jitter (ps)	Incremental Loss (dB)	Cumulative Loss (dB)
1	De-emphasized driver	Behavioral				
2	System board wire (12 inches)	Lossy t-line				
3	Add-in card wire (4 inches)	Lossy t-line				
4	TX and RX packages (1 inch)	Lossy t-line				
5	Impedance tolerance (±10%)	Lossy t-line				
6	RX capacitance (2 pF)	Capacitor				
7	RX termination (+20%)	Resistor				
8	BGA solder balls	CST MWS				
9	Card and board vias	CST MWS				
10	Card edge connector	Vendor				
11	AC coupling capacitors	None				
12	Crosstalk: card wires	Lossy t-line				
13	Crosstalk: BGA	CST MWS				
14	Crosstalk: vias	CST MWS				
15	Crosstalk: connector	Vendor				
16	**Totals**					
17	**PCI Express specification**			90		11.95
18	**Margin**					

Figures 9-2 and 9-3 show physical and schematic views of the complete end-to-end channel topology used to simulate loss and jitter for PCI Express compliance.

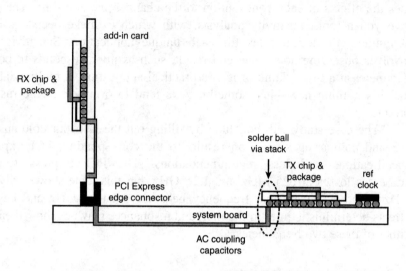

Figure 9-2   System board and add-in card.

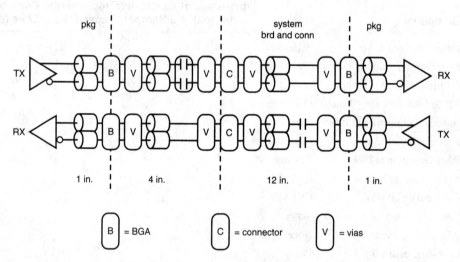

Figure 9-3   Simulation net topology for PCI Express sensitivity analysis.

9.3   IDEAL DRIVER AND LOSSY TRANSMISSION LINE

The PCI Express Base Specification calls for a de-emphasized differential driver with 100 ohm nominal output impedance and 50 ps minimum rise time. Comparing a simple CMOS push-pull driver (see Figure 2-21) with a de-emphasized driver of the same impedance and rise time gives a good indication of how much de-emphasis improves jitter. Assuming the stack-up from Figure 5-1 and a 12-inch system board, the simplest interesting load is 16 inches of 4 mil wire with an ideal 100 ohm termination at the far end. The push-pull driver is easy to construct in IBIS and will run in any behavioral simulator.

As the waveforms shown in Figure 9-4 demonstrate, the step response of a lossy transmission line has three general regions. In the first region, the waveform rises rapidly but more slowly than the response of an ideal transmission line with identical impedance and propagation delay. This portion of the waveform has the appearance of an RC roll-off. A "corner" located somewhere between half- and full-signal swing marks the beginning of the second region. The shape of the waveform changes dramatically beyond the corner to a much lower slope. Finally, the waveform dribbles up to its steady-state solution whose voltage is determined by the DC current flowing on the wire and the DC resistance. The full swing at the receiver input is always less than the full swing at the driver output.

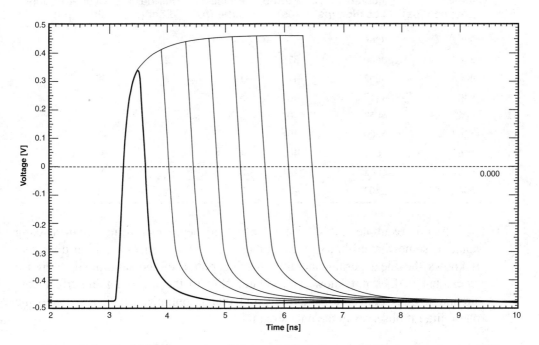

Figure 9-4   Ideal 100 ohm push-pull driver, 16-inch line, and 100 ohm termination PCI.

Note that the DDR2 net in Figure 4-16 achieved full amplitude in the middle of the second bit while the PCI Express net takes 6 or 7 bits.

A time interval analyzer (TIA) measures the zero crossing points of the differential waveform very precisely, stores a deep array of these crossing points, and then generates statistics that help quantify the link's operating margins. Table 9-3 shows the results of a simulated TIA using the waveforms from Figure 9-4.

- Ideal crossing: integer multiples of the ideal unit interval (400 ps) subtracted from the falling edge zero crossing point of the last bit—the "best" bit
- Actual crossing: the zero crossing point of the differential waveform relative to simulation time zero
- Period: the difference between two successive actual crossings
- Phase jitter: the difference between actual and ideal crossings
- Period jitter: the difference between the period and ideal unit interval
- Cycle-to-cycle jitter: the difference between two PCI successive periods

Table 9-3   Jitter with a Push-Pull Driver

Edge	Ideal Location (ps)	Actual Location (ps)	Period (ps)	Phase Jitter (ps)	Period Jitter (ps)	Cycle-to-Cycle Jitter (ps)
1	3262	3265		−3		
2	3662	3640	375	22	−25	
3	4062	4052	412	10	12	37
4	4462	4457	405	5	5	−7
5	4862	4859	402	3	2	−3
6	5262	5260	401	2	1	−1
7	5662	5661	401	1	1	0
8	6062	6062	401	0	1	0
9	6462	6462	400	0	0	−1

In this example, the choice of where to begin measuring ideal crossing points is somewhat arbitrary. A clock data recovery unit faces a similar dilemma: It knows the ideal period and the crossing points of the multiplied reference clock, but it does not know their relationship to the incoming imperfect data stream. It must choose a sampling point that is optimized with respect to the phase jitter it observes in the data stream PCI.

9.4   DIFFERENTIAL DRIVER WITH DE-EMPHASIS

The combination of a generic 100 ohm driver and a 16-inch net consumes one quarter of the interconnect jitter budget; the other remaining effects have yet to be accounted for. If the eye is closing because attenuation prevents a 1-bit signal from achieving full amplitude, why not give the signal a little boost? The additional energy would have to be conditional on the data pattern history, because boosting a signal that is already strong will only worsen the jitter problem in the opposite direction. The PCI Express driver accomplishes this function in Figure 9-5 by adding a second and higher-impedance output stage in parallel with the first. A comparison between the current data bit and the previous data bit controls the output enable of the second stage.

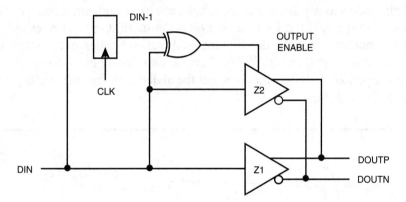

Figure 9-5   De-emphasized driver functional schematic.

This common technique is called "de-emphasis," a reference to signal processing theory that often is used interchangeably with pre-emphasis. The exclusive NOR gate compares the current and previous bits and enables the second stage (Z2) only when two consecutive bits are the same.

If the current data bit and the previous data bit are in opposite logical states, the signal at the receiver input has only one unit interval to travel up the attenuated waveform shown in Figure 9-4. Consequently, it doesn't take much energy to reverse direction. One output stage (Z1) is enough. If the current data bit is the same as the previous data bit, the de-emphasized driver enables its second output stage (Z2) and drives the net harder because the attenuated signal at the receiver input has had two unit intervals to switch. More energy is required to pull it back in the other direction.

Table 9-4   De-emphasized Driver Impedance

DIN	DIN−1	OE	ZOUT
0	0	1	Z1 + Z2
0	1	0	Z1
1	0	0	Z1
1	1	1	Z1 + Z2

The next round of simulations swaps out the push-pull driver for a PCI Express compliant de-emphasized driver; the net topology remains the same. Note the characteristic overshoot on the first bit in Figure 9-6 caused by the additional parallel transistors in the output stage (lower impedance) followed by a slight undershoot at the driver switches back to its higher impedance state. In the case of the push-pull driver, the first bit is the shortest one, whereas the second bit is the shortest one in the de-emphasized driver—although not as short as the first bit of the push-pull driver. Table 9-5 proves that de-emphasis does well at preserving the period of all the other bits, but the undershoot jogs the starting point of the second falling edge.

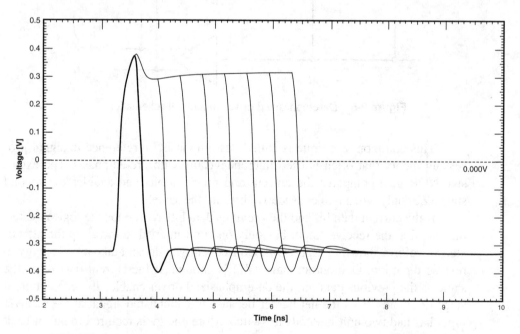

Figure 9-6   De-emphasized driver, 16-inch line, and 100 ohm termination.

Table 9-5 Jitter with a De-emphasized Driver

Edge	Ideal Location (ps)	Actual Location (ps)	Period (ps)	Phase Jitter (ps)	Period Jitter (ps)	Cycle-to-Cycle Jitter (ps)
1	3256	3258		-2		
2	3656	3661	403	-5	3	
3	4056	4051	390	5	-10	-13
4	4456	4453	402	3	2	12
5	4856	4855	402	1	2	0
6	5256	5255	400	1	0	-2
7	5656	5655	400	1	0	0
8	6056	6056	401	0	1	1
9	6456	6456	400	0	0	-1

For all three types of jitter listed in Table 9-6, the de-emphasized driver outperforms the simpler push-pull driver. In high-speed serial interfaces, phase jitter is the most important kind of jitter. Why is this? Consider what a clock recovery unit is trying to accomplish. It already knows what the period should be, because it has a multiplied copy of the reference clock, which is good to ±300 parts per million. Therefore, period jitter is not too important. Furthermore, it has been compiling a long history of zero crossing points and trying to decide where to place the sample clock relative to the boundaries of the "ideal" period established by the reference clock. This comparison of actual crossing points to ideal crossing points fits the definition of phase jitter. The more phase jitter, the harder the clock recovery unit has to work to find an optimum sample point. At some point, the phase jitter encroaches on the internal sample point, and the bit error rate climbs high enough that the link ceases to function. For this simple 8-bit stream of 1s, de-emphasis buys 17 ps less phase jitter. Recall that the PCI Express specification allows 90 ps of total jitter for the interconnect.

Table 9-6 Worst-Case Jitter Comparison

Jitter Type	Push-Pull (ps)	De-emphasis (ps)
Phase jitter	22	5
Period jitter	25	10
Cycle-to-cycle jitter	37	13

Next, let's apply a more complex data pattern to the driver and observe the associated jitter. Three general types of signal integrity simulators available on the market can accomplish this. Some of the traditional SPICE-based simulation engines now combine lossy transmission lines and s-parameter models with full transistor-level simulation, making them suitable for signal integrity analysis. They often can import behavioral IO circuit models in the IBIS format as well. The next class of simulators uses behavioral IO circuit models exclusively and does not support transistor-level simulation. These simulators are tailored toward signal integrity applications and can offer improved performance and other useful features. Recently a new class of simulators has appeared on the scene. They linearize the IO circuits, solve the entire network in the frequency domain, and then transform the results into the time domain. They are a natural fit for s-parameter descriptions of interconnect and can process hundreds of thousands of bits in a few minutes.

Figures 9-7, 9-8, and 9-9 plot the results of simulating 16 inches of 5 mil microstrip and an ideal 100 ohm differential termination in ADS from Agilent Technologies, a high-frequency SPICE signal integrity simulator. The simple 8-bit pattern caused 5 ps of phase jitter, and a 250-bit pseudo-random pattern increased the jitter to 13 ps.

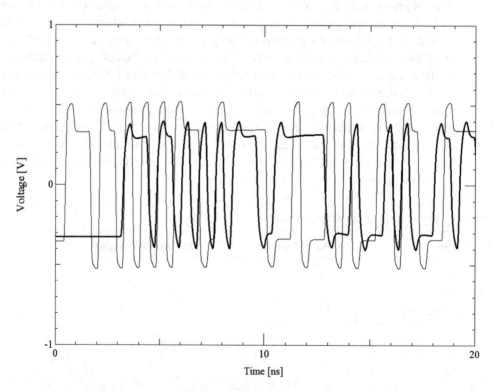

Figure 9-7   Near-end and far-end time domain waveforms.

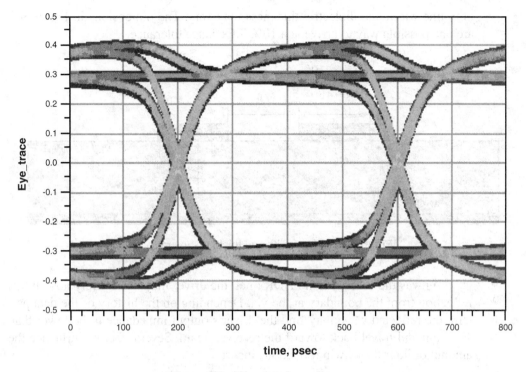

Figure 9-8   Simulated eye diagram.

Note the bifurcation or splitting of waveforms in Figure 9-8. The leading edge of the split comes from the normal signal amplitude, and the trailing edge comes from the de-emphasized amplitude. This signature is responsible for the majority of the jitter in a simple net topology with only lossy transmission lines and no 3D discontinuities. It appears in simulations of the package alone, the add-in card alone, and system board alone. When building up the net one element at a time, loss increases, the bifurcation becomes more fuzzy, and jitter rises slightly.

9.5   CARD IMPEDANCE TOLERANCE

The channel between driver and receiver comprises both 2D and 3D discontinuities that prevent the energy traveling in the electromagnetic wave from reaching its destination or trap that energy between two nodes. When the signal reaches a via and transitions from one layer of the printed circuit board to another, it may encounter a 2D discontinuity if both layers are at opposite extremes of their manufacturing tolerances. Assume that the printed circuit board manufacturer can control the differential impedance to a 10% tolerance. For simplicity, assume that only line width and dielectric thickness may vary. The dimensions in Figure 9-9 are one possible way to arrive at a 10% impedance tolerance.

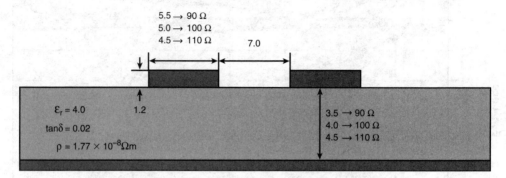

Figure 9-9 Line width and thickness tolerances.

As was the case with the DDR2 net, the driver may not entirely absorb this reflection from the boundary at the via. Depending on the history of the data pattern, the reflected wave may find the driver's output impedance to be lower than 100 ohm and travel back toward the receiver again. Several factors influence the amount of jitter these twin reflections incur:

- Reflection coefficient at the discontinuity
- Location of the discontinuity with respect to the driver and receiver
- Attenuation along the entire path
- Reflection coefficient at the driver

When the incident wave hits the discontinuity, some of it returns to the driver. The loss of this fraction of the original incident wave slightly lowers the edge rate of the wave that continues toward the receiver—on the order of 2 ps. Suppose the sum of the distance between the driver and the discontinuity and the total length of the net is an integer multiple of the unit interval, as is the case in Figure 9-10. Then the secondary reflection off the driver will arrive while the input of the receiver is switching. This effect is about the same size—3 ps. Sweeping the length of the line between the via and the receiver around 11.5 inches finds the sweet spot where the secondary reflection does the worst damage. The size of both these effects pushes the simulator's precision limits, yet we must consider numbers on the order of a handful of picoseconds when the overall budget is 90 ps. This should place simulator accuracy at the top of our lists of things that could impact system reliability.

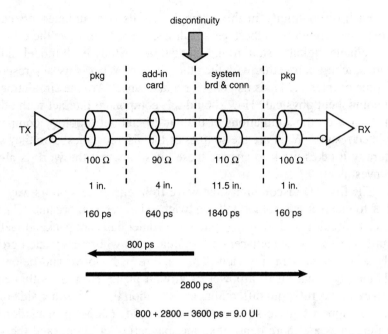

Figure 9-10   Reflections from impedance discontinuity and the driver.

9.6   3D DISCONTINUITIES

Every time an electromagnetic wave encounters a change in conductor geometry or material properties, there is a possibility that some of the energy in the wave will not continue toward the receiver. To make a digital square wave out of sine waves, we need full amplitude at the fundamental frequency, one-third amplitude at three times the fundamental, one-fifth amplitude at five times the fundamental, and so on. If the losses from the reflected energy do not preserve the amplitude relationship between the Fourier components of the signal edge, the shape of the edge will change, and the zero differential crossing point will shift in time. Jitter will occur.

Each of the five 2D transmission line segments in Figure 9-3 lies between two 3D discontinuities: solder balls, vias, edge connector, or AC coupling capacitors. We model these structures using a 3D field solver, export the results into some sort of model such as s-parameters, and then import this model into a time- or frequency-domain engine. The accuracy of the signal and jitter analysis hinges not only on the accuracy of the 3D field solution and associated model but also on our ability to understand the nature of the discontinuity and model its interaction with the rest of the network.

Each discontinuity in this example has its own unique personality. If we become acquainted with these personalities, we will improve the odds of delivering a reliable digital system to the marketplace. Study each model independently before stitching it together with the other models. What is its step response for the rise time of interest? Does its behavior make sense? Are the simulation results for that component physical? How should this component interact with other components? How do the simulations of the end-to-end net topology compare with the expected results? What are the sources of the discrepancies? Are they significant? Naturally it takes work to answer these questions, but the work is always worth the investment.

The first 3D discontinuity the wave front encounters on its way from transmitter to receiver is a solder ball sandwiched between an organic IC package and a printed circuit board. At 1 mm pitch, the solder balls are packed together tightly. Figure 9-11 shows two adjacent differential pairs whose associated ground solder balls do not appear in this view. The capture pads above and below the solder balls may be adjacent to a power or ground plane. How does this environment compare to the 100 ohm differential transmission lines on either side of the solder ball that support transverse electromagnetic mode (TEM) propagation? The electric field lines are more dense than the magnetic field lines, and the solder balls reflect a small burst of energy back toward the driver that opposes the energy in the incident wave. The reflection appears as a negative blip in the voltage waveform at the driver. The properties of this compact differential solder ball structure depend strongly on their proximity to neighboring solder balls and to the structures above and below them. Any 3D electromagnetic model therefore must be large enough to capture the complexity of the physical structure. After the energy has passed through this discontinuity, another TEM wave with slightly less energy forms in the printed circuit board.

In this case study, a set of differential vias lies directly beneath the solder balls. If the transmit nets are routed in a microstrip layer on the bottom side of the system board, the signal must traverse two differential vias to get there. If the receive nets are in a microstrip layer on the top side, there may still be stub vias to facilitate testability. Electromagnetic propagation through vias is not a simple affair. When the wave hits the planes, the electric and magnetic field vectors pick up a component that is parallel to the via barrels. This means the Poynting vector (power flux) is perpendicular to the via barrels, and some of the energy squirts out between the planes and travels away from the vias. If the vias are situated in a field of vias, as they will be under a ball grid array (BGA) package, the neighboring vias act to suppress the parallel plane mode and concentrate the energy around the differential vias. The spacing between the via barrel and the planes (antipad) strongly influences the amount of energy that passes through the via. If the planes are too close to the barrel, the wave has a difficult time passing through this small gap.

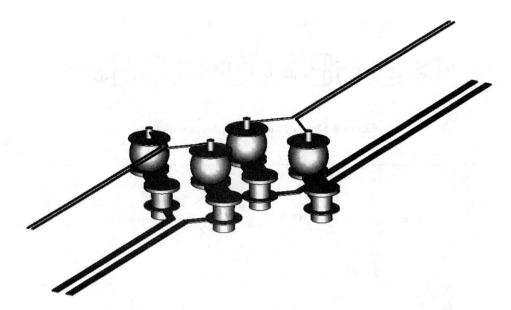

Figure 9-11   A 3D electromagnetic model of the solder ball array.

Some connectors are the antithesis of a solder ball array. For example, a DIMM connector has two rows of tiny springs that wipe against the gold fingers at the edge of the DIMM card. Whereas the solder balls are short and wide, the DIMM connector pins are long and narrow. In this electromagnetic environment, the magnetic field lines are more concentrated than the electric field lines, whereas the opposite is true for the solder ball array. A PCI Express connector is similar to a DIMM connector, except that the pins are shorter and more flat, causing a lower concentration of magnetic field lines when compared to a DIMM connector. When combined with the surface features on the PCI Express add-in card and system board, the PCI Express connector pins actually look more like a solder ball or via than a DIMM connector. If all discontinuities tend to be lower than 100 ohm at 100 ps rise times, why not lower the impedance of card and board wires, too?

9.7   CHANNEL STEP RESPONSE

The channel's step response correlates with the fundamental behavior of these three discontinuities: BGA solder balls, vias, and edge connector. Figure 9-12 points out their locations in space, and Figure 9-13 shows their locations in time.

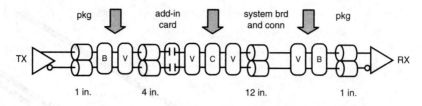

Figure 9-12   Location of 3D discontinuities.

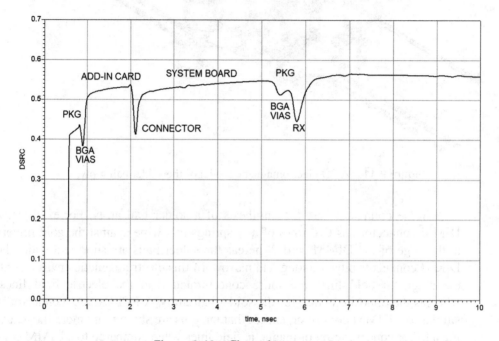

Figure 9-13   Channel step response.

Equation 9-1 shows how to extract the impedance of the first discontinuity from the step response.

$$Z_{3D} = \frac{Z_0 V}{2V_{INCIDENT} - V}$$

Equation 9-1

Z_{3D} is the impedance of the 3D discontinuity, Z_o is the impedance of the source and the transmission line between the source and the discontinuity, VINCIDENT is the step voltage, and V is the voltage at the input of the transmission line after the step is done switching. Strictly speaking, this equation holds for only the first discontinuity. The wave that travels toward the second discontinuity has lost some of its amplitude, which invalidates the assumptions used to derive the equation. Things get even more complicated when reflections begin to bounce between discontinuities and losses come into play. This is still a useful means for calculating the impedance of a single interconnect component in isolation.

Reflections induce jitter in two ways. Each discontinuity takes a little bit of energy from the incident wave (on the order of 5% or 10%), and this slightly changes the signal's edge rate. This effect is not as strong as you might think—on the order of a few ps. Furthermore, suppose the reflected wave reflects off a second boundary, such as a driver, and heads back toward the receiver. It may arrive at the zero volt crossing point and budge the waveform slightly to the left or right of its original location, depending on the sign of the reflection (positive or negative) with respect to the direction of the edge (rising or falling). As we saw in the example of the reflection on the DDR2 net, when a reflection of similar edge rate arrives in the middle of an edge, its effect on the waveform is difficult to observe.

How can we predict and control multiple reflections? After all, it is unlikely that the designer of an add-in card has any direct insight into the size and locations of the 3D discontinuities on the system board, and vice versa. The PCI Express specification addresses losses at 625 MHz and 1.25 GHz, but the combination of add-in card, system board, and two IC packages may result in notches in the s-parameters at other frequencies. If two discontinuities happen to be 400 ps apart, energy can become trapped between them. The PCI Express committee addressed some of these concerns in their design guidelines. For example, they recommend placing AC coupling capacitors near a connector or IC package. Stub vias on a 0.062-inch add-in card resonate around 15 GHz— well beyond the band of interest for 2.5 Gbps signaling. However, a system board may have longer vias that resonate at lower frequencies. It makes good sense to study the resonant behavior of the entire channel in both the time and frequency domains. The specification alone does not guarantee that any given implementation will be trouble-free.

In the final tally, rows 6 through 10 of Table 9-10 list the incremental jitter caused by adding each discontinuity to the simulation net topology shown in Figure 9-3 one at a time. The net is slightly sensitive to variations in length of package, card, and board segments, but only on the order of a few ps. The numbers in Table 9-10 incorporate the effects of pathological lengths.

9.8   CROSSTALK PATHOLOGY

Thanks to Dr. Howard Johnson's *High-Speed Digital Design: A Handbook of Black Magic*, nearly every signal integrity engineer has had some exposure to the fundamental concepts of crosstalk. When two conductors are in close proximity, they can exchange energy through the coupling of electric and magnetic fields in the case of forward crosstalk. Reverse crosstalk depends on amplitude and rise time. Two primary factors determine the amount of energy exchanged: the geometric configuration of the conductors and surrounding material, and the peak instantaneous time derivative of the electric and magnetic fields. Eventually that energy makes its way to a receiver input. It may bounce around on the net a few times, and when it finally reaches the receiver, it may be significantly weaker than it originally was at the point of coupling.

What happens next at the receiver input depends on the other activity on that net. Assume that both nets have the similar peak time derivatives. If the crosstalk arrives at the midpoint of the waveform while the receiver is transitioning from one state to the other, it either speeds up or slows down the edge, depending on the direction of crosstalk with respect to the direction of the edge. If the crosstalk arrives slightly earlier or later than the edge, it distorts the shape of the waveform, causing a flat spot or plateau.

Just how much crosstalk can a digital interface tolerate? This is a difficult question that many intelligent people have sought to answer over the years. In the early days of signal integrity engineering, people counted mV of crosstalk, reflections, and simultaneous switching noise, added the sum to the logic low voltage (or subtracted it from the logic high voltage), and then compared the final number to the receiver threshold voltage. This approach covers the case in which all noise sources superimpose simultaneously on a quiet net and the receive chip happens to be sampling data when the noise is at its peak.

Contemporary digital interfaces are more sensitive to crosstalk-induced jitter. ISI caused by frequency-dependent losses and reflections can cause the zero crossing point to drift to the left or right of its ideal location. A packet of crosstalk energy that is coincident with a zero volt differential crossing point can have a similar effect. Therefore, it is important to pay close attention to the arrival time of the crosstalk pulse with respect to the data edge. When the total jitter from ISI, crosstalk, and simultaneous switching noise approaches the proverbial edge of the cliff, the receiver's output no longer tracks the driver's input. Because most crosstalk simulators report their results in mV, the challenge is to translate these results to ps of jitter.

Each unique physical structure that lies between the IO circuits represents an opportunity for coupling to occur: package wires, package vias, BGA solder balls, printed circuit board vias, printed circuit board wires, edge connector, AC coupling capacitors—even the on-chip metal between the C4 ball (or wire bond pad) and the IO circuit. Each physical structure also represents a unique electromagnetic environment to understand and quantify. Over which structures can we exercise control? If the project involves concurrent ASIC and printed circuit board development, we can influence the package wires, package vias, assignment of BGA signal and return pins, and the artwork that lies on either side of the BGA solder ball. Package vias and wires clearly fall within the realm of the transmit or receive device characteristics, whose boundaries lie at the package pin according to the PCI Express specification. The package solder ball is not such a clear case, because artwork in the package and the artwork in the card both strongly influence its electrical characteristics. In fact, there is a convincing argument that the solder balls and the printed circuit board vias function as one continuous electromagnetic entity.

That leaves the AC coupling capacitors, printed circuit board wires, and edge connector. The size of the smallest AC coupling capacitors (0402) dictates a spacing between adjacent differential pairs that is large enough to preclude any significant coupling. As for the printed circuit board, the PCI Special Interest Group (PCI SIG) recommends 20 mil spacing between differential pairs in add-in cards and system boards for both stripline and microstrip transmission lines.

9.9 CROSSTALK-INDUCED JITTER

What is the motivation for using the recommended 20 mil spacing shown in Figure 9-14? How much crosstalk-induced jitter occurs if both the add-in card and the system board use microstrip differential routing with 20 mils between pairs? As a reference point, assume that 4 inches of add-in card wire and 12 inches of system board wire couple to the victim on one side only. Then vary the spacing between the coupled differential pairs from 5 mils to 20 mils in 5 mil increments. Table 9-7 lists the resulting far-end crosstalk (FEXT) and deterministic jitter (DJ) for these spacing numbers, assuming perfect alignment between the crosstalk pulse and the waveform at the victim receiver. When the total jitter budget is only 90 ps, crosstalk-induced jitter of 52 ps becomes problematic.

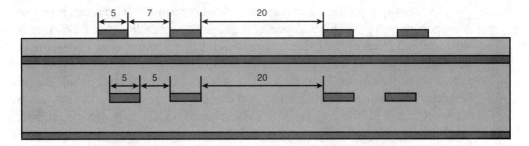

Figure 9-14 Recommended stripline and microstrip spacing.

Table 9-7   Crosstalk and Jitter Versus Spacing for a 16-Inch Coupled Microstrip

Spacing (mils)	FEXT (mV)	DJ (ps)
5	157	26
10	93	16
15	52	9
20	30	5

For each additional 5 mils of spacing between differential pairs, FEXT decreases by roughly 40%. Do these numbers make sense? How much jitter should we expect if 30 mV of crosstalk coincides with the zero volt differential crossing point? This depends on the time derivative of the victim waveform, which is roughly 2.5 V/ns using the 20% and 80% points to calculate the slope.

$$\Delta t = \frac{\Delta V}{dV/dt} = \frac{0.030\ V}{2.5\ V/ns} = 12ps$$

Equation 9-2

There is nearly a factor of 3 difference between calculated and simulated crosstalk-induced jitter! Why? A closer look at the victim waveform reveals the actual derivative to be 4.0 V/ns around the zero volt differential crossing point. This answer is closer to the simulated jitter, which is bumping into the simulator's precision limits.

$$\Delta t = \frac{\Delta V}{dV/dt} = \frac{0.030\ V}{4.0\ V/ns} = 7ps$$

Equation 9-3

This example sheds some light on the common misconception that forward crosstalk is somehow a function of the amplitude of the aggressor waveform. After all, many component vendors quote crosstalk as a percentage of aggressor amplitude. In fact, forward crosstalk is a function of the peak instantaneous time derivative of the aggressor waveform. This number can vary appreciably from a simple calculation using the 20% and 80% points. This explains why crosstalk waveforms have a Gaussian-like signature. The lab data in Figure 9-15 shows how the derivative of a realistic digital waveform peaks near the midpoint and falls off to zero on either side. A 1 V signal generates the same forward crosstalk as a 2 V signal if the values of their peak instantaneous time derivatives are identical.

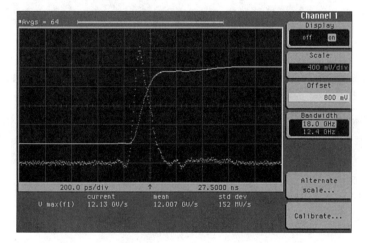

Figure 9-15   Instantaneous dV/dt.

After the crosstalk pulse leaves the region of coupling, conductor and dielectric losses attenuate the pulse on its way to the receiver. The amount of attenuation depends on the pulse's frequency content, the cross-sectional area of the wire, the loss tangent of the dielectric material, and the distance between the region of coupling and the receiver. Gaussian noise pulses have more energy at higher frequencies than do the edges from which they originated, because high frequencies are required to form that sharp peak. Recall how many harmonics were necessary to form the sharp corners in the Fourier construction of a square wave.

To get a sense of how much attenuation affects the size of the crosstalk pulse, run the simulation shown in Figure 9-16. Coupling exists in the add-in card only. Pair-to-pair spacing is 20 mils. Hold the length of the upper system board wire constant, and vary the length of the lower system board wire from 0 to 12 inches. This amount of attenuation spans the range seen in a typical PCI Express application. As shown in Table 9-8, the forward crosstalk pulse shrinks by 30% traveling down 12 inches of 5 mil microstrip wire. Near-end crosstalk (NEXT) saturates at a low level somewhere between 1 inch and 2 inches. Even if the driver reflected all near-end crosstalk toward the receiver, it would not be large enough to count. The driver, being close to 100 ohm differential, absorbs most of the near-end crosstalk—but not all of it. The driver may be in a low-impedance state, and the capacitance of the output transistors and electrostatic discharge (ESD) protection devices lowers the effective impedance even further.

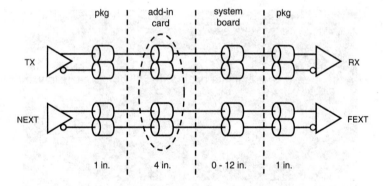

Figure 9-16   Add-in card crosstalk simulation.

Table 9-8   Crosstalk Attenuation Versus System Board Length

Length (inches)	NEXT (mV)	FEXT (mV)
0	2	9
4	2	8
8	2	7
12	2	6

By inserting coupled models one at a time into the simulation shown in Figure 9-16, you can generate Table 9-9. Even if each differential pair were surrounded on all sides by six other differential pairs in the BGA pin field (an extremely poor signal-to-return ratio), it would hardly be enough crosstalk to measure. The differential via model from Chapter 5 is a bit pessimistic with so few other vias in the vicinity. For the final tally, let's assume two neighboring via pairs at the TX chip and two at the RX chip for a total of 16 mV of crosstalk. According to Table 9-7, 16 mV will generate 2.5 ps of jitter. In the PCI Express edge card connector, two ground pins surround each differential pair on the left and two more on the right. It will serve well in the generations to come.

Table 9-9 Component Crosstalk

Component	NEXT (mV)	FEXT (mV)
Card and board wires	2	30
BGA	1	0.2
Card and board vias	9	–4
Edge card connector	1	0.2

How much routing skew between victim and aggressor does it take to decrease the crosstalk to half its peak value? The crosstalk pulse drops to half its amplitude in about 75 ps, which is approximately 0.4 inches. In other words, the worst-case assumption of simultaneous aggressors and victims is valid only if each of the three differential pairs is routed within a 0.2-inch spread. So we see that the assumption of absolute worst-case crosstalk is somewhat pessimistic—even for an engineer.

Crosstalk analysis depends strongly on implementation details. Although the mechanisms are familiar to most signal integrity engineers, crosstalk behavior is more complex than it may first appear. Transmission lines attenuate crosstalk. Discontinuities reflect crosstalk. Crosstalk pulses can interfere constructively and destructively with each other.

A thorough and accurate crosstalk analysis must account for these elements:

- Geometric configuration of coupled conductors and surrounding material
- Peak instantaneous time derivative of voltage waveforms at
  - Coupling location
  - Receiver input
- Cumulative attenuation between location of coupling and receiver
- Timing relationship between victim and each aggressor

- Complex reflection coefficients at
  - Victim driver (depends on driver state)
  - Victim receiver
- Bandwidth of crosstalk pulse
- Bandwidth of all models

Resist the temptation to run one or two simple simulations and be done.

9.10 CHANNEL CHARACTERISTICS

In theory, a set of s-parameters contains all the information necessary to describe a channel's response to any stimulus. The thin curve in Figure 9-17 represents s21 for the lossy transmission lines in the add-in card and system board. The bold curve represents s21 after adding the discontinuities: impedance tolerance, BGA solder ball, vias, and edge connector. Altogether, it is a rather well-behaved channel—no deep notches where energy can get trapped. That explains the exceptional time-domain performance: only 48 ps of deterministic jitter for the entire channel, including crosstalk.

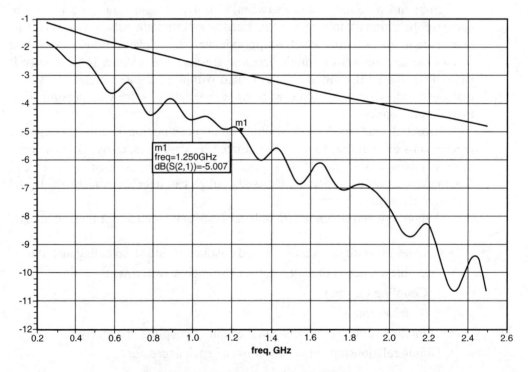

m1
freq=1.250GHz
dB(S(2,1))=-5.007

Figure 9-17   Simulated channel s-parameters.

So far we have considered only jitter, but the PCI Express standard also recommends an eye mask to cover both the time and voltage axes. The dimensions of the eye mask are 175 mV by 260 ps (0.65 UI). The eye diagram shown in Figure 9-18 meets the specification with margin to spare.

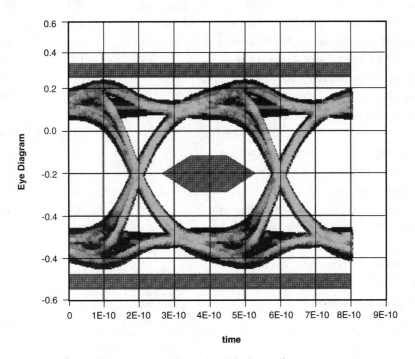

Figure 9-18 Final simulated eye diagram.

9.11 SENSITIVITY ANALYSIS RESULTS

We are now in a position to compile the results of all the previous simulations and see how the dBs and ps stack up against the PCI Express specification. This case study has constructed a typical PCI Express net one component at a time. Table 9-10 lists the cumulative effects on loss and jitter of adding each component. Incremental loss and jitter are simply the differences between the current and previous rows.

Table 9-10 Sensitivity Analysis Results

	Component	Model	Incremental Jitter (ps)	Cumulative Jitter (ps)	Incremental Loss (dB)	Cumulative Loss (dB)
1	De-emphasized driver	Behavioral	11	11	N/A	N/A
2	System board wire (12 inches)	Lossy t-line	3	14	2.2	2.2
3	Add-in card wire (4 inches)	Lossy t-line	2	16	0.8	3.0
4	TX and RX packages (1 inch)	Lossy t-line	2	18	1.1	4.1
5	Impedance tolerance (±10%)	Lossy t-line	0	18	0.2	4.3
6	RX capacitance (2 pF)	Capacitor	6	24	0.6	4.9
7	RX termination (+20%)	Resistor	−1	23	0.3	5.2
8	BGA solder balls	CST MWS	0	23	−0.2	5.0
9	Card and board vias	CST MWS	2	25	−0.1	4.9
10	Card edge connector	Vendor	5	30	0.1	5.0
11	AC coupling capacitors	None	0	30	0.0	5.0
12	Crosstalk: card wires	Lossy t-line	10	40	N/A	N/A
13	Crosstalk: BGA	CST MWS	0	40	N/A	N/A
14	Crosstalk: vias	CST MWS	2	42	N/A	N/A
15	Crosstalk: connector	Vendor	0	42	N/A	N/A
16	**Totals**			**42**		**5.0**
17	**PCI Express specification**			**90**		**11.95**
18	**Margin**			**48**		**6.95**

There is a reason why Table 9-10 contains no entries for the AC coupling capacitors. The spacing between the many planes inside a ceramic capacitor is much smaller than any other dimension in the structure: solder fillet, surface pads, surface traces, vias, and planes. A 3D electromagnetic field solver would have extreme difficulty solving such a problem, because the mesh cell count would get too large. Therefore, AC coupling capacitors are better suited to measurement-based modeling, which is beyond the scope of this book. By inference from add-in card jitter measurements, 0402 AC coupling capacitors contribute on the order of 1 ps to the total jitter.

The dielectric weave effect is also conspicuously missing from the table. When a printed circuit board trace runs nearly parallel to a glass fiber in the dielectric material, its impedance can vary outside the specified tolerance window. This effect is significant only when the angle between the trace and the fiber is only a few degrees. Clearly there is enough margin to absorb this effect if it is on the same order as the 2D discontinuity due to impedance tolerance.

The value of this sensitivity analysis lies in the ability to realistically assess operating margins—to strike a balance between extreme conservatism and ignorance—and to use this knowledge to make decisions in the best interests of product cost, schedule, manufacturability, and reliability.

Table 9-11 lists the initial design rule assumptions for this example. Knowing the sensitivity of the PCI Express interface to each of these parameters, you can confidently make the trade-offs that are so much a part of everyday interactions between engineers. For instance, if the thermal engineer were to ask for an extra inch of add-in card trace length to optimize placement of a heat sink in the airflow, you could refer to Figure 9-19 and respond with a resounding yes. Or if the manufacturing engineer asked for another via for in-circuit test, you could say, "That via doesn't resonate until 15 GHz. I have margin for that." This approach allows you to be both cooperative and responsible. The alternative is to stonewall your colleagues because you have insufficient information to make the trade-offs, and you don't want them to know that.

Table 9-11   Design Rule Assumptions

Design Rule	Value
Dielectric material	FR4
Transmission line construction	microstrip
Add-in card line length	4 inches
System board line length	12 inches
Line width	5 mils
Spacing between true and complement	7 mils
Spacing between diff pairs	20
Via count	2
Via diameter	8
Via pad diameter	20
Via antipad diameter	30

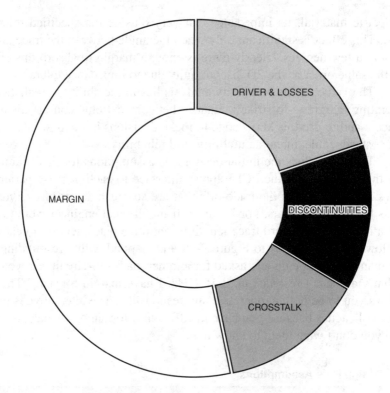

Figure 9-19   Interconnect jitter allocation.

9.12   MODEL-TO-HARDWARE CORRELATION

On the project mentioned at the beginning of this chapter, we had the advantage that the circuit, ASIC, package, and card designers all worked for the same company. Our model-to-hardware correlation began with a 64-page lab report written by the people who tested the high-speed serial macro. They established PCI Express compliance of the TX and RX circuits by measuring nominal, fast, and slow silicon at temperature and voltage corners. One could not ask for a more thorough characterization effort.

The impedance coupon for our prototype board included two PCI Express nets pinned out to probe pads for s-parameter extraction using Agilent's E8364 PNA. We found very little difference between TX and RX nets. In other words, losses due to stub vias, through vias, and 0402 capacitors were small. Both nets measured around –1.5 dB at 1.25 GHz and –3.0 dB at 2.5 GHz.

Add-in card compliance measurements went smoothly. We used the Agilent DSO 81204A loaded with the PCI Express compliance software and the compliance base board from the PCI-SIG. Table 9-12 lists the data from the compliance test report, and Figure 9-20 shows the eye mask test. When comparing lab and simulation results, bear in mind that the compliance base board includes 2 inches of etch, an SMA connector, and cabling that is probably not in your simulation. The specifications account for these effects. Still, it's remarkable how clean the eye looked. Note that simulations do not account for TX jitter, but measurements do. That's why the eye mask in Figure 9-20 is diamond-shaped, but the eye mask in Figure 9-18 includes a rectangle to represent TX jitter.

Table 9-12   Compliance Test Results

Description	Specification	Measurement
Unit interval	±0.12 ps	–0.01 ps
Template test (eye mask failures)	0	0
Median to maximum jitter	56.5 ps	26 ps
Eye width	287 ps	364 ps
Peak differential output voltage	0.36 to 1.2 V	0.998 V

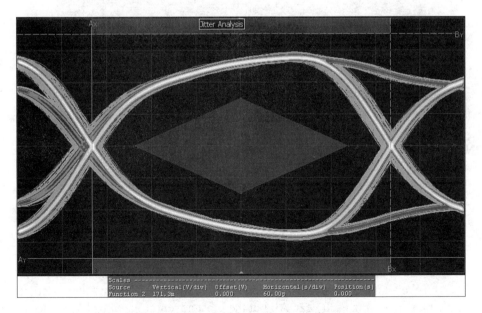

Figure 9-20   Add-in card eye mask test.

For reference, we also measured jitter using the BERTScope BSA 12500A/PL from SyntheSys. The plot shown in Figure 9-21 allowed us to distinguish between random and deterministic jitter. If we assume that most of the random jitter (RJ) comes from the silicon and that deterministic jitter (DJ) is due to the interconnect, we can compare 9.3 ps to rows 1 and 3 of Table 9-10. The results are within a few picoseconds of each other—acceptable accuracy for a 400 ps unit interval.

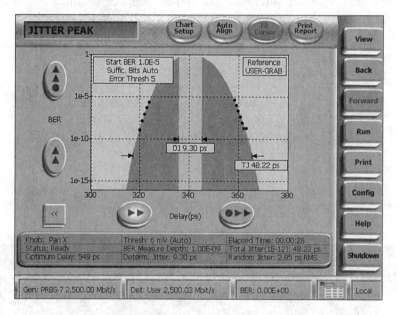

Figure 9-21   Add-in card jitter versus bit error rate.

CONCLUSION

The topic of margins has always seemed to border on taboo, yet it is common to every engineering discipline. We live in fear of people discovering how much conservatism there really is in our part of the budget. If they knew, they might want to take it away. And when things break, guess who gets to spend twelve-hour days in the lab? Guess who gets to explain to the VP why a big pile of scrap is sitting at the end of the assembly line? These are natural fears, to be sure.

We all have physical limitations that we must respect, and they become a little more daunting each year. All but the most adventurous of us have at least some curiosity about where the edge of the cliff lies. Only when the numbers are on the table can an engineering team engage in a rational discussion about their relevance. We may not always agree on the accounting, but the discussion will result in more reliable digital interfaces.

The Wireless Signal

Wireless signals are an integral part of many of today's embedded system designs. Mobile computer providers talk of media convergence, where consumers will be able to browse the Internet or watch live sports on a wireless computer, mobile telephone, portable digital television, or personal digital assistant (PDA). Put simply, the media will be transparent to the wireless technology. Nevertheless, media convergence is the precursor to a myriad of complex technological issues, such as enhanced data compression, interoperability, propagation, and interference. Numerous other wireless uncertainties, such as the large number of international standards and media formats, deserve a book of their own. This chapter, in keeping with signal integrity engineering, is less concerned with media, standards, and the peculiarities of wireless propagation; it focuses on measuring and analyzing wireless signals. Wireless signals and spectrum analysis are wide-ranging subjects with several specialist areas, and it could be argued that such topics are better suited to dedicated wireless books. However, because wireless is becoming so prevalent in embedded system design and there are so many fresh wireless issues, the wireless environment deserves valuable thinking time from the signal integrity engineer. Consequently, this book would be incomplete without an explanation of modern wireless signals and their measurement. There-fore, it is the aim of this chapter to help you understand some of the new tech-niques in wireless signal measurement. This chapter also offers a few thought-provoking ideas about signal analysis in the modern wireless environment.

With such a rich and diverse subject as wireless signals and their measure-ment, it will always be debatable which wireless instruments and applications to include in a broad SI book. Nevertheless, this topic is somewhat straightforward, because it can be argued that the spectrum analyzer (SA) is the principal tool for evaluating radio frequency (RF) signal characteristics. Moreover, spectrum analy-sis is the dominant test setting for a wide range of wireless systems and device designs. Also, spectrum analysis currently supports research and development applications ranging from low-power radio frequency identification (RFID) sys-tems to high-power radar and RF transmitter measurements.

10.1   RADIO FREQUENCY SIGNALS

An RF carrier signal is like a blank piece of paper on which a message can be written and dispatched. RF carriers can transport information in many ways based on variations in the carrier's amplitude or phase, where modulation is simply a change in the shape of a wireless carrier signal. In practice, we talk of amplitude modulation (AM) and frequency modulation (FM), but to be pedantic, frequency modulation is the time derivative of phase modulation (PM). Combinations of AM and PM lead to numerous variations of modulation schemes, such as Quadra-ture Phase Shift Keying (QPSK), a digital modulation format in which the symbol decision points occur at multiples of 90 degrees of phase. Quadrature Amplitude Modulation (QAM) is a high-order modulation format in which both amplitude and phase are varied simultaneously to provide multiple states. Even highly com-plex modulation formats such as Orthogonal Frequency Division Multiplexing (OFDM) can be decomposed into magnitude and phase components. Most ele-mentary texts on wireless provide comprehensive illustrative examples that make plain the methods used to modulate a carrier signal. In the case of understanding modulation, a picture really is worth a thousand words.

However, to understand the digital representation of a modulated wireless carrier, you must be familiar with the vector model that is commonly used to rep-resent a signal's amplitude and phase, as shown in Figure 10-1. A signal vector can be thought of as representing the instantaneous value of the magnitude (amplitude) and phase of a signal as the length and angle of a vector, respectively. In a polar coordinate system, the same point could be expressed on a graph in tra-ditional Cartesian coordinates, or rectangular horizontal X and vertical Y coordi-nates. In a digital representation of RF signals, an in-phase (I) and quadrature (Q) format of time samples are commonly used. These are mathematically equivalent to Cartesian coordinates, with I representing the horizontal or X component and Q the vertical or Y component. Figure 10-2 illustrates the magnitude and phase of a vector, along with corresponding I and Q components.

$$Vector\ magnitude = \sqrt{I^2 + Q^2}\qquad Vector\ phase = tan^{-1}(Q/I)$$

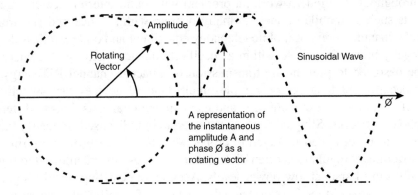

Figure 10-1   A vector model representing a signal's amplitude and phase.

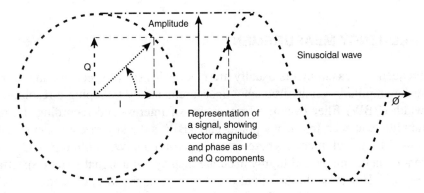

Figure 10-2   A digital representation of a rotating vector with the I (in-phase) and Q (quadrature) components.

For example, consider the demodulation of an AM signal that is to be represented in I and Q components. This basically requires computing the instantaneous carrier signal magnitude for each I and Q sample, where each result is digitally represented and stored in memory. A plot of the stored data (amplitudes) over time would give a representation of the original modulating signal. However, PM demodulation is more complex and consists of computing the phase angle of

I and Q samples, storing the results in memory, and performing some trigonometric computations to correct the data. Then the data is used to reconstruct the original modulating signal. Understanding quadrature I and Q signals appears difficult, but in reality it's similar to understanding the representation of the value of a sinusoid at any point in time using a vector's X and Y coordinates.

However, the ideal signals represented in Figures 10-1 and 10-2 are seldom realized in practice. As mobile telephony and various wireless systems extend throughout the modern world, a problem with radio interference emerges. Products such as mobile phones normally operate within a licensed spectrum. As a rule, manufacturers of mobile telephony equipment and other wireless devices are legally bound to operate within a specific frequency band. Such devices have to be designed to prevent the transmission of adjacent channel RF energy. This is especially challenging for complex multistandard devices that switch between different modes of transmission and maintain simultaneous links to different telephone networks. Simpler devices that operate in unlicensed frequency bands also have to be designed to function properly in the presence of interfering signals. Government regulations often dictate that these devices are allowed to transmit in only short bursts at low power levels. Accurately detecting, measuring, and analyzing "bursty"-type wireless signals are tasks of significant concern in SI engineering.

10.2   FREQUENCY MEASUREMENT

Frequency measurements usually are made with swept spectrum analyzers. They make amplitude-versus-frequency measurements by sweeping a resolution bandwidth (RBW) filter over the frequencies of interest and recording the resultant amplitude at each frequency point. The RBW is the key measurement method for most RF measurements, where swept spectrum analyzers often provide an excellent dynamic range and highly accurate displays of a signal's static spectral components. However, the principal disadvantage of the swept spectrum analyzer is that it records amplitude data at only one frequency at a particular moment in time. This is a weakness because new RF applications are emerging with RF signals that have complex time-related characteristics. Modern RF signals, especially those in the open-access Industrial, Scientific, and Medical (ISM) band, often employ spread spectrum techniques, such as Bluetooth and WiFi, that are more intermittent or "bursty" in nature. Such short-duration wireless signals are significantly more variable in frequency than radio signals of the past. As a result, today's RF signals are more difficult to analyze with a traditional swept spectrum analyzer, which is limited in its digital modulation analysis and multidomain capabilities. Even the vector signal analyzer (VSA), which emerged to specifically address digitally modulated signals, has limited capabilities for analyzing transient signals over time in the frequency modulation domains.

Spectrum monitoring today often involves detecting elementary temporal events in the presence of nonstationary signals and uncorrelated noise. Put simply, transients, predictable and unpredictable frequency shifts, and complex modulation schemes are the norm in many of today's RF and wireless disciplines and applications. Common examples are spread-spectrum and RFID, which communicate via brief bursts of information. Although the traditional swept frequency spectrum analyzer and vector signal analyzer remain the instruments of choice for mainstream RF test and measurement applications, this chapter focuses on innovative real-time spectrum analysis (RTSA). We discuss RTSA because of the migration of today's RF applications to transient signal behavior. SI engineers now need to trigger and capture signals of interest simultaneously in the time and frequency domains. Often SI engineers need to capture a continuous record of signal fluctuations, including transients and frequency deviations, and they need to analyze the signal for changes in frequency, amplitude, and modulation. Moreover, all these operations often need to be carried out over a lengthy period of time. For example, an SI engineer could wait a significant amount of time for a swept-spectrum analyzer to detect a transient event in a modern RF system. Even then there would be a limited means of determining when the event occurred or whether the engineer had actually missed the capture of one or more events.

The common thread that runs through many emerging RF applications is the time-varying nature of the wireless signal. This characteristic, coupled with the factors already discussed, calls for a new type of analysis. As a result, SI engineers and designers are using real-time spectrum analysis in increasing numbers. Although real-time spectrum analysis is not a new concept, and it is very similar in concept to the VSA, the relevance of RTSA to SI engineering is significant. Consequently, today's SI engineer is encouraged to consider both conventional frequency-domain information and RTSA. Moreover, given the trend in SI engineering toward RTSA as more systems require simultaneous time and frequency information to reveal underlying RF signal behavior, there is a reasoned argument for focusing on RTSA in this chapter.

10.2.1 The Swept Spectrum Analyzer

The swept-tuned, superheterodyne spectrum analyzer is the traditional architecture that first enabled engineers to make frequency domain measurements several decades ago. Originally built with purely analog components, the swept SA has since evolved along with the applications it serves. Current-generation swept SAs now include advanced digital elements such as analog-to-digital conversion

(ADC), digital signal processing (DSP), and microprocessors. However, the basic swept approach remains largely the same, and the instrument retains its role as the primary measurement tool for observing controlled RF signals. A clear advantage of a modern swept SA is its excellent dynamic range, whereby it can capture and detect a broad array of RF data.

The swept SA makes power-versus-frequency measurements by down-converting the signal of interest and sweeping it through the passband of an RBW filter. The RBW filter is followed by a detector that calculates the amplitude at each frequency point in the selected passband, as shown in Figure 10-3.

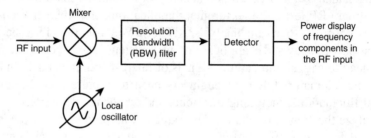

Figure 10-3   A block diagram of a traditional RBW spectrum analyzer.

Figure 10-3 shows the measurement trade-off between frequency resolution and time. The local oscillator sweeps through a "span" of frequencies feeding the mixer. Each sweep produces sum and difference frequencies at the mixer output. The resolution filter has a bandwidth that is set to a user-selected frequency, the RBW. The narrower the filter bandwidth, the higher the resolution of the measurement, and the greater the exclusion of unwanted instrument-generated noise. The RBW filter feeds the detector, which measures the spectral power at each instant in time to produce a frequency-domain display plotting spectral power against frequency. While this method can provide high dynamic range, its principal disadvantage is that it can calculate the amplitude data for only one frequency point at a time. If the RBW filters are made too narrow, the time taken to complete a sweep of the RF input is too long, and any changes in the RF input go undetected. Sweeping the analyzer over a span of frequencies or number of passbands can take considerable time. This measurement technique is based on the assumption that the analyzer can complete several sweeps without any significant changes to the signal being measured. Consequently, a relatively stable, unchanging input signal is required. If the signal changes rapidly, the change probably will be missed.

For example, the left part of Figure 10-4 shows an RBW logic analyzer sweep that is looking at frequency segment Fa while a momentary spectral event occurs at Fb. By the time the sweep arrives at segment Fb, the event has vanished and goes undetected. The RBW spectrum analyzer sweep fails to provide a trigger for the transient signal at Fb and fails to store a comprehensive record of signal behavior over time. This is an example of the classic trade-off between frequency resolution and measurement time, which is the Achilles' heel of the traditional RBW spectrum analyzer.

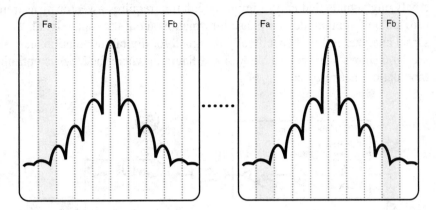

Figure 10-4 The sweep on the left is looking at frequency segment Fa while a momentary spectral event occurs at Fb. The right side of the figure shows the Fb measurement taking place, with the missing transient.

However, a modern swept SA is significantly faster than its traditional purely analog predecessor. Figure 10-5 shows a classic modern swept SA architecture. Traditional wide analog RBW filters have been enhanced with digital techniques to facilitate fast and accurate narrower band filtering. Nevertheless, the filtering, mixing, and amplification preceding the ADC are analog processes. When exacting narrow band-pass filters are needed, they are implemented by DSP in the stage following the ADC. However, the job of the ADC and DSP is rather demanding. In particular, nonlinearity and noise in the ADC can be a concern, and there remains a place for the analog spectrum analyzer, which is an ideal tool for eliminating these problems.

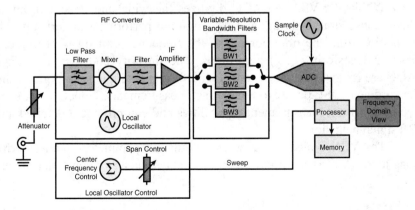

Figure 10-5 A modern swept SA architecture with digital circuitry.

10.2.2   The Vector Signal Analyzer

Traditional swept spectrum analysis enables scalar measurements to provide information about the magnitude of each frequency component in an input signal. Analyzing signals that carry digital modulation requires a vector measurement to provide both magnitude and phase information about the input signal. The vector signal analyzer is a tool specifically designed for digital modulation analysis. Figure 10-6 shows a simplified VSA block diagram, where the VSA is optimized for modulation measurements.

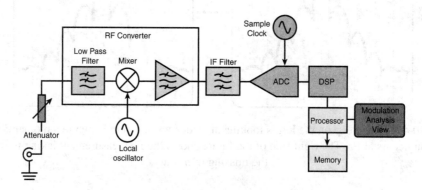

Figure 10-6   A simplified block diagram of the VSA.

Like the Real-Time Spectrum Analyzer (RTSA) described in the next section, a VSA digitizes all the RF energy within the instrument's passband to extract the magnitude and phase information required for digital modulation measurements. However, most, but not all, VSAs are designed to take snapshots of the input signal at arbitrary points in time. This makes it difficult or impossible to store a long record of successive acquisitions for a cumulative history of how a signal behaves over time. Like a swept SA, the triggering capabilities typically are limited to an intermediate frequency (IF) level trigger and an external trigger.

Within the VSA, an ADC digitizes the wideband IF signal, and then the down-conversion, filtering, and detection are performed numerically. The transformation from the time domain to the frequency domain is processed by Fast Fourier Transform (FFT) algorithms. The linearity and dynamic range of the ADC are critical to the instrument's performance. Equally important, there must be sufficient DSP power to enable fast measurements. Nonetheless, some VSAs have significant measurement capabilities and can provide a detailed analysis of RF signal behavior.

The VSA typically measures modulation parameters such as error vector magnitude (EVM) and provides other displays, such as a constellation diagram. A stand-alone VSA is often used to supplement the capabilities of a traditional swept SA. In addition, many modern instruments have architectures that can perform both swept SA and VSA functions. These instruments provide frequency and modulation domain measurements within one instrument; however, both measurements are not always correlated in time.

10.3   OVERVIEW OF THE REAL-TIME SPECTRUM ANALYZER

The RTSA is designed to capture and analyze RF signals that have transient and dynamic characteristics. The fundamental concept of real-time spectrum analysis is to trigger on an RF signal event, seamlessly capture the digitized data into memory, and then provide a built-in analysis of the data in multiple domains. Figure 10-7 is a simplified block diagram of the RTSA architecture. The RF front end can be tuned across the instrument's entire frequency range, and it downconverts the RF input signal to a fixed IF that is related to the RTSA's maximum real-time bandwidth. The signal is then filtered, digitized by the ADC, and passed to the DSP engine that manages the instrument's triggering, memory, and analysis functions.

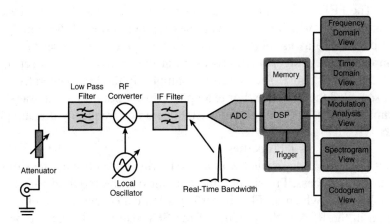

Figure 10-7   A simplified block diagram of the RTSA architecture.

While elements of the RTSA block diagram and acquisition process are similar to those of the VSA architecture, the RTSA is optimized to deliver real-time triggering, seamless signal capture, and time-correlated multidomain analysis. An important consideration for the RTSA is the need to incorporate advanced ADC technology to enable a conversion with high dynamic range and low noise for the measurement to be comparable to a swept SA. For measurement spans that are less than or equal to the instrument's real-time passband, the RTSA architecture lets you seamlessly capture the input signal with no gaps in time by digitizing the RF signal and storing the time-contiguous samples in memory. RTSA has several advantages over the acquisition process of a swept spectrum analyzer for measuring modern "bursty" wireless signals. This makes it an exciting modern instrument to discover and is why RTSA is the focus of this chapter. Nonetheless, the SI engineer must bear in mind that the classic RBW spectrum analyzer remains the instrument of choice for a significant number of RF measurements.

10.3.1   Fast Fourier Transform Analysis

The Fast Fourier Transform is at the heart of real-time spectrum analysis. In the RTSA, FFT algorithms are employed to transform time-domain signals into their frequency-domain spectral components. Conceptually, FFT processing can be considered as passing an RF signal through a bank of parallel filters where each filter has an equal frequency resolution and bandwidth but detects a different component. The FFT output generally is complex-valued, which means that it represents each frequency component as a vector with a particular phase and magnitude. For spectrum analysis, the amplitude of the complex component is usually of most interest.

The FFT process starts with properly decimated and filtered baseband I and Q components, which form the complex representation of the signal, with I as its real part and Q as its imaginary part. In FFT processing, a set of samples of the complex I and Q signals are processed at the same time. This set of samples is called the FFT frame. The number of samples in the FFT, generally a power of 2, is also called the FFT size. For example, a 1,024-point FFT can transform 1,024 I samples and 1,024 Q samples into 1,024 complex frequency-domain points.

10.3.1.1   FFT Properties

A lot of jargon is associated with an FFT, which can obscure the simple concepts of an FFT process. Put simply, the amount of time required to acquire a set of samples upon which an FFT is performed is called the frame length. The frame length in the RSA is the product of the FFT size and the sample period; it's the time taken to perform a single measurement. Any changes, or temporal events, that occur in the measured signal during the frame length cannot be resolved. Therefore, the frame length is simply the time resolution of the FFT process or a single measurement period for an RTSA. The frequency domain points of FFT processing are often called FFT bins. Therefore, the FFT size is equal to the number of bins in one FFT frame. Those bins are equivalent to the individual filter output of an RTSA parallel filter. All bins are spaced equally in frequency. Two spectral lines closer than the bin width cannot be resolved. The FFT frequency resolution therefore is the width of each frequency bin, which is equal to the sample frequency divided by the FFT size. Therefore, for the same sample rate, a larger FFT size yields a finer frequency resolution. For example, an RTSA with a sample rate of 25.6 MHz and an FFT size of 1,024 gives a frequency resolution of 25 kHz.

Frequency resolution can be improved by increasing the FFT size or by reducing the sampling frequency. The RTSA typically uses a digital down-converter (DDC) and decimator to reduce the effective sampling rate as the frequency span is narrowed. This effectively trades time resolution for frequency resolution while keeping the FFT size and computational complexity at manageable levels. This approach allows fine resolution on narrow spans without excessive computation time on wide spans, where coarser frequency resolution is sufficient. The practical limit on FFT size is often display resolution, because an FFT with resolution much higher than the number of display points does not provide any additional information on the instrument's screen.

10.3.2   Windowing

An assumption inherent in the mathematics of Discrete Fourier Transforms (DFTs) and FFT analysis says that the data to be processed is a single period of a periodically repeating signal. Figure 10-8 depicts the FFT of a series of time domain frames. For example, when FFT processing is applied to Frame 2, the transform process assumes that the signal is periodic. However, the process generally produces discontinuities between successive frames, as shown in Figure 10-9. These artificial discontinuities generate spurious responses not present in the original signal, which can make it impossible to detect small signals in the presence of nearby large ones. This effect is called spectral leakage.

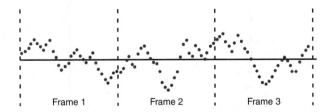

Figure 10-8   Three frames of a sampled time domain signal.

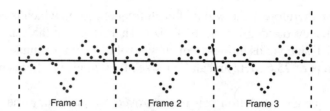

Figure 10-9 Discontinuities between frames.

Typically an RSA applies a windowing technique to the FFT frame before FFT processing is performed to reduce the effects of spectral leakage. The window functions usually have a bell shape. Numerous window functions are available. Figure 10-10 shows the popular Blackman-Harris 4B (BH4B) profile. The Blackman-Harris 4B windowing function shown in Figure 10-10 has a value of 0 for the first and last samples and a continuous curve in between. Multiplying the FFT frame by the window function reduces the discontinuities at the ends of the frame. In the case of the Blackman-Harris window, we can eliminate discontinuities. The effect of windowing is to place a greater weight on the samples in the center of the window than those away from the center, bringing the value to 0 at the ends. This can be thought of as effectively reducing the time over which the FFT is calculated. It should be noted that time and frequency are reciprocal quantities and a smaller time sample implies poorer (wider) frequency resolution. For Blackman-Harris 4B windows, the effective frequency resolution is approximately twice as wide as the value achieved without windowing.

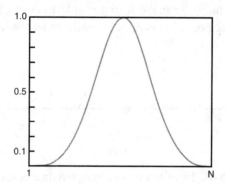

Figure 10-10 A BH4B window profile.

Another implication of windowing is that the time-domain data modified by this window produces an FFT output spectrum that is most sensitive to behavior in the center of the frame and is insensitive to behavior at the beginning and end of the frame. Transient signals appearing close to either end of the FFT frame are de-emphasized and can be missed. This problem can be resolved by using over-lapping frames, a complex technique involving trade-offs between computation time and time-domain flatness to achieve the desired performance.

10.3.3   Key Concepts of Real-Time Spectrum Analysis

As you have seen, the measurements performed by an RTSA are implemented using DSP techniques. To understand how an RF signal can be analyzed in the time, frequency, and modulation domains, first you must examine how the instrument acquires and stores a signal. After the ADC digitizes an RF input, the signal is represented by time domain data, from which all frequency and modulation parameters can be calculated using DSP. Three terms samples, frames, and blocks describe the hierarchy of data stored when an RTSA contiguously captures a signal using real-time acquisition. Figure 10-11 illustrates the sample-frame-block structure.

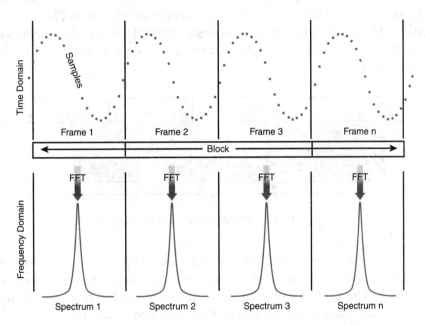

Figure 10-11   The sample-frame-block structure of an RTSA acquisition.

The lowest level of the data hierarchy is the sample, which represents a discrete time-domain data point. This construct is familiar from other applications of digital sampling, such as real-time oscilloscopes and PC-based digitizers. The effective sample rate that determines the time interval between adjacent samples depends on the selected span. In the RTSA, each sample is stored in memory as an IQ pair, where I is the signal's in-phase magnitude component and Q is the signal's quadrature component.

The next in the hierarchy is the frame. A frame consists of an integer number of contiguous samples and is the basic unit to which the FFT can be applied to convert time domain data into the frequency domain. In this process, each frame yields one frequency domain spectrum.

The highest level in the acquisition hierarchy is the block, which is made up of many adjacent frames that are captured seamlessly in time. The block length, also called the acquisition length, is the total amount of time that is represented by one continuous acquisition. Within a block, the input signal is represented with no gaps in time.

In the real-time measurement modes of the RTSA, each block is seamlessly acquired and stored in memory. It is then post-processed using DSP techniques to analyze the signal's frequency, time, and modulation behavior. In standard SA modes, the RTSA can emulate a swept SA by stepping the RF front end across frequency spans that exceed the maximum real-time bandwidth. Figure 10-12 shows block acquisition mode. Although each acquisition is seamless in time for all the frames within a block, remember that data capture is not contiguous between blocks.

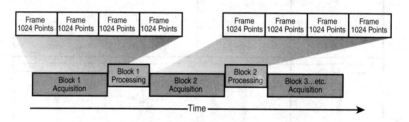

Figure 10-12   RTSA block acquisition mode.

After the signal processing of one acquisition block is complete, the next block is acquired. As soon as the block is stored in memory, real-time measurements can be applied. For example, a signal captured in real-time SA mode can then be analyzed in demodulation mode and time mode. The number of frames acquired within a block can be determined by dividing the acquisition length by the frame length. Normally the user enters the acquisition length; it is rounded so that the block contains an integer number of frames. The maximum acquisition length ranges from seconds to days. It is limited by the selected measurement span and the instrument's memory depth.

10.4   HOW A REAL-TIME SPECTRUM ANALYZER WORKS

A modern real-time spectrum analyzer can acquire a passband, or span, anywhere within the analyzer's input frequency range. At the heart of this capability is an RF down-converter followed by a wideband IF section. An ADC digitizes the IF signal, and the system performs all further steps digitally. An FFT algorithm implements the transformation from time domain to frequency domain, where subsequent analysis produces displays such as spectrograms, codograms, and more.

Several key characteristics distinguish a successful real-time architecture:

- An ADC system that can digitize the entire real-time bandwidth with sufficient fidelity to support the desired measurements.
- An integrated signal analysis system that provides multiple analysis views of the signal under test, all correlated in time.
- Sufficient capture memory and DSP power to enable continuous real-time acquisition over the desired time measurement period.
- DSP power to enable real-time triggering in the frequency domain.

This section contains several architectural diagrams of the main acquisition and analysis blocks of a proprietary RTSA. The principles of operation that are described typically apply to real-time spectrum analysis in general. However, some ancillary functions, such as minor triggering-related blocks and display and keyboard controllers, have been omitted to clarify the discussion.

10.4.1   Digital Signal Processing in Real-Time Spectrum Analysis

A modern RSA uses a combination of analog and digital signal processing to convert RF signals into calibrated, time-correlated multidomain measurements. This section deals with the digital portion of the RSA signal processing flow. Figure 10-13 illustrates the major digital signal processing blocks used in a modern RSA. An analog IF signal is band-pass filtered and digitized. A digital down-conversion and decimation process converts the analog to digital converted samples into streams of in-phase (I) and quadrature (Q) baseband signals. A triggering block detects signal conditions to control acquisition and timing. The baseband I and Q signals as well as triggering information are used by a baseband DSP system to perform spectrum analysis by means of FFT, modulation analysis, power measurements, timing measurements, and statistical analysis.

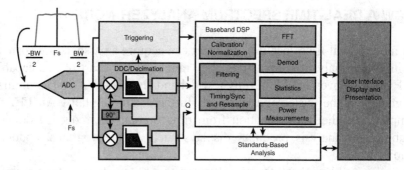

Figure 10-13   The major digital signal processing blocks used in a modern RSA.

10.4.2   IF Digitizer

An RSA typically digitizes a band of frequencies centered on an IF. This band or span of frequencies is the widest frequency for which real-time analysis can be performed. Digitizing at a high IF rather than at DC or baseband has several signal processing advantages, such as overcoming spurious performance, DC rejection, and improved dynamic range. But this can result in excessive computation to filter and analyze the data if processed directly. An alternative technique, shown in Figure 10-13, is to employ a DDC and a decimator to convert the digitized IF into I and Q baseband signals at an effective sampling rate just high enough for the selected span.

10.4.3   Digital Down-Converter

As shown in Figure 10-13, the IF signal in an RSA is digitized with a sampling rate designated as FS. The digitized IF is then sent to a DDC. A numeric oscillator in the DDC generates sine and cosine signals at the centered frequency of the band of interest. The sine and cosine signals are numerically multiplied with the digitized IF to generate streams of I and Q baseband samples that contain all the information present in the original IF signal. The I and Q streams then pass through variable-bandwidth low-pass filters where the cut-off frequency of the low-pass filters is varied according to the selected span.

10.4.4   I and Q Baseband Signals

Figure 10-14 illustrates the process of taking a frequency band and converting it to baseband using quadrature down-conversion. The original IF signal is contained in the space between three halves of the sampling frequency and the sampling frequency itself.

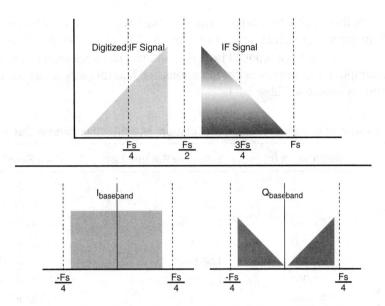

Figure 10-14   Information in the passband is maintained in I and Q at half the sample rate.

Sampling produces an image of this signal between zero and one-half the sampling frequency. The signal is then multiplied with coherent sine and cosine signals at the center of the passband of interest, generating I and Q baseband signals. The baseband signals are real-valued and symmetric about the origin. The same information is contained in each so-called sideband; therefore, it is possible to decimate by 2.

10.4.5   Decimation

The Nyquist theorem states that for baseband signals, you only need to sample at a rate equal to twice the highest frequency of interest. Remembering that time and frequency are reciprocal quantities, it is evident that to study low frequencies with an RSA, you must observe a long time record. For this reason, among others, decimation is used to balance span, processing time, record length, and memory usage.

Consider a practical example. An RTSA uses a 51.2 million samples per second (MSs^{-1}) sampling rate at an ADC to digitize a span of 15 MHz bandwidth. Then the I and Q records that result after DDC, filtering, and decimation for the 15 MHz span are at an effective sampling rate of half the original sample rate, which is 25.6 MSs^{-1}. The total number of samples is unchanged, but we are left with two sets of samples, each at an effective rate of 25.6 MSs^{-1} instead of a single set at 51.2 MSs^{-1}.

Further decimation can be made for narrower spans, resulting in longer time records for an equivalent number of samples. The disadvantage of the lower effective sampling rate is a reduced time resolution. The advantages of the lower effective sampling rate are fewer computations and less memory usage for a given time record, as shown in Table 10-1.

Table 10-1   Examples of Selected Span, Decimation, and Effective Sample Rates

Span	Decimation (n)	Effective Sample Rate	Time Resolution
15 MHz	2	25.6 MSs^{-1}	39.0625 ns
10 MHz	4	12.8 MSs^{-1}	78.1250 ns
1 MHz	40	1.28 MSs^{-1}	781.250 ns
100 kHz	400	128 kSs^{-1}	7.81250 ns
10 kHz	4000	12.8 kSs^{-1}	78.1250 ns
1 kHz	40000	1.28 kSs^{-1}	781.250 ns
100 Hz	400000	128 Ss^{-1}	7.81250 ms

10.4.6   Time and Frequency Domain Effects on the Sampling Rate

Using decimation to reduce the effective sampling rate has several consequences for important time and frequency domain measurement parameters. Figures 10-15 and 10-16 show an example contrasting a wide span and a narrow span. A wide-capture bandwidth displays a broad span of frequencies with relatively low frequency domain resolution. Compared to narrower-capture bandwidths, the sample rate is higher, and the resolution bandwidth is wider. In the time domain, frame length is shorter, and time resolution is finer. Record length is the same in terms of the number of stored samples, but the amount of time represented by these samples is shorter.

In contrast, a narrow-capture bandwidth displays a small span of frequencies with higher frequency domain resolution. In a wide-capture bandwidth, the sample rate is lower and the frequency domain resolution is lower. In the time domain, the frame length is longer and the time resolution is coarser, although the available record length encompasses more time. This can be seen in Figure 10-15, which illustrates a narrow bandwidth capture. Table 10-1 provides some real-world examples. It is important to note the scale of the numbers such as frequency resolution, which are several orders of magnitude different from the wideband capture. This is significant when you're using an RSA.

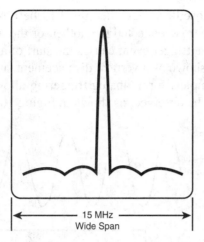

15 MHz
Wide Span

Figure 10-15   Wide bandwidth capture.

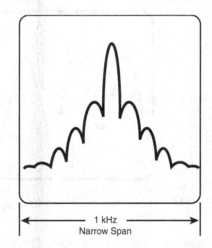

1 kHz
Narrow Span

Figure 10-16   Narrow bandwidth capture.

10.5   APPLYING REAL-TIME SPECTRUM ANALYSIS

Real-time spectrum analysis adds the time domain to spectrum and modulation analysis. Triggering is critical to capturing time domain information, and trigger functionality provides the ability to detect and correct signal integrity errors in common with the oscilloscope and logic analyzer. The most common trigger system used in benchtop instrumentation is the one found in most oscilloscopes. In traditional analog oscilloscopes, the signal to be observed is connected to one input, and the trigger is connected to another, or the trigger is extracted from the captured signal. The trigger event causes the start of a horizontal sweep. The signal's amplitude is shown as a vertical displacement superimposed on a calibrated graticule. In its simplest form, analog triggering allows events that happen after the trigger point to be observed, as shown in Figure 10-17.

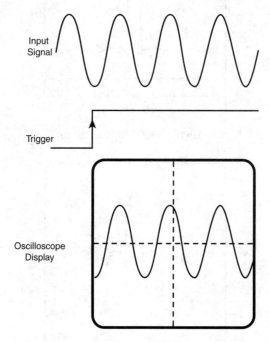

Figure 10-17   Traditional oscilloscope triggering.

The ability to represent and process signals digitally and with a large memory capacity allows the capture of events before and after the trigger point. Digital acquisition systems of the type used in a typical RSA use an ADC to fill a deep memory with time samples of the received signal. Conceptually, new samples are continuously fed to the memory while the oldest samples fall off. Figure 10-18 shows a memory configured to store N samples. The arrival of a trigger stops the acquisition, freezing the contents of the memory. The addition of a variable delay in the path of the trigger signal allows events that happen before a trigger as well as those that come after it to be captured.

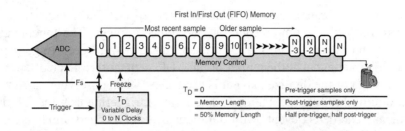

Figure 10-18 Triggering in digital acquisition systems.

Consider a case in which there is no delay. The trigger event causes the memory to freeze immediately at the trigger point and store a sample concurrent with the trigger point. The memory then contains the sample at the time of the trigger, as well as N samples that occurred before the trigger. Only pre-trigger events are stored. Now consider the case in which the delay is set to match exactly the memory's length. N samples are then allowed to come into the memory after the trigger occurrence before the memory is frozen. The memory now contains N samples of signal activity after the trigger. Only post-trigger events are stored. Both post- and pre-trigger events can be captured if the delay is set to a fraction of the memory length. If the delay is set to half of the memory depth, half of the stored samples are those that preceded the trigger and half of the stored samples that followed it. This concept is similar to a trigger delay used in the zero span mode of a conventional swept SA. However, an RSA typically can capture much longer time records; this signal data can subsequently be analyzed in the frequency, time, and modulation domains. This is an important measurement technique for applications that require signal monitoring and SI troubleshooting.

10.5.1 RTSA Trigger Sources and Data-Capture Techniques

An RTSA normally provides several methods of internal and external triggering. Various real-time trigger sources exist. Each is suited to a particular application, setting, and required time resolution:

- External triggering typically allows an external transistor-transistor logic (TTL) signal to control data acquisition. This is usually a control signal such as a frequency switching command from the system under test. This external signal prompts the acquisition of an event in the system under test.
- Internal triggering depends on the characteristics of the signal being tested. An RSA normally can trigger on the level or amplitude of the digitized signal, on the power of the signal after filtering and decimation, or on the occurrence of specific spectral components using a frequency mask trigger.

Each of the trigger sources and modes offers specific advantages in terms of frequency selectivity, time resolution, and dynamic range. The functional elements that support these features are shown in Figure 10-19. Level triggering compares the digitized signal at the output of the ADC with a user-selected setting. The full bandwidth of the digitized signal is used, even when observing narrow spans that require further filtering and decimation. Level triggering uses the full digitization rate and can detect events with durations as brief as one sample at the full sampling rate. The time resolution of the downstream analyses, however, is limited to the decimated effective sampling rate. The trigger level is set as a percentage of the ADC clip level—that is, its maximum binary value, all 1s. This is a linear quantity not to be confused with the logarithmic display, which is expressed in dB.

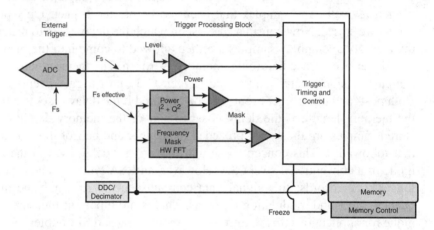

Figure 10-19   A simplified block diagram of a digital SA trigger processing system.

Power triggering calculates the power of the signal after filtering and decimation. The power of each filtered pair of I and Q samples ($I^2 + Q^2$) is compared with a user-selected power setting. The setting is in dB relative to full scale (dBfs), as shown on the logarithmic display screen. A setting of 0 dBfs places the trigger level at the top graticule of the display and generates a trigger when the total power contained in the span exceeds that trigger level. Likewise, a setting of –10 dBfs triggers when the total power in the span reaches a level 10 dB below the display's top graticule. It should be noted that the total power in the span generates a trigger. For example, two carrier wave signals, each at a level of –3 dBm, would have an aggregate power of 0 dBm.

Frequency mask triggering compares the spectrum shape to a user-defined mask. This technique allows changes in a spectrum shape to trigger an acquisition. Frequency mask triggers typically are used to detect signals that are significantly below full-scale measurement, even in the presence of other signals at much higher levels. This ability to trigger on weak signals in the presence of strong ones is critical for detecting intermittent signals in the presence of intermodulation products and transient spectrum containment violations. However, a full FFT is required to compare a signal to a mask, and this takes a complete frame. Therefore, the time resolution for a frequency mask trigger is roughly one FFT frame, typically 1,024 samples at the effective sampling rate. Trigger events normally are determined in the frequency domain using a dedicated hardware FFT processor, as shown in Figure 10-19.

10.5.2 Constructing a Frequency Mask

Like other forms of mask testing in oscilloscopes and logic analyzers, the RSA frequency mask trigger, also called the frequency domain trigger, starts with a definition of an on-screen mask. The mask typically is defined with a set of frequency points and their amplitudes. The mask normally is defined point by point or graphically by drawing it with a mouse or other computer pointing device. Triggers generally are set to occur when a signal outside the mask boundary "breaks in" or when a signal inside the mask boundary "breaks out." Figure 10-20 shows a frequency mask defined to allow the passage of a signal's normal spectrum but not momentary aberrations. Figure 10-21 shows a spectrogram display for an acquisition that was triggered when the signal momentarily exceeded the mask.

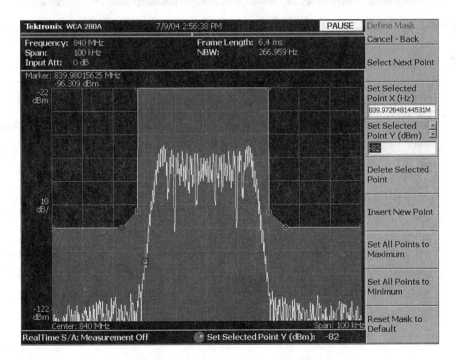

Figure 10-20   A frequency mask definition.

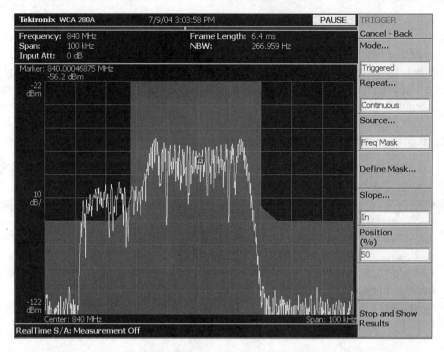

Figure 10-21   An acquisition that was triggered when the signal momentarily exceeded the
frequency mask.

10.5.3   Frequency Domain Measurements

Real-time spectrum analysis lets you analyze captured time domain data using a
power-versus-frequency spectrogram display. A spectrogram has three axes:

- The horizontal axis represents frequency.
- The vertical axis represents time.
- The gray scale or color represents amplitude.

The spectrogram is an important measurement that provides an intuitive display
of how frequency and amplitude behavior change over time. Figure 10-22 shows

the spectrogram of a dynamic signal. The horizontal axis represents the same range of frequencies that a traditional spectrum analyzer shows on the power-versus-frequency display. In the spectrogram, the vertical axis represents time, and amplitude is represented by the trace's gray scale or color. In an RTSA, each "slice" of the spectrogram corresponds to a single frequency spectrum calculated from one frame of captured time domain data. Figure 10-23 shows the power-versus-frequency and spectrogram displays for the signal shown in the figure. On the spectrogram, the oldest frame is shown at the top of the display, and the most recent frame is shown at the bottom. This measurement shows an RF signal whose frequency is changing over time. It also reveals a low-level transient signal that appears and disappears near the end of the time block. Since the data is stored in memory, a marker can be used to scroll "back in time" through the spectrogram. In Figure 10-23 a marker has been placed on the transient event on the spectrogram display. This causes the spectrum corresponding to that particular point in time to be shown on the power-versus-frequency display.

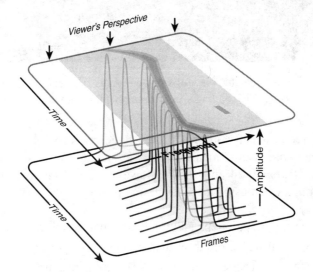

Figure 10-22 A real-time spectrogram dynamic signal.

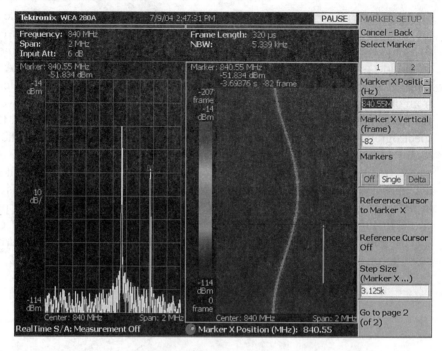

Figure 10-23 A time-correlated view of a power-versus-frequency display, shown on the left, and the corresponding spectrogram display, shown on the right.

When combined with real-time triggering capabilities, the spectrogram becomes a more useful measurement, especially for dynamic RF signals, as shown in Figures 10-24 and 10-25. However, the SI engineer must remember a few key points when using a spectrogram display:

- Frame time is span-dependent; a wider span equals a shorter time.
- One vertical step-through of the spectrogram equals one real-time frame.
- One real-time frame equals 1,024 time domain samples.
- The oldest frame is at the top of the screen; the most recent frame is at the bottom.
- The data within a block is contiguous in time.
- Horizontal black lines on the spectrogram represent boundaries between blocks; these are the gaps in time that occur between acquisitions.
- The white bar on the left side of a spectrogram display denotes post-trigger data.

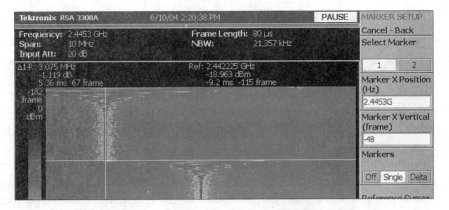

Figure 10-24 Real-Time SA mode showing a spectrogram of a frequency-hopping signal.

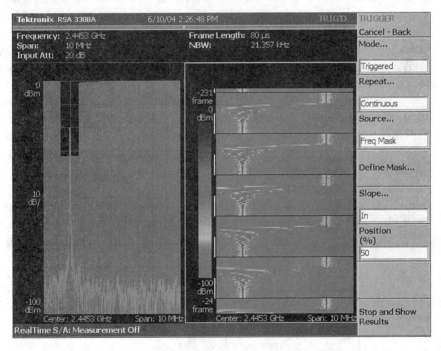

Figure 10-25 Real-Time SA mode showing several blocks acquired using a frequency mask trigger to measure the repeatability of frequency switching transients.

10.5.4   Time Domain Measurements

The frequency-versus-time measurement displays frequency on the vertical axis and time on the horizontal axis. It provides a result similar to what is shown on the spectrogram display, with two important differences. First, the frequency-versus-time view has much better time domain resolution than the spectrogram. Second, this measurement calculates a single average frequency value for every point in time, which means that it is unable to display multiple RF signals like the spectrogram. The spectrogram is a compilation of frames and has a line-by-line time resolution equal to the length of one frame. In contrast, the frequency-versus-time view has a time resolution of one sample interval. For example, if a frame has 1,024 samples, the resolution of the frequency-versus-time measurement would be 1,024 times finer than that of a spectrogram. This makes it easy for a frequency-versus-time measurement to show small or short frequency shifts in greater detail. The frequency-versus-time measurement acts almost like a very fast frequency counter, where each of the 1,024 sample points represents a frequency value, whether the span is a few hundred hertz or many megahertz. Constant-frequency signals such as a carrier wave or amplitude modulation produce a flat, level display.

The frequency-versus-time view provides the best results when there is a relatively strong signal at one unique frequency. Figure 10-26 is a simplified illustration contrasting the frequency-versus-time display with a spectrogram. The frequency-versus-time display is in some ways a zoomed-in view that magnifies a portion of the spectrogram. This is very useful for examining transient events such as frequency overshoot or ringing. When there are multiple signals in the measured environment, or one signal with an elevated noise level, or intermittent spurs, the spectrogram remains the preferred view. Put simply, it provides a visualization of all the frequency and amplitude activity across the chosen span.

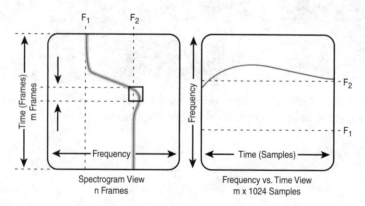

Figure 10-26   A comparison of spectrogram and frequency-versus-time views.

Figures 10-27, 10-28, and 10-29 compare the frequency-versus-time and spectrogram displays by showing three different analyses or views of the same acquisition. In Figure 10-27, the frequency mask trigger was used to capture a transient signal coming from a transmitter having occasional problems with frequency stability during turn-on. Since the oscillator was not tuned to the frequency at the center of the screen, the RF signal broke the frequency mask, as shown on the left of Figure 10-27, and caused a trigger. The spectrogram plot on the right of Figure 10-27 shows a 35 ms time segment of the device's frequency settling behavior.

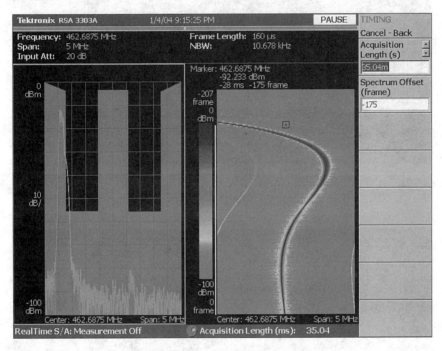

Figure 10-27   A spectrogram showing a 5 MHz signal settling over 35 ms of time.

The next two figures show frequency-versus-time displays of the same signal. Figure 10-28 shows the same frequency settling behavior of the 5 MHz signal from Figure 10-27, but the frequency-versus-time display allows a shorter 25 ms analysis length. Figure 10-29 shows that you can zoom in to a frequency-versus-time display to give an analysis length of 1 ms, which shows the changes in frequency over time with a much finer time domain resolution. This reveals a residual oscillation on the signal even after it has settled to the correct frequency.

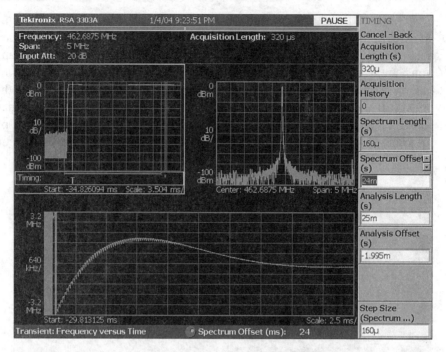

Figure 10-28   A frequency-versus-time view of the frequency settling behavior of the 5 MHz signal over 25 ms of time.

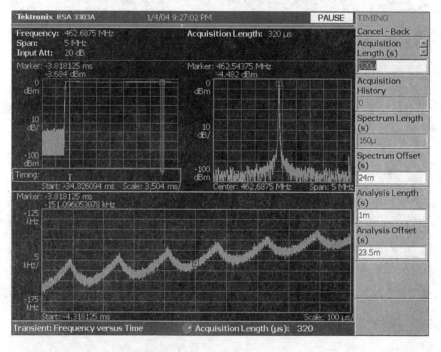

Figure 10-29   The same 5 MHz signal with a zoomed-in view to give a 50 kHz frequency versus 1 ms time display of the signal's settling behavior.

Another time-domain measurement is the magnitude of both I and Q as a function of time. The amplitude-versus-time measurement shows the raw I and Q output signals coming from the digital down-converter. This measurement is a useful troubleshooting tool for expert users, especially in terms of providing insight into irregular frequencies, phase errors, and instabilities.

10.5.5   Power Measurements

A power-versus-time measurement shows how a signal's power changes on a sample-by-sample basis. This is shown in Figure 10-30, where the signal's amplitude is plotted in dBm on a logarithmic scale. The power-versus-time display is similar to an oscilloscope time-domain view in that the horizontal axis represents time. However, in contrast to an oscilloscope display, the vertical axis of a power-versus-time display shows power on a logarithmic scale instead of voltage on a linear scale. In addition, the power-versus-time display gives a measurement that represents the total power detected within the measurement span. For example, a constant power signal yields a flat trace display because there is no average power change per cycle. For each time domain sample point, a signal's logarithmic power ratio is calculated with reference to a 1mW signal, as shown in Equation 10-1.

$$Power = \frac{10log\ (I^2 + Q^2)}{1mW}$$

Equation 10-1

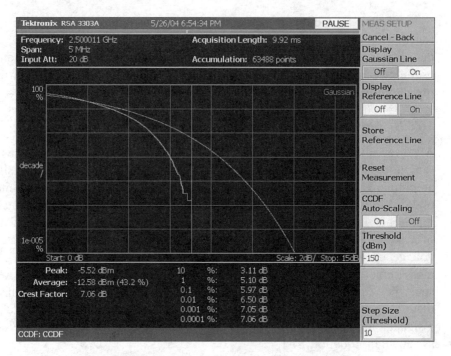

Figure 10-30   A power-versus-time display.

10.5.6   Complementary Cumulative Distribution Measurements

An experienced SI engineer typically uses the Complementary Cumulative Distribution Function (CCDF) measurement to determine the probability of a peak power signal component exceeding the average power of a measured signal. In other words, it displays the percentage probability that a part of the signal will momentarily exceed the average power amplitude that is displayed on the horizontal scale. The probability is displayed as a percentage value on the vertical scale, where the vertical axis is a logarithmic scale.

   CCDF analysis actually measures the time-varying crest factor, which is an important measurement for many digital signals, especially those that use Code Division Multiple Access (CDMA) and Orthogonal Frequency Division Multiplexing (OFDM). The crest factor measurement is the calculated ratio of a signal's peak voltage divided by its average voltage with the result expressed in dB, as shown in Equation 10-2.

$$Crest\ factor = 20\ log\left\{\frac{Vpeak}{Vrms}\right\}$$

Equation 10-2

The signal's crest factor determines how linear a transmitter or receiver must be to avoid unacceptable levels of signal distortion. The CCDF curves in Figure 10-31 show the measured signal against a Gaussian reference trace. The CCDF and crest factor measurements are especially interesting to designers, because they must balance power consumption and distortion performance of devices such as RF amplifiers.

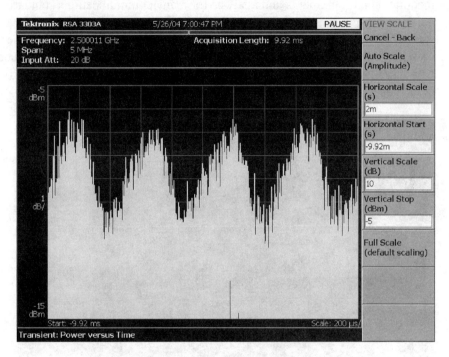

Figure 10-31   A CCDF measurement.

10.5.7   Modulation Domain Measurements

Apart from signal capture and automated measurement, the trend today is for a modern instrument to automatically analyze the captured signal. In the case of the RTSA, the instrument provides automated demodulation analysis of amplitude modulation, frequency modulation, and phase modulation. Just like time domain measurements, these analytical tools are based on the concept of multidomain analysis.

The digital demodulation mode can demodulate and analyze many common digital signals based on phase shift keying (PSK), frequency shift keying (FSK), and QAM. An RSA typically provides a wide range of measurements, including constellation, error vector magnitude (EVM), magnitude error and phase error, demodulated IQ versus time, symbol table, and eye diagrams. The SI engineer has to properly configure the instrument to enable accurate demodulation measurements and a number of variables such as the modulation type, symbol rate, and measurement filter types. This is a complex task that requires patience and practice. However, today's instruments, such as VSAs and RTSAs, provide built-in modulation measurements and analysis for many communications standards. Figure 10-32 shows the modulation analysis of a Wideband Code Division Multiple Access (W-CDMA) mobile telephone handset signal that is under closed-loop power control. The lower-right constellation display shows the error associated with large glitches that occur during level transitions. The transitions can be seen in the upper-left correlated power-versus-time display.

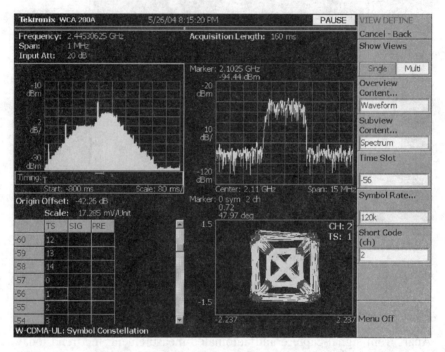

Figure 10-32 Modulation analysis of a W-CDMA handset under closed-loop power control. The lower right is a constellation display, and the upper left is a correlated power-versus-time measurement.

10.5.8 The Codogram Display

Modern mobile telephony provides voice, text messaging, and data communication, often including Internet access. Such systems naturally make use of complex modulation and signaling techniques that require complex measurement and analysis. The codogram illustrated in Figure 10-33 adds a time axis to code domain power measurements for the complex CDMA-based communications standards. Like the spectrogram, the codogram intuitively shows changes over time. Figure 10-34 is an actual W-CDMA codogram display showing a simulated W-CDMA compressed mode handoff in which the data rate is momentarily increased to make room for temporary dual-mode operations.

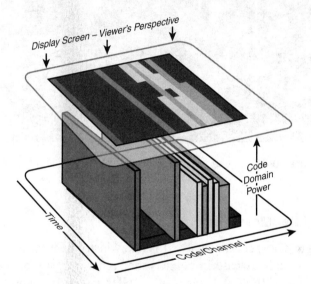

Figure 10-33 The codogram display.

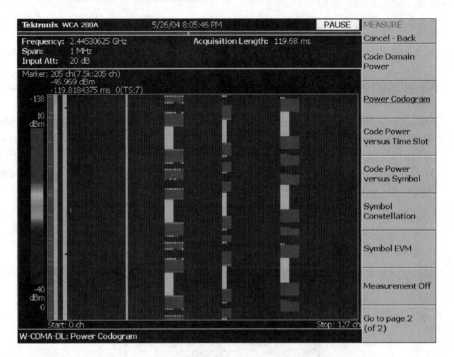

Figure 10-34   A codogram measurement of a W-CDMA signal.

10.5.9   Jitter Measurements

Interestingly, the spectrum analyzer is a key instrument for measuring and testing bound RF signals. A case in point is the signal integrity of the new high-speed serial buses. An RF digital signal in the GHz region is bound to a printed circuit board and, unlike a wireless signal, its electromagnetic radiation is deliberately kept to a minimum. A primary measurement is random jitter and maximum periodic jitter or deterministic jitter, which you need to consider when evaluating a new high-speed serial bus design. Nonetheless, radio signals, particularly wireless data, often require jitter analysis.

Combined with a precise frequency source, an RF signal can be translated to a lower frequency to allow minute changes in signal phase to be accurately measured with a spectrum analyzer. After you measure signal phase deviation, a direct conversion to time is possible where, for example, a signal that changes phase by 1 degree has a 1/360 change in period. When jitter is measured with a spectrum analyzer, it is called phase noise and is measured relative to the correct signal in dBc. The signal under test is considered the fundamental frequency or carrier frequency. Its phase noise results are normalized to a unit Hertz value based on several aspects of the spectrum analyzer measurement system, such as filter bandwidths and sweep rates. The advantage of using a spectrum analyzer for a jitter measurement is that normally there is a very good phase measurement resolution, and the instruments themselves typically have very low noise floors. This provides a platform that can measure frequency and phase change well below what is possible in conventional time-domain instrumentation. For example, several common spectrum analyzers have noise floors below −105 dB at a 10 kHz offset. This is compared to a current real-time oscilloscope, which comes in with −50 dB at a 10 kHz offset. Spectrum analyzers can be quite effective at finding small sources of phase noise and can give a clear analysis with reference to diverse types of signal modulation. Unfortunately, swept mode spectrum analyzers are unable to measure cycle-to-cycle changes in a signal. Also, a swept mode spectrum analyzer cannot be used to measure data signals. However, some of the newer real-time spectrum analyzers can measure cycle-to-cycle changes and, like a VSA, are quite useful for measuring small phase variations in the range between DC and 500 GHz. A general requirement for jitter analysis in a real-time spectrum analyzer is the need to set the analyzer's start and stop offset frequencies to define the integration bandwidth of interest. Put simply, the instrument has to be set to sum the average phase noise between two user-defined points to calculate quantitative values of random jitter and maximum periodic jitter. Phase noise is just jitter by a different name. Consider, for example, Figure 10-35, which shows a sinusoidal clock signal at 1 GHz. The signal's period is 1 ns. If the signal has 0.1 unit of peak jitter, 10% of one unit or signal interval, the phase error is calculated by determining the phase angle subtended by 0.1 unit. Therefore, as shown in Figure 10-35, if one cycle of rotation of the sinusoidal signal is 360 degrees, or a complete phase change of 360 degrees, 2π radians, the signal's phase error is 0.1×360 degrees = 36 degrees peak, $0.1 \times 2\pi$ radians = 0.628 radians—and that's the straightforward view.

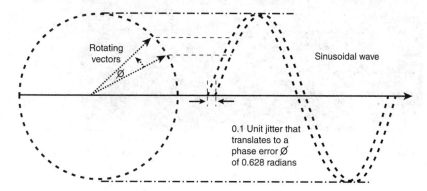

Figure 10-35   Phase noise is simply jitter by a different name.

Sometimes it is necessary to measure nominal jitter levels in the midst of noise peaks that could distort the results. A solution to this problem is to set a maximum threshold value to exclude unwanted peaks above the median. Figure 10-36 is a multidomain display of a high-speed serial bus, showing jitter measurements and other RF signal behavior. Figure 10-37 is a display taken from a proprietary RTSA. It shows the user-defined Start and End offset frequencies that range in this instrument from 10 Hz to 100 MHz. After they are set, the offset frequencies become the phase noise integration bandwidth for the built-in jitter calculations. A single correlated display is used to show the time relationship between phase noise behavior and jitter. The maximum periodic jitter threshold can be set so as to ignore large spurious responses or noise peaks while median jitter values are measured. Although jitter is a complex topic, its measurement and analysis are becoming somewhat easier as instrument manufacturers provide automated built-in jitter analysis and advanced software for straightforward jitter compliance tests.

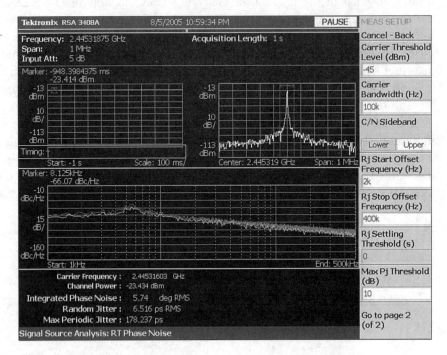

Figure 10-36   A multidomain display of a high-speed serial bus showing jitter.

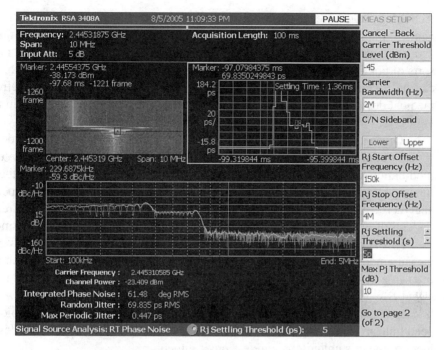

Figure to the right of the image, on the instrument display:

Tektronix RSA 3408A	8/5/2005 11:09:33 PM

Frequency: 2.44531875 GHz
Span: 10 MHz
Input Att: 5 dB

Acquisition Length: 100 ms

PAUSE | MEAS SETUP

Cancel - Back

Carrier Threshold Level (dBm)
-45

Marker: 2.44554375 GHz
-38.173 dBm
-97.68 ms -1221 frame
-1260 frame

Marker: -97.07984375 ms
69.8350249843 ps
184.2 ps   Settling Time : 1.36ms
20 ps/
-15.8 ps

Carrier Bandwidth (Hz)
2M

C/N Sideband

Lower   Upper

Rj Start Offset Frequency (Hz)
150k

-1200 frame

Center: 2.445319 GHz   Span: 10 MHz
Marker: 229.6875kHz
-59.3 dBc/Hz
-99.319844 ms   -95.399844 ms

Rj Stop Offset Frequency (Hz)
4M

-10 dBc/Hz

15 dB/

-160 dBc/Hz

Start: 100kHz   End: 5MHz

Rj Settling Threshold (s)
50

Max Pj Threshold (dB)
10

Carrier Frequency :   2.445310585 GHz
Channel Power : -23.409 dBm
Integrated Phase Noise :   61.48   deg RMS
Random Jitter :   69.835 ps RMS
Max Periodic Jitter :   0.447 ps

Go to page 2 (of 2)

Signal Source Analysis: RT Phase Noise   ● Rj Settling Threshold (ps):   5

Figure 10-37   The relationship between jitter settling time and phase noise offset.

A unique feature of a frequency mask trigger is the ability to trigger on transient signals and capture a contiguous sequence of spectral data. Moreover, complex triggering provides the means to time-correlated views and markers that show the relationship between a jitter settling period and phase noise offset, as shown in Figure 10-37.

CONCLUSION

Along with the sustained growth in the military and mobile telephony industries, the technological innovation in RF has accelerated steadily. In particular, increased RF system development has occurred in the past two decades, and currently we are experiencing a period of intense activity in the RF industry. Consequently, to avoid interference, improve security, and advance communication capacity, modern military and commercial wireless networks have become complex. Current wireless systems typically employ sophisticated combinations of RF techniques, such as bursting, frequency hopping, CDMA, and adaptive modulation. Designing today's advanced RF equipment and successfully integrating it into working systems are complicated tasks. Moreover, the success of cellular technology and wireless data networks has caused the cost of basic RF components to plummet. This has enabled manufacturers outside the traditional military and communications markets to embed relatively complex RF devices into all sorts of commodity products. RF transmitters have become so pervasive that they can be found in almost any imaginable location, such as consumer electronics, medical devices, and industrial control systems. They can even be found in tracking devices implanted under the skin of livestock and pets.

Given the challenge of characterizing the behavior of today's RF devices, we set out in this chapter to help you understand how wireless parameters such as modulation and jitter are measured. Although this chapter has discussed the real-time spectrum analyzer (RTSA), it could have focused on the swept spectrum analyzer (SA) or the vector signal analyzer (VSA). Even though we have reached the end of our discussion, we still haven't decided what is the best instrument to use. This is because no single RF analyzer is ever the best solution for every RF measurement. In fact, many common measurements, such as the measurement of jitter in a new high-speed bus, can be performed with equal effectiveness using a swept SA, VSA, or RTSA. Today, however, the SI engineer is provided with RF benchtop instrumentation with built-in real-time analysis tools that would have been unheard of a decade ago. This adds significantly to the support and development of today's ubiquitous wireless environment.

Index

Symbols

A

C

cable absolute differential impedance, 340

cable losses, TDR (time domain reflectometry), 107-108

calculations

output high impedance, 64

output low impedance, 64

capturing

hold violations, logic analyzers, 240-241

setup violations, logic analyzers, 240-241

card impedance tolerance, PCI Express, 379-381

card-to-cable lane, 305

card-to-card lane, 305

case studies

DDR2 interface, 117-120, 155-157

conductor losses, 135-138

contents, 121

dielectric losses, 135-138

DIMM connector crosstalk, 143-147

final read timing budgets, 153

final write timing budgets, 149-152

functional diagram, 122

impedance tolerance, 138-142

interconnect sensitivity analysis, 132-135

IO circuit, 127-128

length variation within a byte lane, 142

off-chip drivers, 128-129

on-die termination, 129-131

pin-to-pin capacitance variation, 142

read timing, 125-127, 154

resistor tolerance, 147-149

signaling, 121-123

slope derating factor, 149

sources of conservatism, 154-155

voltage margins, 154

Vref AC noise, 147-149

waveforms, 131-132

write timing, 123-125

PCI Express

3D discontinuities, 381-384

card impedance tolerance, 379-381

channel step responses, 383-385

channels, 392-393

crosstalk pathology, 386-387

crosstalk-induced jitter, 387-392

de-emphasized differential drivers, 375-379

high-speed serial interfaces, 368-370

ideal drivers, 373-374

lossy transmission lines, 373-374

model-to-hardware correlation, 396-398

sensitivity analysis, 371-372

sensitivity analysis results, 393-396

CCDF (Complimentary Cumulative Distribution Function) measurements, RTSA, 430-431

channel count, logic analyzers, 235

channels, PCI Express, 392-393

step respoonses, 383-385

characteristic impedance, 92-96

impedance, 97, 99

chip-to-chip lanes, 304-305

chip-to-chip networks, simulation, 2-5, 7-9

circuits

behavior, controlling and monitoring, 256-259

digital modulation, analyzing, 258

H

I

U–V

BOOKS ONLINE
ENABLED

THIS BOOK IS SAFARI ENABLED

INCLUDES FREE 45-DAY ACCESS TO THE ONLINE EDITION

The Safari® Enabled icon on the cover of your favorite technology book means the book is available through Safari Bookshelf. When you buy this book, you get free access to the online edition for 45 days.

Safari Bookshelf is an electronic reference library that lets you easily search thousands of technical books, find code samples, download chapters, and access technical information whenever and wherever you need it.

TO GAIN 45-DAY SAFARI ENABLED ACCESS TO THIS BOOK:

- Go to **informit.com/safarienabled**
- Complete the brief registration form
- Enter the coupon code found in the front of this book on the "Copyright" page

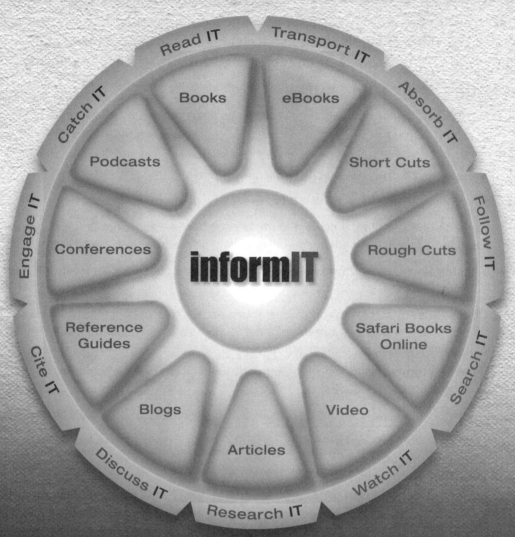